Share—wing KO

Techniques
and Applications of
Plasma Chemistry

CONTRIBUTORS

ALEXIS T. BELL, University of California, Department of Chemical Engineering, Berkeley, Ca.

JOHN R. HOLLAHAN, Tegal Corporation, Richmond, Ca.

MARTIN HUDIS, General Electric, Corporate Research and Development Center, Schenectady, N.Y.

RALPH W. KIRK, Materials Research Laboratories, Motorola Semiconductor Products Division, Phoenix, Ar.

MERLE MILLARD, U.S. Department of Agriculture, Western Regional Laboratory, Albany, Ca.

ATTILA E. PAVLATH, U.S. Department of Agriculture, Western Regional Laboratory, Albany, Ca.

HARALD SUHR, University of Tübingen, Department of Chemistry, Tübingen, West Germany.

RICHARD S. THOMAS, U.S. Department of Agriculture, Western Regional Laboratory, Albany, Ca.

THEODORE WYDEVEN, National Aeronautics and Space Administration, Ames Research Center, Moffett Field, Ca.

Techniques and Applications of Plasma Chemistry

edited by

John R. Hollahan

Tegal Corporation
Richmond, Ca.

Alexis T. Bell

University of California, Berkeley

A Wiley-Interscience Publication

JOHN WILEY & SONS

New York • London • Sydney • Toronto

Library of Congress Cataloging in Publication Data:

Hollahan, John R.   1936–
Techniques and applications of plasma chemistry.

"A Wiley-Interscience publication."
Includes bibliographies.
1. Plasma chemistry—Industrial applications.
I. Bell, Alexis T., 1942–   joint author.   II. Title.
[DNLM: 1. Gases.   QC718 H733t 1974]

TP156.P5H64   660'.04'4   74-5122
ISBN 0-471-40628-7

Printed in the United States of America

10 9 8 7 6 5 4 3 2 1

Preface

Plasma chemistry as a field of research has had a long history, dating as far back as 1796 (see Chapter 2). The work carried out over the years rapidly demonstrated the potential of plasmas for promoting a wide variety of reactions but the development of practical applications proceeded more slowly. The last several decades, however, have seen the introduction of a number of such applications in the fields of organic chemistry, polymer chemistry, biology, solid-state electronics, and metallurgy. Although descriptions of these advances have appeared in the literature, they have been dispersed over a wide range of journals, making it extremely difficult for both the initiated and uninitiated reader to keep up with this work. Several books on plasma chemistry have appeared recently but these have focused on fundamental or research aspects of the field. Consequently the editors felt that it would be appropriate to assemble a volume outlining the current applications for plasma chemistry.

In selecting the contributions for this volume, the editors have sought to represent a broad spectrum of applications giving special attention to those in which the plasma is used because of its unique physical characteristics rather than as a source of heat. For this reason the contributions have been restricted to those in which a nonequilibrium plasma ($T_e \gg T_g$) is used. Applications of high-temperature equilibrium plasmas ($T_g > 10^{3}°K$) have been excluded since the principal use of such plasmas is as a source of high-temperature heat.

It was the intention of the editors that this volume should also serve as guide to the methods for applying plasma chemistry in addition to being a review of the most recent work in each subject area. For this reason, Chapter 1 has been included to serve as an introduction to the physical and chemical fundamentals of plasma chemistry and a portion of each of the other chapters has been devoted to an extended discussion of experimental technique. The editors would also like to bring to the reader's attention the Appendix which describes the construction of radio-frequency equipment, which can be used to sustain an electric discharge.

The preparation of this volume was initiated three years ago and has come to its completion mainly through the efforts of the contributors. The Editors are indebted to B. D. Blaustein of the U.S. Bureau of Mines for his helpful suggestions and discussions concerning the initial organization of this book. Support for the production and editing of the manuscripts was provided by the Boeing Company and the NASA Ames Research Center (JRH) as well as the Department of Chemical Engineering of the University of California at Berkeley (ATB). Finally, the editors would like to acknowledge the assistance of Linda Sorensen who helped to type a number of the manuscripts and aided in bringing the book to its completion.

Richmond, California   *J. R. Hollahan*
Berkeley, California   *A. T. Bell*

Contents

Techniques
and Applications of
Plasma Chemistry

Chapter 1

Fundamentals of
Plasma Chemistry

Alexis T. Bell

1.1. INTRODUCTION

The field of plasma chemistry deals with the occurrence of chemical reactions in a partially ionized gas composed of ions, electrons, and neutral species. This state of matter can be produced through the action of either very high temperatures or strong electric or magnetic fields. The present book focuses attention on ionized gas produced by gaseous electric discharges. In a discharge, free electrons gain energy from an imposed electric field and lose this energy through collisions with neutral gas molecules. The transfer of energy to the molecules leads to the formation of a variety of new species including metastables, atoms, free radicals, and ions. These products are all active chemically and thus can serve as precursors to the formation of new stable compounds.

For the ionized gas produced in a discharge to be properly termed a plasma, it must satisfy the requirement that the concentrations of positive and negative charge carriers are approximately equal. This criterion is satisfied when the dimensions of the discharged gas volume characterized by Λ are significantly larger than the Debye length

$$\lambda_D = \left(\frac{\varepsilon_0 k T_e}{ne^2}\right)^{1/2} \tag{1.1}$$

which defines the distance over which a charge imbalance can exist. In (1.1) ε_0 is the permittivity of free space, k is the Boltzmann constant, T_e is the electron temperature, n is the electron density, and e is the charge on the electron.

Many types of electric discharge have been described in the physical literature and the properties of the plasma produced in them can differ widely. Figure 1.1 illustrates the characteristics for a number of man-made as well as naturally occurring plasmas in terms of the electron temperature and density. The two regions of greatest interest to plasma chemistry are

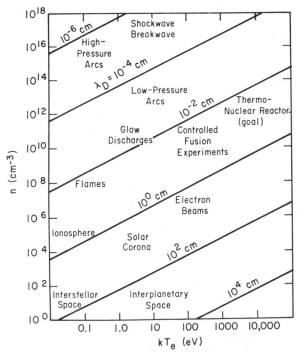

FIGURE 1.1. Typical plasmas characterized by their electron energy and density.

those labeled "glow discharges" and "arcs." The plasma produced in a glow discharge is characterized by average electron energies of 1–10 eV and electron densities of 10^9–10^{12} cm^{-3}. An additional characteristic of such plasmas is the lack of equilibrium between the electron temperature T_e and the gas temperature T_g. Typical ratios for T_e/T_g lie in the range of 10–10^2. The absence of thermal equilibrium makes it possible to obtain a plasma in which the gas temperature may be near ambient values at the same time that the electrons are sufficiently energetic to cause the rupture of molecular bonds. It is this characteristic which makes glow-discharge plasmas well suited for the promotion of chemical reactions involving thermally sensitive materials.

By contrast, the conditions found in the plasmas produced by arcs or plasma jets lead to an equilibrium situation in which the electron and gas temperatures are nearly identical. The very high gas temperatures ($> 5 \times 10^3$ °K) measured in arcs and plasma jets makes them suitable for processing inorganic materials and organic compounds with very simple structures. More complex organic materials and polymers cannot be treated under these conditions since they would be rapidly degraded.

Since the applications described in this book deal exclusively with plasmas produced in glow discharges, the balance of this chapter will be devoted to a discussion of their physical and chemical properties. This discussion will begin with an examination of the factors controlling the electron-velocity distribution and in turn the average electron energy. The determination of the plasma transport properties are discussed next. This is followed by an examination of the types of chemical reactions which occur in plasmas and the means available for describing their kinetics. The material of the previous sections is then brought together to give a complete physical characterization of the plasma as a function of its operating conditions. Finally, using the dissociation of oxygen as an example, a model of the reaction kinetics in a discharge is assembled and tested against experimental data.

1.2. PLASMA KINETIC THEORY

Electron-Velocity Distribution Functions

The electron-velocity distribution plays a central role in defining the physical properties of a plasma. From it are derived the electron-energy distribution, the average electron energy, the electron-transport properties, and the rate constants for reactions involving electron–molecule collisions. As we shall see from the development presented below, the shape of the distribution function depends upon the magnitude of the applied electric field and the nature of the elastic and inelastic interactions which the electrons undergo.

By definition, the electron-velocity distribution function f represents the density of electrons in both position and velocity space. If the vector $\mathbf{r}$ is used to describe a given point in position space and the vector $\mathbf{v}$ a given point in velocity space, f may be expressed as $f(\mathbf{r}, \mathbf{v})$. The product $f\, d\mathbf{r}\, d\mathbf{v}$, in which $d\mathbf{r} = dx\, dy\, dz$ and $d\mathbf{v} = dv_x\, dv_y\, dv_z$, denotes the number of electrons whose positions are located within the volume element $d\mathbf{r}$ and whose velocities lie within the volume element $d\mathbf{v}$. Dividing $f\, d\mathbf{r}\, d\mathbf{v}$ by $d\mathbf{r}$ gives us $f\, d\mathbf{v}$, the number of electrons per unit volume at $\mathbf{r}$ with velocities in $d\mathbf{v}$. If the product $f\, d\mathbf{v}$ is integrated, we will obtain the electron density at the point $\mathbf{r}$. Thus

$$n = \int_{-\infty}^{\infty} f\, d\mathbf{v} \tag{1.2}$$

where the single integral sign is used to denote a triple integration extending over the three coordinates of velocity space.

If $\phi(\mathbf{r}, \mathbf{v})$ defines some property of the electrons which might depend on both position and velocity, the average value of this property weighted against the distribution of velocities can be expressed as

$$n\langle \phi \rangle = \int_{-\infty}^{\infty} \phi f\, d\mathbf{v} \tag{1.3}$$

Thus, for instance, to evaluate the average translational velocity $\langle v_x \rangle$, let $\phi = v_x$. Then

$$\langle v_x \rangle = \frac{1}{n} \int_{-\infty}^{\infty} v_x f\, d\mathbf{v} \tag{1.4}$$

Equation (1.3) can also be used to determine the average electron energy. We begin by noting that the velocity $\mathbf{v}$ can be expressed as $\mathbf{v} = \mathbf{V} + \langle \mathbf{v} \rangle$, the sum of the random velocity plus the translational velocity of the whole electron cloud. Since positive and negative values of $\mathbf{V}$ are equally probable, the average value of $\mathbf{V}$ will equal zero. The total translational kinetic energy can then be expressed as

$$\varepsilon = \tfrac{1}{2}m\langle v^2 \rangle = \frac{1}{n} \int_{-\infty}^{\infty} \tfrac{1}{2}m\mathbf{v} \cdot \mathbf{v} f\, d\mathbf{v}$$

$$= \tfrac{1}{2}m\langle (\mathbf{V} + \langle \mathbf{v} \rangle) \cdot (\mathbf{V} + \langle \mathbf{v} \rangle) \rangle$$

$$= \tfrac{1}{2}m(\langle \mathbf{V} \cdot \mathbf{V} \rangle + 2\langle \mathbf{V} \rangle \cdot \langle \mathbf{v} \rangle + \langle \mathbf{v} \rangle \cdot \langle \mathbf{v} \rangle)$$

$$= \tfrac{1}{2}m\langle V^2 \rangle + \tfrac{1}{2}m\langle v \rangle^2 \tag{1.5}$$

where v and V are the magnitudes of the velocities $\mathbf{v}$ and $\mathbf{V}$, respectively. The first term in (1.5) represents the kinetic energy associated with the random motion of the electrons, and the second term represents the kinetic energy associated with the translation of the electron cloud as a whole. In a system in which the velocity distribution is Maxwellian, the random kinetic energy can be related to the electron temperature by

$$\tfrac{1}{2}m\langle V^2\rangle = \tfrac{3}{2}kT_e \tag{1.6}$$

The Boltzmann Equation

The exact form of the velocity distribution function can be derived through consideration of the gain and loss of electrons from an incremental volume in phase space defined by $d\mathbf{r}\,d\mathbf{v}$. The equation summarizing the net rate of transfer of electrons from this volume is known as the Boltzmann equation and is expressed as

$$\frac{\partial f}{\partial t} + \mathbf{v}\cdot\mathbf{\nabla}_r f + \frac{e\mathbf{E}}{m}\cdot\mathbf{\nabla}_v f = \left(\frac{\partial f}{\partial t}\right)_{\text{coll}} \tag{1.7}$$

The first term on the left-hand side of (1.7) gives the local variation of the distribution function with time. The second term describes the variation in the distribution function resulting from electrons streaming in and out of a given volume element. This term is closely related to the description of diffusion. The third term is the variation of the distribution function resulting from an applied electric field $\mathbf{E}$ acting on the electrons. The single term on the right-hand side of (1.7) accounts for the net transfer of electrons from the differential volume by the mechanism of binary collisions between electrons and molecules, ions, and other electrons.

The net rate at which electrons are removed from the differential volume $d\mathbf{v}$ by binary collisions is given by

$$\left(\frac{\partial f}{\partial t}\right)_{\text{coll}} = \int_{-\infty}^{\infty}\int_0^{\pi}(\tilde{f}\tilde{f}_T - ff_T)g\sigma(\chi, v)2\pi\sin\chi\,d\chi\,d\mathbf{v}_T \tag{1.8}$$

In this expression f and f_T are the velocity distribution functions of the electrons and target paricltes, respectively; g is the relative velocity; and $\sigma(\chi, v)$ is the collision cross-section which depends on the scattering angle χ and the electron speed v. The second term on the right-hand side of (1.8) describes the rate at which electrons are removed from $d\mathbf{v}$ by collisions. The product ff_T associated with this term is evaluated at the velocities v and v_T which hold just before collision. The first term on the right-hand side of

(1.8) describes the rate at which electrons are scattered into $d\mathbf{v}$ by collisions which are the reverse of those removing electrons from $d\mathbf{v}$. In this case the product $\tilde{f}\tilde{f}_T$ must be evaluated at the velocities $\tilde{v}$ and $\tilde{v}_T$ which describe the motion of the electron and target immediately after collision. The two sets of velocities are related to each other by the conservation of momentum and energy.

Many types of binary collisions can occur within a plasma. These can most conveniently be broken down into the categories of elastic and inelastic collisions. The former involves electron collisions with either neutral or charged targets such that there is no excitation of the target particle. If the target is a molecule or ion, then there is only a very small transfer of energy from the electrons. By contrast, interactions between two electrons can lead to a significant transfer of energy from one electron to another and, as we shall see, play an important role in shaping the electron-velocity distribution function. Inelastic collisions differ from elastic collisions in that the target particle is left in an excited state after collision. The nature of these excitations ranges across the energy spectrum from rotational excitations in which 0.01–0.1 eV is absorbed to ionizations in which more than 10 eV is absorbed.

Solutions to the Boltzmann Equation

The Boltzmann equation must be solved in order to obtain the velocity distribution function. Since no exact solution to this equation is known, an approximate solution must be sought. A technique which has been used extensively considers that the solution can be expressed as the sum of an isotropic portion plus a small anisotropic perturbation. The basis for this form is the observation that, in the absence of any fields or gradients in electron concentration, the velocity distribution becomes completely isotropic. The effect of spatial gradients and external forces is thus assumed to produce a perturbation onto the isotropic solution. Under this assumption, the distribution function f can be expressed as

$$f = f^\circ + \phi(\mathbf{v}) \tag{1.9}$$

where f° is the isotropic distribution and $\phi(\mathbf{v})$ the anisotropic distribution. It may be further assumed that the anisotropic contribution is largest when $\mathbf{v}$ is in the direction of the gradient or force causing the perturbation and smallest when it is perpendicular to it. By this reasoning we can express the anisotropic distribution as

$$\phi(\mathbf{v}) = \frac{\mathbf{v}}{v} \cdot \mathbf{f}' \tag{1.10}$$

in which the vector $\mathbf{f}'$ points in the direction in which the electrons drift as the result of the external fields and spatial gradients. The total distribution can then be written as

$$f = f^\circ + \frac{\mathbf{v}}{v} \cdot \mathbf{f}' \qquad (1.11)$$

Substitution of the approximate solution (1.11) into the Boltzmann equation leads to the identification of two differential equations which can be solved for f° and $\mathbf{f}'$. The procedure by which this is accomplished is quite complex and it is suggested that the reader who is interested in the details consult the discussion of this problem given by Allis [1]. The two equations which are the final result can be expressed as

$$\frac{\partial f^\circ}{\partial t} + \frac{v}{3} \nabla_r \cdot \mathbf{f}' - \frac{e\mathbf{E}}{3mv^2} \cdot \frac{\partial}{\partial v}(v^2 \mathbf{f}') = \left(\frac{\partial f^\circ}{\partial t}\right)_{cc} + \left(\frac{\partial f^\circ}{\partial t}\right)_{cm} + \left(\frac{\partial f^\circ}{\partial t}\right)_{cx} \qquad (1.12)$$

$$\frac{\partial \mathbf{f}'}{\partial t} + v\nabla_r f^\circ - \frac{e\mathbf{E}}{m}\frac{\partial f^\circ}{\partial v} = \left(\frac{\partial \mathbf{f}'}{\partial t}\right)_{cc} + \left(\frac{\partial \mathbf{f}'}{\partial t}\right)_{cm} + \left(\frac{\partial \mathbf{f}'}{\partial t}\right)_{cx} \qquad (1.13)$$

The three terms appearing on the right-hand side of these equations describe the effects of coulombic, elastic, and inelastic collisions, respectively.

The elastic collision contributions to the right-hand sides of (1.12) and (1.13) are given by

$$\left(\frac{\partial f^\circ}{\partial t}\right)_{cm} = \frac{m}{M}\frac{1}{v^2}\frac{\partial}{\partial v}(v_m v^3 f^\circ) + \frac{1}{v^2}\frac{\partial}{\partial v}\left(v_m v^2 \frac{kT_g}{M}\frac{\partial f^\circ}{\partial v}\right) \qquad (1.14)$$

$$\left(\frac{\partial \mathbf{f}'}{\partial t}\right)_{cm} = -v_m \mathbf{f}' \qquad (1.15)$$

where m and M are the mass of the electron and target molecule, respectively. The momentum transfer collision frequency v_m is defined by

$$v_m = 2\pi N v \int_0^\pi (1 - \cos \chi)\sigma_e(\chi, v) \sin \chi \, d\chi \qquad (1.16)$$

in which N is the total gas density and σ_e is the elastic collision cross-section.

Inelastic collisions contribute principally to the distribution of velocities and relatively little to the transport properties associated with the electrons. As a result, the appropriate terms appearing in (1.12) and (1.13) will be given by

$$\left(\frac{\partial f^\circ}{\partial t}\right)_{cx} = \sum_i \left(\frac{\tilde{v}}{v}\tilde{f}^\circ(\tilde{v})\nu_i(\tilde{v}) - f^\circ(v)\nu_i(v)\right) \qquad (1.17)$$

$$\left(\frac{\partial \mathbf{f}'}{\partial t}\right)_{cx} = 0 \qquad (1.18)$$

where $1/2\, mv^2 = 1/2\, m\tilde{v}^2 + \varepsilon_i$. The collision frequency ν_i for each inelastic process is related to the cross-section for that process by

$$\nu_i = N v \sigma_i(v) \qquad (1.19)$$

The physical significance of the two terms appearing in (1.17) will be discussed below.

Coulombic collisions arising from the interaction of electrons with either ions or other electrons become significant only when the fractional degree of ionization is substantial [2]. In most high-frequency glow discharges, the degree of ionization is small enough ($<10^{-5}$) for coulombic collisions to be neglected. Thus the terms $(\partial f^\circ/\partial t)_{cc}$ and $(\partial \mathbf{f}'/\partial t)_{cc}$ appearing in (1.12) and (1.13) may be set equal to zero.

The final form of (1.12) and (1.13) can be obtained by introducing the expressions for the collision terms. Thus

$$\frac{\partial f^\circ}{\partial t} + \frac{v}{3}\nabla_r \cdot \mathbf{f}' - \frac{e\mathbf{E}}{3mv^2} \cdot \frac{\partial}{\partial v}(v^2\mathbf{f}')$$

$$= \frac{m}{M}\frac{1}{v^2}\frac{\partial}{\partial v}\left(\nu_m v^3 f^\circ + \frac{\nu_m k T_g}{m}v^2\frac{\partial f^\circ}{\partial v}\right) + \sum_i \left(\frac{\tilde{v}}{v}\tilde{f}^\circ\tilde{\nu}_i - f^\circ\nu_i\right) \qquad (1.20)$$

$$\frac{\partial \mathbf{f}'}{\partial t} + v\nabla_r f^\circ - \frac{e}{m}\mathbf{E}\frac{\partial f^\circ}{\partial v} = -\nu_m\mathbf{f}' \qquad (1.21)$$

The solution of these two equations for f° and $\mathbf{f}'$ will now be illustrated by several simple examples.

Consider a homogeneous, isotropic plasma in the presence of an alternating electric field $\mathbf{E}_0 e^{-i\omega t}$. Then, assuming that the time dependence of $\mathbf{f}'$ is given by $e^{-i\omega t}$, the solution for $\mathbf{f}'$ may be obtained from (1.21) as

$$\mathbf{f}' = \frac{e\mathbf{E}}{m(\nu_m - i\omega)}\frac{\partial f^\circ}{\partial v} \qquad (1.22)$$

Neglecting spatial gradients, (1.20) can be written as

$$\frac{e\mathbf{E}}{3mv^2}\cdot\frac{\partial}{\partial v}(v^2\mathbf{f}') + \frac{m}{M}\frac{1}{v^2}\frac{\partial}{\partial v}\left(\nu_m v^3 f^\circ + \frac{\nu_m kT_g}{m}v^2\frac{\partial f^\circ}{\partial v}\right) + \sum_i\left(\frac{\tilde{v}}{v}\tilde{f}^\circ\tilde{\nu}_i - f^\circ\nu_i\right) = 0$$

(1.23)

The time dependence of f° has been neglected in (1.23) since it has been shown by Margenau [3] that it is small in comparison to the steady-state value of f° for $\omega \gg (2m/M)\nu_m$.

Introduction of the solution for $\mathbf{f}'$ into (1.23) requires that the product $\mathbf{E}\cdot\mathbf{f}'$ be interpreted as the time average of the product of the real parts of $\mathbf{E}$ and $\mathbf{f}'$. With this interpretation, (1.23) can be rewritten as

$$\frac{1}{2v^2}\frac{\partial}{\partial v}\left[\frac{e^2 E_0^2\nu_m v^2}{3m^2(\nu_m^2+\omega^2)}\frac{\partial f^\circ}{\partial v}\right.$$

$$\left. + \frac{2m}{M}\nu_m v^2\left(vf^\circ + \frac{kT_g}{M}\frac{\partial f^\circ}{\partial v}\right)\right] + \sum_i\left(\frac{\tilde{v}}{v}\tilde{f}^\circ\tilde{\nu}_i - f^\circ\nu_i\right) = 0 \quad (1.24)$$

It should be noted that (1.24) could also have been obtained for the case of a dc field with the exception that $E_0^2\nu_m^2/2(\nu_m^2+\omega^2)$ appearing in the first term would have been replaced by E_{dc}^2. If we define an effective field E_e by $E_0\sqrt{\nu_m^2/2(\nu_M^2+\omega^2)}$, then identical distribution functions will result when $E_e = E_{\mathrm{dc}}$.

The physical significance of (1.24) can be illustrated if we multiply each term by $1/2\, mv^2$ and integrate over all speeds. Thus

$$\frac{(eE_e)^2}{m\nu_m}\int_0^\infty f^\circ 4\pi v^2\,dv$$

$$= \frac{2m}{M}\nu_m\int_0^\infty \frac{mv^2}{2}f^\circ 4\pi v^2\,dv - \frac{2m}{M}\frac{3kT_g}{2}\nu_m\int_0^\infty f^\circ 4\pi v^2\,dv$$

$$- \sum_i\int_0^\infty\left(\frac{\tilde{v}}{v}\tilde{f}^\circ\tilde{\nu}_i - f^\circ\nu_i\right)\frac{mv^2}{2}4\pi v^2\,dv \quad (1.25)$$

where, for the sake of simplicity, ν_m is assumed to be independent of v. The left-hand side of (1.25) represents the rate at which energy is transferred from the electric field to the electrons. The first term on the right-hand side gives the rate at which the electrons lose energy by elastic collisions. The second term gives the energy gained by slow electrons colliding with faster moving molecules. The final term represents the net loss of energy due to inelastic collisions.

If the value of E_0 is set to zero and we neglect inelastic losses, (1.24) reduces to

$$vf° + \frac{kT_g}{m}\frac{\partial f°}{\partial v} = 0 \tag{1.26}$$

For this case $f°$ is given by

$$f° = Ce^{-mv^2/2kT_g} \tag{1.27}$$

and is therefore Maxwellian. Under these conditions the electrons are in equilibrium with the gas molecules and are characterized by the gas temperature T_g.

At low electric field strengths, the loss of energy by inelastic collisions will be negligible since very few electrons will possess sufficient energy to bring about such excitations. Under these conditions (1.24) takes the form

$$\frac{1}{v^2}\frac{\partial}{\partial v}\left[\frac{e^2E_e^2v^2}{3m^2v_m}\frac{\partial f°}{\partial v}\right] + \frac{m}{M}v^2v_m\left(vf° + \frac{kT_g}{m}\frac{\partial f°}{\partial v}\right) = 0 \tag{1.28}$$

Integrating once, we obtain

$$v^2\left[\frac{e^2E_e^2}{3m^2v_m}\frac{\partial f°}{\partial v} + \frac{m}{M}v_m\left(vf° + \frac{kT_g}{m}\frac{\partial f°}{\partial v}\right)\right] = 0 \tag{1.29}$$

where the constant of integration is zero since $f°$ must be Maxwellian when $E_e = 0$.

Equation (1.29) can be rewritten as

$$\frac{\partial f°}{\partial v}\left(\frac{e^2E_e^2M}{3\,m^2v_m^2} + kT_g\right) + mvf° = 0 \tag{1.30}$$

The solution of (1.30) is then

$$f° = C\exp-\left(\int_0^v\frac{mv\,dv}{kT_g + e^2E_e^2M/3m^2v_m^2}\right) \tag{1.31}$$

where the constant C is determined by the normalization condition

$$n = 4\pi\int_0^\infty f°v^2\,dv \tag{1.32}$$

The form of $f°$ given by (1.31) is known as the Margenau distribution.

We may now consider the case in which the applied electric field is sufficiently large to cause the second term in the denominator of the integral appearing in (1.31) to be much greater than kT_g. In addition, we will assume a sufficiently low frequency for the applied field such that $\omega^2 \ll \nu^2$. The Margenau distribution can then be approximated as

$$f^\circ = C \exp -\left(\int_0^v \frac{mv\,dv}{e^2 E_0^2 M/6m^2 \nu_m^2}\right) \tag{1.33}$$

The integral appearing in (1.33) can be evaluated in closed form provided we make some assumptions concerning the dependence of ν_m on velocity. From the kinetic theory $\nu_m(v)$ can be expressed as

$$\nu_m = N v \sigma_m(v) \tag{1.34}$$

where σ_m is the momentum collision cross-section. For certain gases such as helium and hydrogen $\sigma_m(v)$ is found to vary as $1/v$, and, for such gases, the momentum collision frequency is therefore independent of velocity. In these cases the Margenau distribution will be given by

$$f^\circ = C \exp -\left(\frac{mv^2/2}{e^2 E_0^2 M/6m^2 \nu_m^2}\right) \tag{1.35}$$

Equation (1.35) has the same form as a Maxwellian distribution in which

$$kT_e = \frac{e^2 E_0^2 M}{6m^2 \nu_m^2} \tag{1.36}$$

When the collision cross-section is independent of velocity, the collision frequency ν_m becomes a linear function of velocity. Under such conditions the Margenau distribution may be expressed as

$$f^\circ = C \exp -\left(\int_0^v \frac{6m^3 N^2 \sigma_m^2 v^3\,dv}{e^2 E_0^2 M}\right)$$

$$= C \exp -\left[\frac{(mv^2/2)^2}{e^2 E_0^2 M/6mN^2\sigma_m^2}\right] \tag{1.37}$$

Equation (1.37) is known as the Druyvesteyn distribution. Since the Druyvesteyn distribution varies as e^{-av^4}, the tail of this distribution decreases more rapidly than the Maxwellian distribution. Figure 1.2 shows that, for a

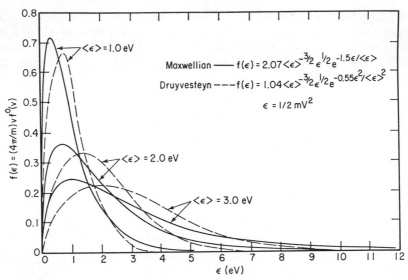

FIGURE 1.2. Comparison of Maxwellian and Druyvesteyn energy distribution functions.

plasma with a given mean energy, the Druyvesteyn distribution predicts fewer high-energy electrons than does the Maxwellian distribution.

An important conclusion which can be derived from (1.36) is that the average electron energy expressed in terms of T_e is a function solely of E_0/p. This result comes about from the linear dependence of ν_m on pressure. A similar conclusion can be drawn for the Druyvesteyn distribution. There it may be noted that the denominator in the exponential portion of (1.37) represents the square of the average electron energy. Since the gas density N is proportional to pressure, the average electron energy is again seen to be a function solely of E_0/p. The effects of electron density on the average electron energy are absent in the two cases examined here. However, as will be discussed below the electron density can have an effect when the fractional degree of ionization becomes large.

The Margenau and Druyvesteyn distributions provide valid solutions to the Boltzmann equation for electric field strengths which are sufficiently low to allow the neglect of inelastic collisions. These conditions are rarely valid in a steady-state discharge. As a result we must return to (1.24) and examine its solution in the presence of inelastic collisions. Since the form of (1.24) becomes quite complex in this case, it is not feasible to provide an analytic solution and one must turn to numerical methods.

An interesting example of the solution of the Boltzmann equation is discussed by Dreicer [2]. In this work the electron-velocity distribution is

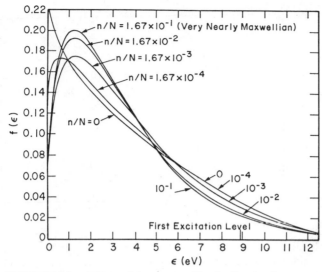

FIGURE 1.3. Effect of the extent of ionization on the energy distribution function for a hydrogen plasma; E/p is held constant at 28.3 V/cm torr [2].

calculated for a hydrogen plasma, taking into account inelastic collisions as well as random two-body coulombic collisions. Elastic collisions between electrons and hydrogen molecules are neglected on the basis that they represent only a small energy loss term. In consideration of the coulombic interactions only those occurring between two electrons are retained since it is shown that electron–ion encounters result in negligible energy losses from the electron.

Figures 1.3–1.5 illustrate the numerical solutions obtained for E/p equal to 28.3 V/cm torr and 48.9 V/cm torr. The function plotted on the ordinate is electron-energy distribution function which is related to the velocity distribution function by

$$f(\varepsilon) = \frac{4\pi}{m} v f^{\circ}(v) \tag{1.38}$$

where $\varepsilon = 1/2\,mv^2$. For each value of E/p, the effects of the extent of ionization are identified by the value of n/N. At very low energies $f(\varepsilon)$ decreases as the extent of ionization increases. In the neighborhood of 1 eV, the distributions cross each other and the situation is reversed. At still higher energies, the distributions cross each other two more times until in the very high-energy tail we find the population increasing with degree of ionization. It should be noted that at the highest degree of ionization the distribution function is very nearly Maxwellian.

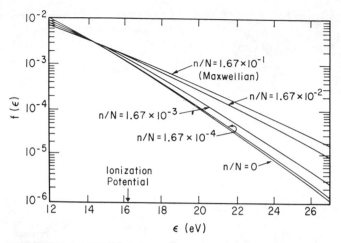

FIGURE 1.4. Effect of the extent of ionization on the energy distribution function for a hydrogen plasma; E/p is held constant at 28.3 V/cm torr [2].

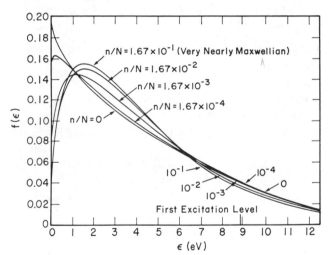

FIGURE 1.5. Effect of the extent of ionization on the energy distribution function for a hydrogen plasma; E/p is held constant at 48.9 V/cm torr [2].

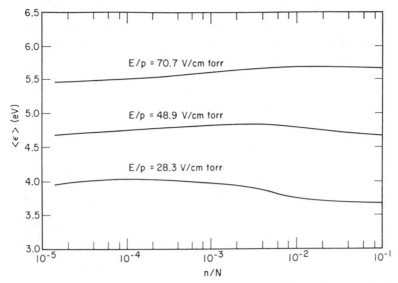

FIGURE 1.6.   Variation of the average electron energy with the extent of ionization [2].

The effects of electron interactions on the average energy are illustrated in Figure 1.6. As the degree of ionization increases, the average energy first rises and then falls. The rise is attributed to the fact that the slow electrons produced in inelastic encounters exchange energy with other electrons, and are quickly redistributed over the body of the distribution. For a fixed value of E/p and N, the rise in average energy is accomplished by the addition to the distribution of electrons whose average energy exceeds the average energy corresponding to $n/N = 0$. The subsequent fall in average energy as n/N increases even further reflects the fact that coulomb encounters have decreased the population in the neighborhood of the first excitation potential and have redistributed these electrons primarily into lower energy regions. A small number of electrons are also redistributed into the very high-energy tail of the distribution. Ultimately, when n/N becomes large enough so that the rate at which electrons exchange energy with each other greatly exceeds the rate at which they gain energy from the field, the distribution function becomes Maxwellian.

1.3.   TRANSPORT PHENOMENA

Electron-Transport Properties

The electron-velocity distribution discussed in the previous section can be used to evaluate the transport properties of electrons. In this section we shall focus on the development of expressions for the electron conductivity and diffusivity. These properties will then be used to determine the spatial distribution of charged particles in a plasma and the rate at which energy is transferred from the sustaining electric field.

A common starting point for the evaluation of both electron conductivity and diffusivity is the flux vector $\mathbf{\Gamma}$ given by

$$\mathbf{\Gamma} = n\langle \mathbf{v} \rangle = \int_{-\infty}^{\infty} \mathbf{v} f \, d\mathbf{v} \tag{1.39}$$

The volume element $d\mathbf{v}$ expressed in spherical coordinates is

$$d\mathbf{v} = v^2 \sin \theta \, d\theta \, d\phi \, dv \tag{1.40}$$

Substituting (1.11) and (1.40) into (1.39), we obtain

$$\mathbf{\Gamma} = \int_0^\infty \int_0^{2\pi} \int_0^\pi \mathbf{v} f^\circ v^2 \sin \theta \, d\theta \, d\phi \, dv + \int_0^\infty \int_0^{2\pi} \int_0^\pi \mathbf{v}(\mathbf{v} \cdot \mathbf{f}')v \sin \theta \, d\theta \, d\phi \, dv$$

$$\tag{1.41}$$

Since f° is isotropic and a function of v alone, the integral over v causes the first term in (1.41) to vanish. The second term can be simplified by taking into account the properties of spherical coordinates. By this means $\mathbf{\Gamma}$ becomes

$$\mathbf{\Gamma} = \frac{4}{3}\pi \int_0^\infty \mathbf{f}' v^3 \, dv \tag{1.42}$$

From (1.42) we thus see that $\mathbf{\Gamma}$ depends on $\mathbf{f}'$ alone. However, since $\mathbf{f}'$ and f° are related, $\mathbf{\Gamma}$ will depend on f° indirectly.

The conductivity of a plasma placed in a uniform ac electric field can be evaluated from (1.42). From the previous section we have seen that the anisotropic portion of the velocity distribution is given by

$$\mathbf{f}' = \frac{e\mathbf{E} \, \partial f^\circ / \partial v}{m(v_m - i\omega)} \tag{1.43}$$

for these conditions. Making use of this result we can express the current density in the plasma as

$$\mathbf{J} = -e\boldsymbol{\Gamma}$$

$$= -\frac{4}{3}e\pi \int_0^\infty f'v^3 \, dv$$

$$= -\frac{4\pi e^2}{3m} \mathbf{E} \int_0^\infty \frac{v^3}{(v_m - i\omega)} \frac{\partial f^\circ}{\partial v} \, dv$$

$$= \sigma\mathbf{E} \tag{1.44}$$

Hence the conductivity is defined by

$$\sigma = -\frac{4\pi e^2}{3m} \int_0^\infty \frac{v^3}{(v_m - i\omega)} \frac{\partial f^\circ}{\partial v} \, dv \tag{1.45}$$

If it is assumed that f° is Maxwellian, then

$$f^\circ = n\left(\frac{m}{2\pi kT_e}\right)^{3/2} e^{-(mv^2/2kT_e)} \tag{1.46}$$

Defining

$$u = \left(\frac{m}{2kT_e}\right)^{1/2} v \tag{1.47}$$

it can be shown that

$$v^3 \frac{\partial f^\circ}{\partial v} \, dv = -2n\pi^{-3/2} u^4 e^{-u^2} \, du \tag{1.48}$$

Substituting (1.49) into (1.45), we obtain

$$\sigma = \frac{8ne^2}{3\sqrt{\pi}\,m} \left(\int_0^\infty \frac{v_m u^4 e^{-u^2}}{\omega^2 + v_m^2} \, du + i\omega \int_0^\infty \frac{u^4 e^{-u^2}}{\omega^2 + v_m^2} \, du \right)$$

$$= \sigma_R + i\sigma_I \tag{1.49}$$

When the collision frequency v_m is independent of the electron speed v, then the real and imaginary parts of σ in (1.49) are given by

$$\sigma_R = \frac{v_m ne^2}{m(\omega^2 + v_m^2)} \tag{1.50}$$

$$\sigma_I = \frac{\omega ne^2}{m(\omega^2 + v_m^2)} \tag{1.51}$$

An expression for the free electron diffusivity can be derived by considering the problem of diffusion in the absence of an electric field. At steady state, (1.21) can then be written as

$$v \nabla_r f^\circ = -\nu_m \mathbf{f'} \tag{1.52}$$

Solving for $\mathbf{f'}$ we obtain

$$\mathbf{f'} = -\frac{v}{\nu_m} \nabla_r f^\circ \tag{1.53}$$

Substituting (1.53) into the expression for the flux vector $\mathbf{\Gamma}$ given by (1.42) we obtain

$$\mathbf{\Gamma} = -\frac{4\pi}{3} \int_0^\infty \frac{v^4}{m} \nabla_r f^\circ \, dv \tag{1.54}$$

The gradient of f° appearing in (1.54) applies to the electron density portion of f° alone, provided the electron speed is not a function of position. Under this restriction $\nabla_r f^\circ$ may be written as

$$\nabla_r f^\circ = \frac{f^\circ}{n} \nabla_r n \tag{1.55}$$

Substituting (1.55) into (1.54) yields

$$\mathbf{\Gamma} = -\frac{4\pi}{3} \frac{1}{n} \nabla_r n \int_0^\infty \frac{v^4}{\nu_m} f^\circ \, dv$$

$$= -D \nabla_r n \tag{1.56}$$

where

$$D = \frac{4\pi}{3} \frac{1}{n} \int_0^\infty \frac{v^4}{\nu_m} f^\circ \, dv \tag{1.57}$$

is the free electron diffusivity.

If f° is Maxwellian, (1.57) can be written as

$$D = \frac{8}{3\sqrt{\pi}} \frac{kT_e}{m} \int^\infty \frac{u^4}{\nu_m} e^{-u^2} \, du \tag{1.58}$$

When the momentum collision frequency can be assumed to be independent of electron speed, (1.58) reduces to

$$D = \frac{kT_e}{m\nu_m} \tag{1.59}$$

Under conditions where the distribution function is Maxwellian or can be approximated as Maxwellian, a useful relationship can be derived between the diffusion coefficient and the electron mobility μ. For this purpose, we must define the mobility by

$$\mu = \frac{\langle \mathbf{v} \rangle}{\mathbf{E}_e}$$

$$= \frac{e}{m\nu_m} \tag{1.60}$$

The ratio of D to μ,

$$\frac{D}{\mu} = \frac{kT_e}{e} \tag{1.61}$$

is known as the Einstein relation. When the distribution is non-Maxwellian, then the ratio D/μ can be expressed as

$$\frac{D}{\mu} = \frac{m\langle v^2 \rangle}{3e} \tag{1.62}$$

when ν_m is constant and

$$\frac{D}{\mu} = \frac{m}{2e} \frac{\langle v \rangle}{\langle 1/v \rangle} \tag{1.63}$$

when σ_m is constant.

The ratio D/μ is a function of E_e/p and is known as the characteristic energy ε_k. Extensive measurements of ε_k for electron swarms have been reported for many gases [4,5,6]. As can be seen from (1.61) and (1.62), ε_k represents the average electron energy, provided that either f° is Maxwellian or ν_m is independent of electron speed. When neither of these conditions is satisfied, the characteristic energy is found to be a poor approximation of the average electron energy corresponding to a given value of E_e/p.

Energy Transfer

The transfer of energy to a plasma occurs principally through the action of the electric field on the free electrons. The instantaneous rate of energy transfer per unit volume is given by

$$\bar{P}(t) = \sigma_r E_0 e^{i\omega t} \tag{1.64}$$

To obtain the net power dissipation, (1.64) is averaged over the period of one cycle of the field. For the case where ν_m is independent of the electron velocity, the time-averaged power density becomes

$$\bar{P} = \frac{ne^2E_0^2}{2m\nu_m}\left(\frac{\nu_m^2}{\nu_m^2 + \omega^2}\right) \tag{1.65}$$

or in terms of the effective field E_e

$$\bar{P} = \frac{ne^2\bar{E}_e^2}{m\nu_m} \tag{1.66}$$

From (1.65) we can observe that ν_m must be finite for energy transfer to occur. Furthermore, the maximum value of $\bar{P}$ is obtained when $\omega^2 \ll \nu_m^2$ for a given value of E_0.

Diffusion of Charged Species

We may turn next to a discussion of the spatial distribution of the charged particles in an ionized gas. At very low charge concentrations, such as those encountered near breakdown, electrons and positive ions will diffuse independently of each other and the flux of each species can be written as:

$$\mathbf{\Gamma}_e = -D_e\nabla n_e \tag{1.67}$$

$$\mathbf{\Gamma}_+ = -D_+\nabla n_+ \tag{1.68}$$

where n_e and n_+ denote the densities of electrons and positive ions, respectively. For steady-state conditions, the charged particles lost from the ionized gas by diffusion must be replaced by ionization. If the rate of ionization is represented by $n_e\nu_i$, where ν_i is the ionization frequency, the steady-state conservation equations for electrons and positive ions will be given by

$$-\nabla \cdot \mathbf{\Gamma}_e + n_e\nu_i = 0 \tag{1.69}$$

$$-\nabla \cdot \mathbf{\Gamma}_+ + n_e\nu_i = 0 \tag{1.70}$$

Substitution of (1.67) and (1.68) into (1.69) and (1.70) gives

$$D_e\nabla^2 n_e + n_e\nu_i = 0 \tag{1.71}$$

$$D_+\nabla^2 n_+ + n_e\nu_i = 0 \tag{1.72}$$

Equation (1.71) can be solved for a variety of plasma geometries in order to obtain the spatial distribution of electron density. We will consider two examples here, that of a rectangular box and that of a cylindrical tube closed at its ends. For the rectangular geometry, (1.71) is expressed as

$$D_e\left(\frac{\partial^2 n_e}{\partial x^2} + \frac{\partial^2 n_e}{\partial y^2} + \frac{\partial^2 n_e}{\partial z^2}\right) + n_e \nu_i = 0 \tag{1.73}$$

The appropriate boundary conditions are given by:

$$\frac{\partial n_e}{\partial x} = 0 \quad \text{at} \quad x = 0 \qquad \text{for all } y \text{ and } z$$

$$\frac{\partial n_e}{\partial y} = 0 \quad \text{at} \quad y = 0 \qquad \text{for all } x \text{ and } z$$

$$\frac{\partial n_e}{\partial z} = 0 \quad \text{at} \quad z = 0 \qquad \text{for all } x \text{ and } y$$

$$\tag{1.74}$$

$$n_e = 0 \quad \text{at} \quad x = \pm\frac{L}{2} \qquad \text{for all } y \text{ and } z$$

$$n_e = 0 \quad \text{at} \quad y = \pm\frac{W}{2} \qquad \text{for all } x \text{ and } z$$

$$n_e = 0 \quad \text{at} \quad z = \pm\frac{H}{2} \qquad \text{for all } x \text{ and } y$$

where L, W, and H represent the length, width, and height of the box. The solution to (1.73) and (1.74) is given by

$$n_e = n_{e0} \cos\left(\frac{\pi x}{L}\right) \cos\left(\frac{\pi y}{W}\right) \cos\left(\frac{\pi z}{H}\right) \tag{1.75}$$

where n_{e0} is the electron density at the center of the box. Substitution of this solution into the original differential equation leads to the constraint that

$$\frac{\nu_i}{D_e} = \frac{1}{\Lambda^2} = \left(\frac{\pi}{L}\right)^2 + \left(\frac{\pi}{W}\right)^2 + \left(\frac{\pi}{H}\right)^2 \tag{1.76}$$

where Λ is defined as the diffusion length.

For a cylindrical geometry, (1.71) takes the form

$$D_e\left[\frac{1}{r}\frac{\partial}{\partial r}\left(r\frac{\partial n_e}{\partial r}\right) + \frac{\partial^2 n_e}{\partial z^2}\right] + n_e \nu_i = 0 \tag{1.77}$$

The boundary conditions in this case are

$$\frac{\partial n_e}{\partial r} = 0 \quad \text{at} \quad r = 0 \qquad \text{for all } z$$

$$\frac{\partial n_e}{\partial z} = 0 \quad \text{at} \quad z = 0 \qquad \text{for all } r$$

$$n_e = 0 \quad \text{at} \quad r = R \qquad \text{for all } z \tag{1.78}$$

$$n_e = 0 \quad \text{at} \quad z = \pm\frac{L}{2} \qquad \text{for all } r$$

where R and L are the radius and length of the cylinder. The solution to (1.77) and (1.78) is given by

$$n_e = n_{e0} J_0\left(\frac{2.405r}{R}\right)\cos\left(\frac{\pi z}{L}\right) \tag{1.79}$$

in which J_0 represents the zeroth order Bessel function. Substitution of the solution for n_e back into (1.77) gives in this case

$$\frac{\nu_i}{D_e} = \frac{1}{\Lambda^2} = \left(\frac{2.405}{R}\right)^2 + \left(\frac{\pi}{L}\right)^2 \tag{1.80}$$

Solutions for the distribution of positive ions can now be obtained from (1.72), recognizing that Γ_e must equal Γ_+ in order to maintain a zero flow of current at the boundaries of the ionized gas. From these considerations we obtain

$$n_+ = \frac{D_e}{D_+} n_e \tag{1.81}$$

Thus, at steady state the process of free diffusion leads to a concentration of positive ions considerably higher than that of electrons. Under these conditions, the ionized gas cannot be considered to be a plasma.

The separation of charge caused by the difference in diffusion coefficients of electrons and ions produces a space charge field. The magnitude of this field will remain small, however, as long as the charge concentrations remain low ($n_e < 10^2$ cm^{-3}) and as a result will not affect the diffusion of either charge carrier.

As the charge concentration is increased, a level is finally reached where $\lambda_D \ll \Lambda$ and the medium may properly be termed a plasma. Under these conditions the space charge field $\mathbf{E}_{sc}$ becomes sufficiently large to affect the transport of both electrons and positive ions. The expressions for charge flux are now given by

$$\mathbf{\Gamma}_e = -D_e \nabla n_e - n_e \mu_e \mathbf{E}_{sc} \tag{1.82}$$

$$\mathbf{\Gamma}_+ = -D_+ \nabla n_+ + n_+ \mu_+ \mathbf{E}_{sc} \tag{1.83}$$

Equations (1.82) and (1.83) can be used to solve for $\mathbf{E}_{sc}$ if we assume $n_e = n_+ = n$ and take $\mathbf{\Gamma}_e = \mathbf{\Gamma}_+ = \mathbf{\Gamma}$ as before. These are the conditions required for ambipolar diffusion and lead to an expression for $\mathbf{E}_{sc}$ as

$$\mathbf{E}_{sc} = -\frac{(D_e - D_+)\nabla n}{(\mu_e + \mu_+)\,n} \tag{1.84}$$

Substitution of (1.84) into either (1.82) or (1.83) gives

$$\mathbf{\Gamma} = -\left(\frac{D_+\mu_e + D_e\mu_+}{\mu_e + \mu_+}\right)\nabla n = -D_a \nabla n \tag{1.85}$$

in which D_a is defined as the ambipolar diffusivity. Since $\mu_e \gg \mu_+$ the magnitude of D_a can be approximated by

$$D_a \simeq D_+\left(1 + \frac{D_e\mu_+}{D_+\mu_e}\right) = D_+\left(1 + \frac{T_e}{T_g}\right) \tag{1.86}$$

Ambipolar diffusion will usually occur when $n_e \simeq n_+ \gtrsim 10^8$ cm^{-3} and the distribution of both charged species will then be governed by

$$D_a \nabla^2 n + n\nu_i = 0 \tag{1.87}$$

Equation (1.87) has the identical form of (1.71). As a result, the solutions for n_e and n_+ will have the same form as (1.75) and (1.79) with the exception that D_a will replace D_e in the expressions for Λ.

In the transition region between free and ambipolar diffusion, the space charge field cannot be represented by (1.84). To solve for the charge distribution it now becomes necessary to consider Poisson's equation in addition

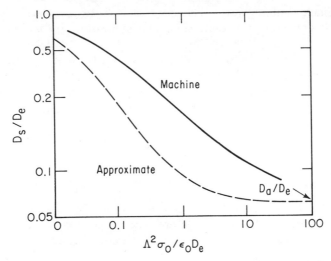

FIGURE 1.7. The effective diffusivity as a function of the plasma conductivity [7].

to the conservation equation for each species. Allis and Rose [7] have solved this set of three equations for the system of a discharge sustained between two infinite parallel planes. In the course of their analysis an effective diffusivity was defined as

$$D_s = D_a\left(1 + \frac{\mu_e \rho}{\sigma}\right) \qquad (1.88)$$

where

$$\rho = (n_+ - n_e)e \qquad (1.89)$$

and

$$\sigma = (\mu_+ n_+ + \mu_e n_e)e \qquad (1.90)$$

Figure 1.7 illustrates the value of D_s at the center of the discharge as a function of the central conductivity σ_0. Also shown in this figure is an approximation for D_s derived from the expression

$$D_s = D_a\frac{D_e + \Lambda^2\sigma_0/\varepsilon_0}{D_a + \Lambda^2\sigma_0/\varepsilon_0} \qquad (1.91)$$

From Figure 1.7 and (1.91), we see that $D_s = D_e$ when $\sigma_0 \approx 0$, and that D_s decreases as the central conductivity increases. For large values of σ_0, D_s approaches the ambipolar value D_a.

When a significant concentration of negative ions is present in addition to the positive ions, the diffusion of electrons must be considered as part of a three-particle system. The appropriate flux equations are now

$$\mathbf{\Gamma}_e = -D_e \nabla n_e - \mu_e n_e \mathbf{E}_{sc} \tag{1.92}$$

$$\mathbf{\Gamma}_- = -D_- \nabla n_- - \mu_- n_- \mathbf{E}_{sc} \tag{1.93}$$

$$\mathbf{\Gamma}_+ = -D_+ \nabla n_+ + \mu_+ n_+ \mathbf{E}_{sc} \tag{1.94}$$

An expression for the space charge field can be obtained by substituting (1.92)–(1.94) into the equation for a current balance

$$\mathbf{\Gamma}_+ = \mathbf{\Gamma}_e + \mathbf{\Gamma}_- \tag{1.95}$$

Thus

$$\mathbf{E}_{sc} = \frac{-D_e \nabla n_e - D_- \nabla n_- + D_+ \nabla n_+}{\mu_e n_e + \mu_- n_- + \mu_+ n_+} \tag{1.96}$$

Substitution of (1.96) together with the charge neutrality condition $n_+ = n_e + n_-$ into (1.92) gives

$$\mathbf{\Gamma}_e = -\left\{ \frac{[D_e(\mu_- + \mu_+)n_- + (D_e\mu_+ + D_+\mu_e)]\nabla n_e + \mu_e n_e(D_+ - D_-)\nabla n_-}{(\mu_e + \mu_+)n_e + (\mu_- + \mu_+)n_-} \right\} \tag{1.97}$$

Since $D_+ \simeq D_-$, the term in the numerator of (1.97) containing the difference between the two diffusion coefficients can be neglected. If the dimensionless groups

$$\beta = \frac{n_-}{n_e}$$

$$\phi = \frac{D_e}{\mu_e} \frac{\mu_+}{D_+} = \frac{D_e}{\mu_e} \frac{\mu_-}{D_-} = \frac{T_e}{T_\pm} \tag{1.98}$$

are introduced, (1.97) can finally be written as

$$\mathbf{\Gamma}_e = -D_+ \left[\frac{1 + \phi + 2\beta\phi}{1 + (1 + \beta)\mu_+/\mu_e + \beta\mu_+/\mu_e} \right] \nabla n_e$$

$$= -D_{a_e} \nabla n_e \tag{1.99}$$

Figure 1.8 shows the behavior of D_{a_e}/D_+ as a function of β and ϕ. As β increases, D_{a_e} increases until it reaches a maximum value of $D_{a_e} = D_e$. At

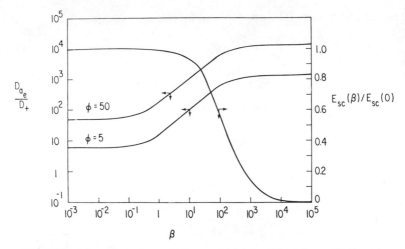

FIGURE 1.8. Variation of the ambipolar diffusivity and the space charge field in an oxygen plasma as a function of the relative concentration of negative ions.

this point the electrons are in free diffusion. The cause for the increase of D_{a_e} with β is the reduction in the space charge field as illustrated by Figure 1.8.

1.4. CHEMICAL REACTIONS OCCURRING IN PLASMAS

A broad spectrum of reactions have been observed to take place in plasmas. These include reactions between electrons and molecules, ions and molecules, ions and ions, and electrons and ions. Due to the very large number of such processes occurring when only one parent gas is present, we shall not attempt to provide a comprehensive discussion of this subject here. Instead, we shall illustrate the diversity of plasma reactions by examining in detail the processes occurring in an oxygen discharge. The reader who is interested in pursuing this subject as it applies to other gases should consult the books by McDaniel [5], McDaniel et al. [8], and Massey, Burhop, and Gilbody [6].

We may begin our discussion of electron–molecule reactions involving oxygen by considering the potential energy diagram for O_2 shown in Figure 1.9. The ground state is designated by the spectroscopic symbol $X^3\Sigma_g^-$. Above this lie four higher states, $a^1\Delta_g$, $b^1\Sigma_g^+$, $C^3\Delta_u$, and $A^3\Sigma_u^+$, all of which dissociate into ground-state atoms. Oxygen in the metastable $a^1\Delta_g$ state is known as "singlet" oxygen. This species is particularly long-lived and participates in

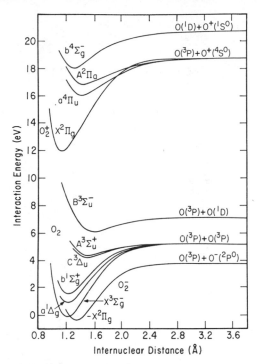

FIGURE 1.9. Potential energy curves for some states of O_2^-, O_2, and O_2^+.

a number of chemical reactions in oxygen discharges (see Table 1.1). The next highest state is the $B^3\Sigma_u^-$. Vertical excitation to this state from the ground state leads to dissociation producing one ground-state atom and one singlet D atom. Further excitation leads to the ionization of the molecule and the formation of O_2^+ in the $X^2\Pi_g$ state.

Electrons colliding with molecules of oxygen undergo either elastic or inelastic collisions. In the former, the molecule is left in an unexcited state but gains $2m/M$ of the original kinetic energy of the electron. This process leads to a slight increase in the kinetic energy of the molecule and, when repeated frequently, is responsible for the increase in the gas temperature. The probability of elastic collisions is characterized by the momentum cross-section σ_m. Figure 1.10 illustrates the magnitude of σ_m for oxygen as a function of electron energy.

At low electron energies, the principal inelastic processes which occur are the excitation of rotational and vibrational levels of ground-state O_2. Very little experimental data has been obtained related to the cross-sections for rotational excitation and the best estimates of their size are the theoretical

TABLE 1.1. Elementary Reactions Occurring in an Oxygen Discharge

Reaction	k	σ_{max} (cm²)	Reference
Ionization			
1. $e + O_2 \rightarrow O_2^+ + 2e$		2.72×10^{-16}	[9]
2. $e + O \rightarrow O^+ + 2e$		1.54×10^{-18}	[10]
Dissociative ionization			
3. $e + O_2 \rightarrow O^+ + O$		1.0×10^{-16}	[11]
Dissociative attachment			
4. $e + O_2 \rightarrow O^- + O$		1.41×10^{-18}	[12]
5. $e + O_2 \rightarrow O^- + O^+ + e$		4.85×10^{-19}	[12]
Dissociation			
6. $e + O_2 \rightarrow 2O + e$		2.25×10^{-18}	[13]
Metastable formation			
7. $e + O_2 \rightarrow O_2(^1\Delta_g) + e$		3.0×10^{-20}	[13]
Charge transfer			
8. $O^+ + O_2 \rightarrow O_2^+ + O$	2×10^{-11} cm³/sec		[14]
9. $O_2^+ + O \rightarrow O^+ + O_2$		8×10^{-16}	[15]
10. $O_2^+ + O_2 \rightarrow O_3^+ + O$		1×10^{-16}	[16]
11. $O_2^+ + 2O_2 \rightarrow O_4^+ + O_2$	2.8×10^{-30} cm⁶/sec		[17]
12. $O^- + O_2 \rightarrow O_2^- + O$	2.5×10^{-14} cm³/sec at $E/p = 20$ V/cm torr		[18]
13. $O^- + O_3 \rightarrow O_3^- + O$	3.4×10^{-12} cm³/sec at $E/p = 45$ V/cm torr		[19]
14. $O^- + 2O_2 \rightarrow O_3^- + O_2$	$1.0 \pm 0.2 \times 10^{-30}$ cm⁶/sec		[18]
15. $O_2^- + O \rightarrow O^- + O_2$	5.3×10^{-10} cm³/sec		[20]
16. $O_2^- + O_2 \rightarrow O_3^- + O$		$<10^{-18}$	[21]
17. $O_2^- + O_3 \rightarrow O_3^- + O_2$	5×10^{-10} cm³/sec		[19]
18. $O_2^- + 2O_2 \rightarrow O_4^- + O_2$	3×10^{-31} cm⁶/sec		[22]
19. $O_3^- + O_2 \rightarrow O_2^- + O_3$	4.0×10^{-10} cm³/sec	4×10^{-17}	[21]

20. $O_4^- + O \to O_3^- + O_2$ 4×10^{-10} cm³/sec   [23]

21. $O_4^- + O_2 \to O_2^- + 2 O_2$ 6×10^{-15} cm³/sec   [22]

Detachment

22. $O^- + O \to O_2 + e$ 3.0×10^{-10} cm³/sec   [20]

23. $O^- + O_2 \to O + O_2 + e$ 7×10^{-16} [24]

24. $O^- + O_2(^1\Delta_g) \to O_3 + e$ $\sim 3 \times 10^{-10}$ cm³/sec   [25]

25. $O_2^- + O \to O_3 + e$ 5.0×10^{-10} cm³/sec   [20]

26. $O_2^- + O_2 \to 2 O_2 + e$ 7×10^{-16} [24]

27. $O_2^- + O_2(^1\Delta_g) \to 2 O_2 + e$ $\sim 2 \times 10^{-10}$ cm³/sec   [25]

Electron–ion recombination

28. $e + \left\{ \begin{array}{c} O^+ \\ O_2^+ \\ O_3^+ \\ O_4^+ \end{array} \right\} \to \left\{ \begin{array}{c} O \\ 2\,O \\ O + O_2 \\ 2\,O_2 \end{array} \right\}$ $\leqslant 10^{-7}$ cm³/sec   [26]

Ion–ion recombination

29. $\left\{ \begin{array}{c} O^- \\ O_2^- \\ O_3^- \\ O_4^- \end{array} \right\} + \left\{ \begin{array}{c} O^+ \\ O_2^+ \\ O_3^+ \\ O_4^+ \end{array} \right\} \to \left\{ \begin{array}{c} O \\ O_2 \end{array} \right\}$ $\sim 10^{-7}$ cm³/sec   [27]

Atom recombination

30. $2 O + O_2 \to 2 O_2$ 2.3×10^{-33} cm⁶/sec   [28]

31. $3 O \to O + O_2$ 1.5×10^{-34} cm⁶/sec   [28]

32. $O + 2 O_2 \to O_2 + O_2$ $1.9 \times 10^{-35} \exp(2100/RT)$ cm⁶/sec   [29]

33. $O + O_3 \to 2 O_2$ $2.0 \times 10^{-11} \exp(-4790/RT)$ cm³/sec   [29]

34. $O \xrightarrow{\text{wall}} \tfrac{1}{2} O_2$ $\gamma = 1.6 \times 10^{-4}$ to 1.4×10^{-2} ($T = 20$–600°C)   [30]

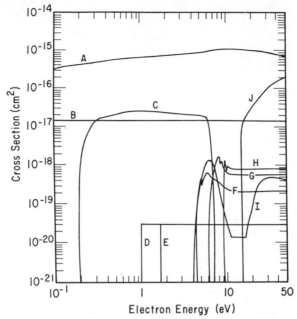

FIGURE 1.10. Elastic and inelastic collision cross-sections for electrons in oxygen; (A) elastic scattering; (B) rotational excitation; (C) vibrational excitation; (D) excitation to the $a^1\Delta_g$ state; (E) excitation of the $b^1\Sigma_u^+$ state; (F) excitation of the $A^3\Sigma_u^+$ state; (G) excitation of the $B^3\Sigma_u^-$ state; (H) excitation of higher electronic states; (I) dissociative attachment; (J) ionization [13].

calculations presented by Gerjuoy and Stein [31]. A curve representing the sum of all rotational excitations is shown in Figure 1.10. Cross-sections for the excitation of individual vibrational levels, calculated by Hake and Phelps [32], are shown in Figure 1.11. The total cross-section for vibrational excitation is shown in Figure 1.10.

For electron energies above 1 eV it becomes possible to excite higher electronic levels. The cross-sections for some of these processes are illustrated in Figure 1.10. In each case the magnitude of the cross-section rises very rapidly near the threshold energy and attains a maximum value at several electron volts above the threshold.

Because of its electronegative nature, oxygen is able to form negative ions either by direct attachment or by dissociative attachment. The former process requires a three-body collision involving an electron and two oxygen molecules. Attachment by this means produces an O_2^- ion and the second molecule carries away the attachment energy. A curve for the potential energy

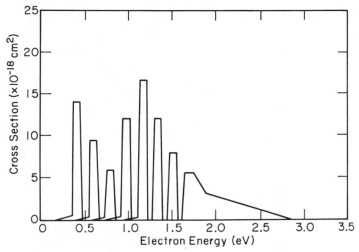

FIGURE 1.11. Vibrational excitation cross-sections for electrons in oxygen [32].

of O_2^- is illustrated in Figure 1.9. By contrast, dissociative attachment is a two-body process leading to the formation of an O^- ion and an oxygen atom. The cross-section for this process is shown in Figure 1.10. It should be noted that the cross-section exhibits two maxima, the first corresponding to the process

$$e + O_2 \rightarrow O^- + O$$

and the second to the process

$$e + O_2 \rightarrow O^- + O^+ + e$$

Reaction rate constants for the processes described above can be derived from the following relationship

$$k_i = \int_0^\infty \left(\frac{\varepsilon}{2m}\right)^{1/2} \sigma_i(\varepsilon) f(\varepsilon) \, d\varepsilon \tag{1.100}$$

where ε is the electron energy, σ_i the cross-section for the ith process, and $f(\varepsilon)$ the electron-energy distribution function. For a rigorous calculation of the rate constant k_i, the exact form of the electron-energy distribution function should be used. Since the determination of this function is difficult (see Section 1.2), it is desirable to determine to what degree a Maxwellian distribution might be used as an approximation.

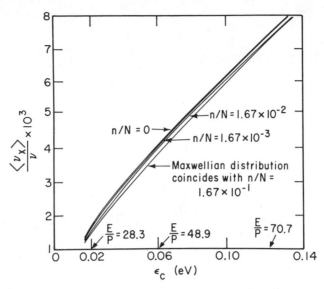

FIGURE 1.12. The ratio of the average inelastic collision frequency to the elastic collision frequency as a function of the energy parameter ε_c and the extent of ionization in a hydrogen plasma [2].

Dreicer [2] has discussed this problem for the case of a hydrogen plasma. In this work, solutions to the Boltzmann equation were obtained which took into account the effects of coulombic interactions between electrons and as well as elastic and inelastic interactions between electrons and molecular hydrogen. The electron-energy distribution functions (see Figs. 1.3–1.5) derived by this means were then used to determine the ratio of the average inelastic collision rate to the elastic collision rate as a function of the average energy transferred upon collision. These results are shown in Figure 1.12 for different values of the degree of ionization. Values of $\langle \nu_x \rangle / \nu$ computed for a Maxwellian distribution function are also shown in this figure. As can be seen, the values of $\langle \nu_x \rangle / \nu$ show very little dependence on either the degree of ionization or the form of the distribution function. The principal cause of these observations is that the true electron-energy distribution differs very little from a Maxwellian form over the range of energies of interest in evaluating $\langle \nu_x \rangle$.

By contrast, Dreicer's computations of the ratio of the average ionization rate to the elastic collision rate show a strong influence of both the degree of ionization and the form of the distribution function. These results are shown in Figure 1.13. In this instance differences in the values of $\langle \nu_i \rangle$ are the result of differences in the tail of the distribution function (see Fig. 1.4).

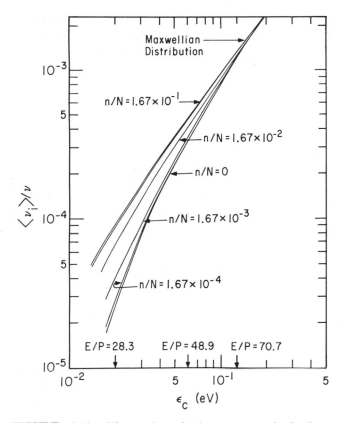

FIGURE 1.13. The ratio of the average ionization frequency to the elastic collision frequency as a function of the energy parameter ε_c and the extent of ionization in a hydrogen plasma [2].

On the basis of Dreicer's work we may conclude that the use of a Maxwellian distribution function to calculate rate constants is justified provided the threshold for the process being considered is not much larger than the average electron energy. Further substantiation of this conclusion can be derived from the work of Myers [33]. In this case, the distribution function was evaluated for an electron swarm in oxygen. Computations of the average dissociative attachment cross-section as a function of the average electron energy showed that very nearly the same values were obtained using the true distribution function and a Maxwellian distribution function.

The elastic and inelastic collisions which electrons undergo in the plasma lead to a loss of energy from the electron cloud. These energy losses are compensated by the gain of energy from the electric field. When the plasma reaches a steady state, the rate of energy gain from the field just balances the loss of energy due to electron–molecule collisions as well as other loss processes. An expression for the energy balance is given by

$$\bar{P} = \frac{2m}{M} \langle \varepsilon \rangle k_m nN + \sum_j \varepsilon_j k_j nN + \langle \varepsilon \rangle k_i nN + \langle \varepsilon \rangle k_d n \qquad (1.101)$$

where $\langle \varepsilon \rangle$ is the average electron energy, ε_j is the energy loss for the jth process, $k_m = \nu_m/N$ is the rate constant for momentum transfer, k_j is the rate constant for the jth inelastic process, k_i is the ionization rate constant, and $k_d = D_a/\Lambda^2$ is the effective diffusion rate constant. The first two terms in (1.101) account for the loss of energy by elastic and inelastic collisions, respectively. The third term accounts for the energy needed to bring each newly formed electron up to the average electron energy. The fourth term describes the loss of energy associated with ambipolar diffusion. The fraction of the total power which is dissipated in any one process can be evaluated by dividing the rate of energy loss for that process by the total power. As can be seen from (1.101), the fractional loss of energy is independent of n and only depends upon the magnitudes of k_m, k_j, and k_d. Since the rate constants for the collisional processes are functions of E_e/p, the fractional loss of energy may also be expressed in terms of this parameter.

Figure 1.14 illustrates the fractional energy losses for a variety of processes occurring in oxygen. At low values of E_e/p, the principal loss of energy occurs through the excitation of vibrational states. As E_e/p is increased, ever larger fractions of the total power are consumed by various electronic excitation processes. With the exception of ionization, the losses for these processes exhibit a maximum. This characteristic is due in part to the shape of the collision cross-sections (see Fig. 1.10) and in part to the very rapid increase in the losses associated with ionization. In contrast to the inelastic processes, the loss of energy through elastic collisions remains small throughout the range of E_e/p, which means that not more than about 1.5% of the total power is directed towards heating the gas.

The positive and negative ions produced by electron collisions with the molecules of the sustaining gas can react further to produce a large number of new ionic species. As an example, we may look at the reactions which occur in an oxygen discharge. The elementary steps involved in the formation and loss of all of the observed species are listed in Table 1.1. Most of the ionic reactions appearing in Table 1.1 are exothermic and hence proceed at a rate which is essentially independent of ion energy. Processes such as the charge transfer reactions 9, 10, 16, and 19, and the collisional detachment reactions 23 and 26 are endothermic and proceed only if the ion energy is sufficient to surpass the activation energy for the reaction. To determine the effective ion temperature as a function of the applied field, it is recommended

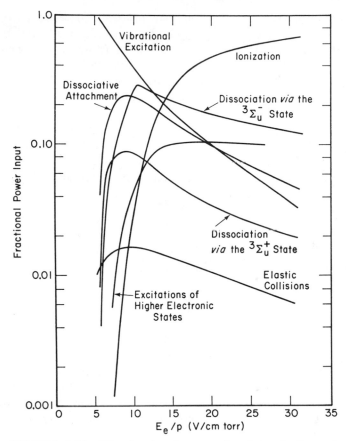

FIGURE 1.14. The fractional power input to elastic and inelastic collisions as a function of E_e/p for oxygen [34].

that Wannier's relationship

$$\tfrac{3}{2}kT_i = \tfrac{1}{2}(M_i + M)v_d^2 + \tfrac{3}{2}kT_g \tag{1.102}$$

be used. In (1.102) T_i is the ion temperature, M_i the mass of the ion, M the mass of the neutral gas molecule, and v_d the drift velocity of the ion, which is a function of E_e/p.

1.5 PHYSICAL CHARACTERISTICS OF HIGH-FREQUENCY DISCHARGES

Breakdown and Steady-State Operating Conditions

As was discussed in the previous sections, the value of E_e/p in a plasma plays an important role in determining the properties of the plasma and the rates at which chemical processes occurring in it proceed. Consequently, we wish to develop methods for predicting E_e/p both at breakdown and under steady-state operating conditions. Since the rates of all processes involving electron–molecule collisions are proportional to the electron density, it will be necessary to determine this quantity as well. To begin our discussion we shall consider the evaluations of E_e/p and n for a plasma in which the field strength is constant throughout the plasma.

The breakdown of a gas in an electric field will occur when the rate of ionization becomes sufficiently large to balance the loss of electrons by various processes. If the principal loss mechanism is considered to be free electron diffusion, then breakdown will be achieved when the condition

$$\nu_i = \frac{D_e}{\Lambda^2} \tag{1.103}$$

is satisfied.

Equation (1.103) can be used to determine the magnitude of the breakdown field. For this purpose it is first rewritten as

$$\frac{\nu_i}{p} = \frac{D_{e_1}}{(p\Lambda)^2} \tag{1.104}$$

where D_{e_1} is the free electron diffusivity evaluated at 1 torr. The quantity ν_i/p essentially represents the ionization constant $k_i = \nu_i(RT/p)$ and is thus a function of E_e/p [see (1.100) in Section 1.4]. As a result, we see that for a fixed value of $p\Lambda$ (1.104) will be satisfied by only one value of E_e/p. An illustration of the relationships between $p\Lambda$, ν_i/p and E_e/p is shown in Figure 1.15.

Figure 1.16 shows a curve of E_e/p versus $p\Lambda$ for the breakdown of hydrogen between two parallel plate electrodes. Notice that, at low values of $p\Lambda$, E_e/p is quite large due to the rapid loss of electrons under these conditions. As $p\Lambda$ increases, E_e/p decreases and tends to level out. These characteristics are due to the reduced loss of electrons by diffusion and the shape of the plot of ν_i/p versus E_e/p. It should be noted that the actual value of the applied field will vary with frequency since E_e is defined as $E_0[\nu_m^2/2(\nu_m^2 + \omega^2)]^{1/2}$. Thus when $\omega^2 \ll \nu_m^2$, $E_e = E_0/\sqrt{2}$ and when $\omega^2 \gg \nu_m^2$, $E_e \simeq E_0\nu_m/\sqrt{2}\omega$.

Under certain conditions, mechanisms other than electron diffusion may control the breakdown-field strength. Thus, we must define the limits within which diffusion-controlled breakdown is valid. In Figure 1.17 these limits are illustrated on a plot of $p\Lambda$ versus $p\lambda$ where λ is the wave length of the applied field. The uniform field line defines the limit at which the field can no longer be considered uniform across the gap separating the electrodes. The mean free-path line describes the limit at which the electron mean free path becomes

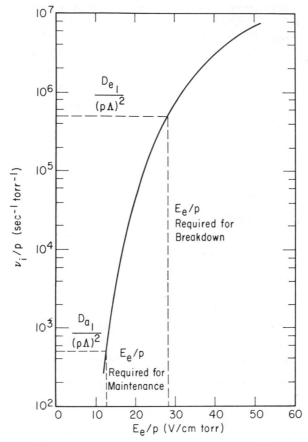

FIGURE 1.15. Illustration of the breakdown and maintenance values of E_e/p on a plot of the reduced ionization frequency versus E_e/p for hydrogen.

comparable to the interelectrode gap. Finally, the oscillation amplitude line describes the limit at which the excursion of an electron during one quarter of a cycle becomes comparable to one half of the interelectrode gap. Within the region defined by the three limiting lines, breakdown is controlled by diffusion.

When breakdown occurs in an electronegative gas, electrons can be lost through attachment to form negative ions in addition to being lost by diffusion. Under such conditions (1.104) must be defined as

$$\frac{\nu_i - \nu_a}{p} = \frac{D_{e_1}}{(p\Lambda)^2} \tag{1.105}$$

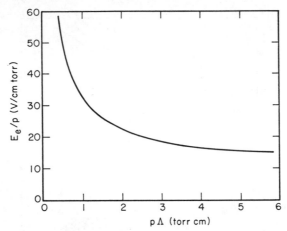

FIGURE 1.16.   The value of E_e/p as a function of $p\Lambda$ required for breakdown in hydrogen.

where ν_a is the electron attachment frequency. The breakdown-field strength can then be determined as explained above, provided information is available on ν_i/p and ν_a/p as functions of E_e/p.

As the electron density is increased from zero near breakdown to higher values, the effective electron diffusivity decreases until it reaches the ambipolar value. This pattern was shown in Figure 1.7. Due to the decrease in

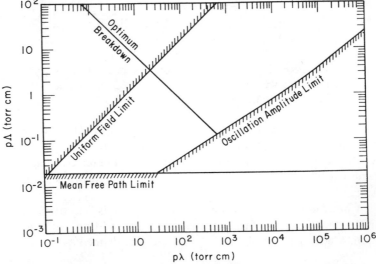

FIGURE 1.17.   A plot of the limits of the diffusion theory for high-frequency breakdown [35].

the diffusion coefficient, the value of ν_i/p needed to sustain a steady-state discharge will be reduced and will approach a lower limit given by

$$\frac{\nu_i}{p} = \frac{D_{a_1}}{(p\Lambda)^2} \tag{1.106}$$

when n_e exceeds 10^8 cm^{-3}. The decrease in ν_i/p in turn causes a decrease in the value of E_e/p as is shown in Figure 1.15 and represents the final consequence of the increase in electron density.

Rose and Brown [36] have demonstrated that the effects of $p\Lambda$ and n on the steady-state value of E_e/p can be summarized on a single plot. Figure 1.18 illustrates their results for a hydrogen discharge maintained between two parallel plate electrodes. The electron density at the midplane is given by n_0. As we can see, for a fixed value of $p\Lambda$, E_e/p decreases from its value at breakdown, corresponding to the point on the curve marked $n_0\Lambda^2 = 0$, to a lower limit achieved when the electron density is very large. This limiting value of E_e/p occurs when $D_s = D_a$. The effects of a change in $p\Lambda$ are also readily observed in Figure 1.18. As the value of $p\Lambda$ is increased, E_e/p is decreased for a given value of n_{e_0}. However, when $n_0\Lambda^2 \gtrsim 10^9$ cm^{-3}, E_e/p becomes independent of the electron density and is then a function of $p\Lambda$ alone.

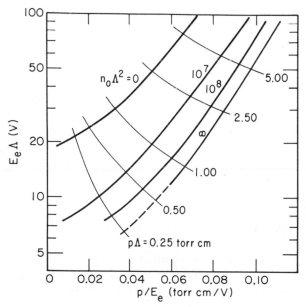

FIGURE 1.18.   Theoretical calculations of E_e/p as a function of $p\Lambda$ and $n_0\Lambda^2$ for hydrogen [36].

The electron density in the high-frequency glow discharges used in most plasma-chemistry studies lie in the range of 10^{10}–10^{12} cm^{-3}. These levels are sufficiently high to make E_e/p independent of the electron density as shown in Figure 1.18.

When the value of E_e/p becomes independent of the electron density, the electron density can be evaluated directly from the expression for the power density

$$\bar{P} = \frac{ne^2E_e^2}{mv_m} \tag{1.107}$$

Equation (1.107) can be rewritten in the following form

$$\frac{n}{\bar{P}\Lambda} = \frac{mv_{m1}}{[e^2p\Lambda(E_e/p)^2]} \tag{1.108}$$

where v_{m_1} is the value of v_m at 1 torr. Since the right-hand side of (1.108) is a function of $p\Lambda$ alone, $n/\bar{P}\Lambda$ can be plotted as a function of $p\Lambda$. Figure 1.19 shows a plot of both E_e/p and $n/\bar{P}\Lambda$ versus $p\Lambda$ for a discharge in oxygen.

From Figure 1.19 we can conclude that, given the discharge dimensions and hence Λ, the gas pressure p, and the power density $\bar{P}$, we can determine E_e/p and n. Since E_e/p defines the electron-energy distribution function, a complete description of the physical properties of the discharge is now possible.

Electric Field and Electron-Density Distributions

The discussion of maintenance fields up to this point has considered the field to be uniform throughout the plasma. This condition is satisfied for electric fields applied in a direction perpendicular to the gradient in electron density or for very high-frequency fields regardless of their orientation. Thus, it is appropriate to discuss the distribution of electric-field strength for intermediate driving frequencies and situations where the field is parallel to the gradient in electron density.

Let us consider a discharge sustained between two infinite parallel plate electrodes placed on the outside of a dielectric container, as shown in Figure 1.20. For this geometry, the width of the discharge zone is the only important linear dimension.

The analysis of the spatial distribution of the electric field can be initiated by recognizing that the sum of conduction and displacement currents must remain constant for the geometry considered. This may be expressed as

$$J = \sigma E + \varepsilon_0 \frac{\partial E}{\partial t} \tag{1.109}$$

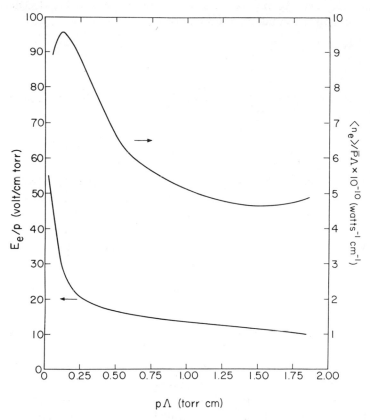

FIGURE 1.19. The values of E_e/p and $n_e/\bar{P}\Lambda$ as a function of $p\Lambda$ for a steady-state discharge in oxygen.

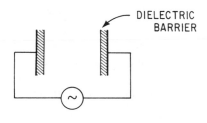

HF VOLTAGE SOURCE

FIGURE 1.20. Schematic of a high-frequency discharge.

in which J is taken to be constant. For a purely sinusoidal electric field, $E = |E| e^{i\omega t}$ and the conductivity σ can be expressed as

$$\sigma = \frac{ne^2(v_m - i\omega)}{m(v_m^2 + \omega^2)} \tag{1.110}$$

Substituting the expressions for the electric field and the conductivity into (1.109), we obtain

$$J = \left\{ \frac{ne^2}{m}\left(\frac{v_m}{v_m^2 + \omega^2}\right) + i\omega\varepsilon_0\left[1 - \frac{ne^2}{m\varepsilon_0}\left(\frac{1}{v_m^2 + \omega^2}\right)\right]\right\}E \tag{1.111}$$

Equation (1.105) can be put in a more convenient form by rewriting it in terms of $r = ne^2/m\varepsilon_0\omega^2$, a dimensionless electron density, and $\alpha = v_m/\omega$, a dimensionless collision frequency. Thus,

$$J = \left\{ \varepsilon_0\omega r\left(\frac{\alpha}{1 + \alpha^2}\right) + i\varepsilon_0\omega\left[1 - r\left(\frac{1}{1 + \alpha^2}\right)\right]\right\}E \tag{1.112}$$

Taking the magnitude of (1.112), we arrive at the desired expression

$$|E| = \frac{|J|}{\varepsilon_0\omega}\left[\frac{(1 + \alpha^2)}{(r - 1)^2 + \alpha^2}\right]^{1/2} \tag{1.113}$$

Equation (1.113) allows us to calculate the magnitude of the field at any position, provided the total current density and the local electron density are known. For conditions in which $r \ll 1$, (1.113) reduces to $|E| = |J|/\varepsilon_0\omega$ which represents the field across a parallel plate capacitor. At lower frequencies or higher electron densities, $r \gg 1$. In these cases (1.113) can be conveniently rewritten as

$$E = E_0\left[\frac{(r_0 - 1)^2 + \alpha^2}{(r - 1)^2 + \alpha^2}\right]^{1/2} \tag{1.114}$$

where the subscript zero denotes conditions at the center of the discharge. It should be noted that from this point on, E refers to the magnitude of the electric field.

In order to make use of (1.114), we must first determine the electron-density distribution. This may be accomplished by solving the diffusion equation

$$D_a\frac{d^2n}{dx^2} + nv_i = 0 \tag{1.115}$$

Since the ionization frequency v_i is a function of E_e/p, the value of v_i will vary with position as long as E_e is not constant across the gap. Thus to

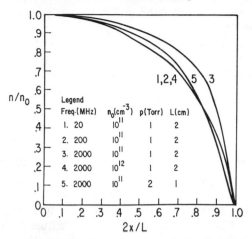

FIGURE 1.21. The spatial distribution of electron density in a helium discharge. Operating conditions for each curve are given in Table 1.2.

determine $n(x)$ we must consider (1.114) and (1.115) together with experimental data on ν_i/p versus E_e/p.

Solutions for the distributions of electron density and E_e/p in a helium discharge have been obtained by Bell [37]. These results are illustrated in Figures 1.21 and 1.22. The conditions for each of the curves are given in Table 1.2. For a midplane electron density of 10^{11} cm^{-3} it can be observed that

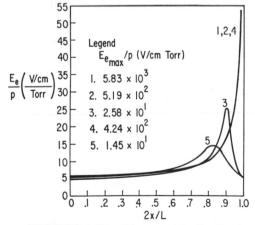

FIGURE 1.22. The spatial distribution of electric field strength in a helium discharge. Operating conditions for each curve are given in Table 1.2.

TABLE 1.2. Operating Conditions for Results Shown in Figs. 1.21, 1.22, and 1.23

f (MHz)	p (torr)	L (cm)	n_0 (cm^{-3})	r	α
20	1	2	10^{11}	2.01×10^4	20.30
200	1	2	10^{11}	2.01×10^2	2.03
2000	1	2	10^{11}	2.01	0.20
2000	1	2	10^{12}	2.01×10^1	0.20
2000	2	1	10^{11}	2.01	0.40

the curves for n^* and E_e/p at 20 and 200 MHz are identical. A noticeable change is produced, however, when the driving frequency is increased to 2000 MHz. Increasing the midplane electron density to 10^{12} cm^{-3} causes the solutions to coincide again with those obtained at the lower frequencies.

The observed behavior can be interpreted in terms of (1.114) which governs the magnitude of E_e as a function of electron density. When $r \gg \alpha$, then

$$E_e \simeq E_{eo}\left(\frac{n_0}{n}\right) \tag{1.116}$$

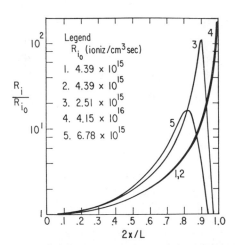

FIGURE 1.23. The spatial distribution of ionization rate in a helium discharge. Operating conditions for each of the curves is given in Table 1.2.

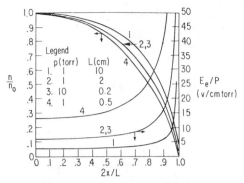

FIGURE 1.24. The spatial distribution of electron density and E_e/p as a function of pressure and gap length in a helium discharge ($r \gg \alpha$).

except for a very narrow region near $x^* = 1$, where $(r - 1)$ approaches the magnitude of α. Where (1.116) is valid, E_e is insensitive to both n_0 and ω. By contrast, when $r \gtrsim \alpha$, both variables affect the field distribution. Since the local value of E_e determines the local value of ν_i, the relative magnitudes of r and α will also influence n^*.

The effects of operating conditions on the distribution of rates of reaction is illustrated in Figure 1.23. The normalized rate of ionization is plotted as a function of position. As can be seen, the distributions of ionization rates for 20 and 200 MHz are identical. For the same value of pL, the distribution at 2000 MHz is different and is, furthermore, sensitive to variations in p and L, even though their product is held constant. When the midplane electron density is raised to 10^{12} cm^{-3}, the distribution of rates for 2000 MHz becomes identical to that observed at the lower frequencies.

Figure 1.24 illustrates the effects of varying p and L under the restriction $r \gg \alpha$. Both the electron density and the electric field distributions are seen to change when either p or L is varied. However, when the product pL is maintained constant, the two distributions are not affected by the individual values of p and L.

1.6. PRODUCTION OF ATOMIC OXYGEN IN A MICROWAVE DISCHARGE

The material discussed in the previous sections can now be used to develop a model for the overall kinetics of a reaction occurring in a high-frequency discharge. As an example we shall consider the dissociation of oxygen in a microwave discharge. This system is chosen because of the availability of

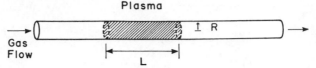

FIGURE 1.25. Schematic of the geometry used for the model of an oxygen discharge.

the necessary physical data and experimental measurements against which the predictions of the model can be tested.

For the model we shall consider the discharge to have the form of a cylindrical plug as shown in Figure 1.25. Within the discharge the electron density and electric field are taken to be constant at the value of the volumetric averages of these quantities. Since we are considering an open system, the gas pressure throughout the discharge tube will be uniform. Consequently, the gas density within the discharge will differ from that elsewhere only if there is a difference in gas temperature.

Dissociation of molecular oxygen can occur by one of two processes involving excitation from the ground state to either the $A^3\Sigma_u^+$ or the $B^3\Sigma_u^-$ excited state by electron collision. These reactions can be expressed as

(1a) $e + O_2 \rightarrow O_2^*(A^3\Sigma_u^+) \rightarrow O(^3P) + O(^3P) + e$

(1b) $e + O_2 \rightarrow O_2^*(B^3\Sigma_u^-) \rightarrow O(^3P) + O(^1D) + e$

The production of atomic oxygen by dissociative attachment

(1c) $e + O_2 \rightarrow O^- + O$

can be neglected since the reverse of reaction (4) is very rapid ($k = 3.0 \times 10^{-10}$ cm³/sec) and would serve to remove atomic oxygen as quickly as it was formed by reaction (1c). The experimentally determined cross-sections for reactions (1a–1c) have already been illustrated in Section 1.4 (see Fig. 1.10).

In the present model the rate of oxygen dissociation is characterized by the sum of reactions (1a) and (1b). A rate constant can then be determined from the expression

$$k_1 = \sqrt{\frac{8}{\pi m}} (kT_e)^{-3/2} \int_0^\infty \varepsilon \sigma_1(\varepsilon) e^{-\varepsilon/kT_e}\, d\varepsilon \qquad (1.117)$$

where T_e is the electron temperature, m is the mass of the electron, ε is the electron energy, and σ_1 is the total dissociation cross-section. In formulating

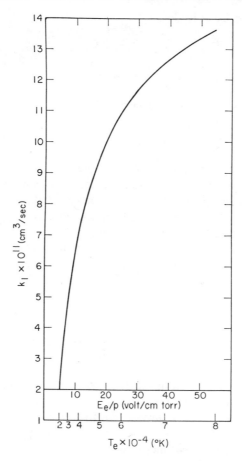

FIGURE 1.26. The rate constant for oxygen dissociation as a function of T_e and E_e/p.

(1.117), it has been assumed that the electron-energy distribution function is Maxwellian. A discussion of the suitability of this form of distribution function has been given in Section 1.4. The value of k_1 computed from (1.117) is shown in Figure 1.26 as a function of T_e. The scale along the abscissa has also been represented in terms of E_e/p. In making this transformation, Myers' [33] computations of the average electron energy versus E_e/p have been used together with the relationship $3/2 \, kT_e = \langle \varepsilon \rangle$.

Loss of atomic oxygen within the discharge proceeds by both homogeneous and heterogeneous processes. Three separate paths for homogeneous recombination can be considered:

$$(2) \qquad 2\,O + O_2 \rightarrow 2\,O_2$$

$$(3) \qquad 3\,O \rightarrow O + O_2$$

$$(4) \qquad O + 2\,O_2 \rightarrow O_3 + O_2$$

$$(5) \qquad O + O_3 \rightarrow 2\,O_2$$

Rate constants for these reactions are listed in Table 1.1. Heterogeneous recombination of atomic oxygen takes place on the walls of the discharge tube, and the efficiency of this process is characterized by the wall recombination coefficient γ. An experimentally determined value of γ for oxygen recombining on silica is also given in Table 1.1.

If we now consider a differential volume of the reactor, a species balance on atomic oxygen can be expressed as

$$\frac{4FN}{(2N - n_1)^2}\frac{dn_1}{dV} = 2k_1 \langle n_e \rangle (N - N_1)$$

$$- \frac{1}{2R} n_1 v_r \gamma - 2k_2 n_1^2 (N - n_1) - 2k_3 n_1^3 - 2k_4 n_1 (N - n_1)^2 \quad (1.118)$$

where n_1 is the concentration of atomic oxygen, N is the total gas concentration, $\langle n_e \rangle$ is the volume-averaged electron density, F is the flow rate of molecular oxygen fed to the reaction, V is the reactor volume, R is the radius of the discharge tube, and v_r is the random velocity of oxygen atoms. In formulating (1.125), a one-dimensional plug flow model of the reactor has been assumed. Radial diffusion has been neglected on the basis of a previous analysis [38] of a similar problem which showed that the rate of diffusion of atomic species in a discharge is sufficiently fast to eliminate transverse concentration gradients. Axial diffusion has also been neglected since the ratio $k_1 \langle n_e \rangle D_{12}/(F/N\pi R^2)^2$ is less than 1 [39] and hence the axial gradients will not be sufficiently high to cause significant axial dispersion.

In order to evaluate the term in (1.118) describing the rate of dissociation, we must first evaluate k_1 and $\langle n_e \rangle$ for a chosen set of operating conditions. If we describe the operating conditions by the gas pressure p, the diffusion length Λ, and the power density $\bar{P}$, then we have seen that the curves given in Figure 1.19 of Section 1.5 will define the corresponding values of E_e/p and n_e. The appropriate value of k_1 may be determined from Figure 1.26.

Equation (1.118) can now be solved by numerical integration in order to determine n_1 for a given set of experimental conditions. The final results are

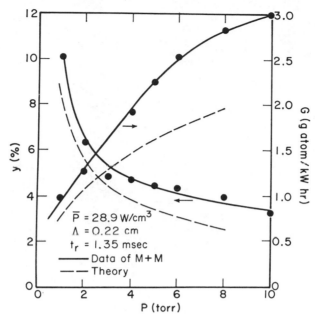

FIGURE 1.27. Conversion and yield versus pressure for reactor A.

then expressed in terms of the conversion y and the yield G defined as

$$y = \frac{n_1}{2N - n_1}$$

$$G = \frac{7.2 \times 10^6 yF/V}{\bar{P}}$$

(1.119)

A comparison between the predicted values of the conversion and yield and the experimental data of Mearns and Morris [40] is given in Figures 1.27–1.32. In examining these figures it should be noted that two reactors were used in the experimental work. These are referred to as reactor A and reactor B. The type of microwave cavity used with each reactor, the dimensions of the discharge tube, and the approximate discharge volume are listed in Table 1.3.

In Figures 1.27–1.32 the experimental data represent the conversion and yield at the exit from the discharge. Since a direct measurement of these quantities could not be made, Mearns and Morris determined them by back-extrapolating from measurements made downstream from the discharge

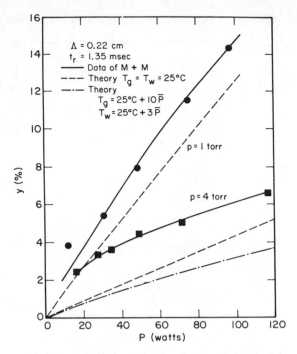

FIGURE 1.28. Conversion versus power for reactor A.

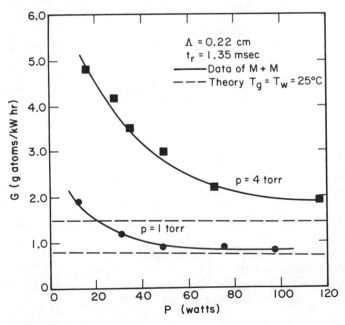

FIGURE 1.29. Yield versus power for reactor A.

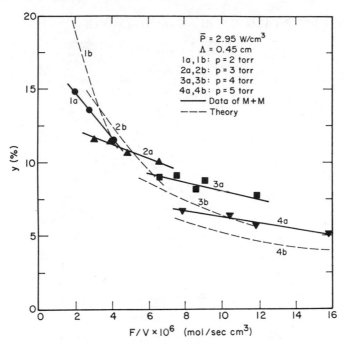

FIGURE 1.30. Conversion versus F/V for reactor B.

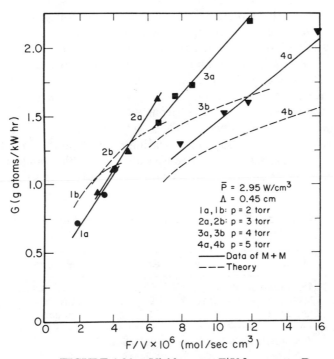

FIGURE 1.31. Yield versus F/V for reactor B.

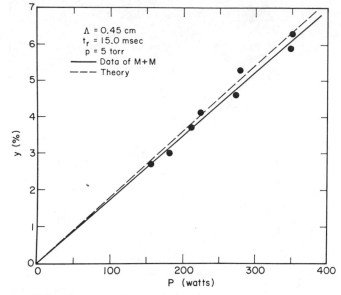

FIGURE 1.32.   Conversion versus power for reactor B.

itself. As a result, the experimental points may contain inaccuracies arising from the method and rate constants used to perform the extrapolations. However, it is not possible to evaluate the significance of these inaccuracies.

The response of conversion and yield to increasing pressure is illustrated in Figure 1.27. The experimental results shown there are for reactor A and were obtained for a constant power of 63 W and a residence time of 1.35 msec. The computed curves derived from the model are seen to follow the trends of the data but lie between 20 and 30% lower than the data.

The decrease in conversion with increasing pressure shown in Figure 1.27 results from the superposition of three separate effects which can be elucidated through an examination of (1.118). For this purpose (1.118) is first rewritten

TABLE 1.3.   Dimensions of the Reactors Used by Mearns and Morris [40]

Reactor characteristic	Reactor A	Reactor B
Microwave cavity	Coaxial	Cylindrical
Discharge tube[a] radius, R	0.53 cm	1.09 cm
Discharge volume	2–3 cm^3	70 cm^3

[a] Both discharge tubes were made of silica.

in terms of atomic mole fraction $x = n_1/N$ as

$$\frac{4F}{N(2-x)^2}\frac{dx}{dV} = 2k_1\langle n_e\rangle(1-x)$$

$$-\frac{1}{2R}v_r\gamma x - 2k_2N^2x^2(1-x) - 2k_3N^2x^3 - 2k_4N^2x(1-x)^2 \quad (1.120)$$

From Figures 1.19 and 1.26 it can be observed that both k_1 and $\langle n_e\rangle$ decrease as the pressure is increased. This has the effect of reducing the dissociation rate term on the right-hand side of (1.120). The second effect of increased pressure is to cause an increase in the rate of homogeneous recombination of atomic oxygen. Finally one must take into account the increase in F which is required in order to maintain a constant residence time when the pressure is increased. This change produces a reduction in the conversion by enhancing the convective loss of atomic oxygen.

The effects of power for a fixed pressure and residence time are shown in Figures 1.28 and 1.29. Here again the experimental data are for reactor A. In this case the model predicts a linear rise in conversion with increased power. The experimental data for 1 torr are essentially consistent with this trend but for 4 torr show a less rapid rise in conversion with power. For both pressures the computed conversions are less than the experimental, the maximum deviation being 14% at 1 torr and 54% at 4 torr. The linear relationship between conversion and power predicted by the model is the result of the proportionality between the electron density and power indicated in Figure 1.19. Since the overall conversion for both pressures is relatively low, nonlinear effects due to homogeneous recombination cannot be observed.

The calculations shown in Figures 1.28 and 1.29 were performed under the assumption that the temperature of the gas and the discharge tube wall remained at 25°C independent of the power dissipated. In order to explore the effects of variations in these temperatures, calculations were performed under the assumption that the gas temperature increased at the rate of $10°C\,cm^3/W$ and the wall temperature at the rate of $3°C\,cm^3/W$. An examination of (1.119) or (1.120) shows that, for a fixed gas pressure, five quantities will respond to a variation in the gas temperature, N, k_1, $\langle n_e\rangle$, v_r, and k_4. The reduction of gas density with increasing temperature leads to a decrease in the value of $p\Lambda$ which in turn causes k_1 and $\langle n_e\rangle$ to increase. The increase in the random velocity enhances the rate of heterogeneous recombination by increasing the flux of atomic oxygen to the wall of the discharge tube. The increase in the discharge tube wall temperature affects only γ which for a silica tube has been shown to vary from 1.6×10^{-4} at 20°C to 1.4×10^{-2} at 600°C [30].

Figure 1.29 illustrates the predicted values of the conversion for 4 torr when both the gas and discharge tube wall temperatures are taken to increase with power. As can be seen, the conversions computed for the elevated temperatures lie below those computed for $T_g = T_w = 25°C$. These results can be explained as follows. With the increase in gas temperature, both k_1 and $\langle n_e \rangle$ increase; however, the product $k_1 \langle n_e \rangle N$ decreases due to the inverse dependence of N on the gas temperature. On the other hand, the heterogeneous rate of recombination increases with power due to the increases in both v_r and γ. This latter effect is only partially offset by the reduction in the rate of reaction (4) due to the decrease in k_4. Thus, for a given power, the effect of assuming an increase in T_g and T_w is to cause the rate of dissociation to decrease and rate of recombination to increase, both of which lead to a reduction in the conversion.

The effects of pressure and flow rate on the conversion and yield obtained from reactor B are illustrated in Figures 1.30 and 1.31. For each pressure the curve of predicted conversion intersects the curve drawn through the data and exhibits a more rapid decrease with flow rate than shown by the data. The maximum deviations between the calculated and measured conversions occur at 2 torr where they reach a level of 22%. As the pressure is increased, the slopes of the pairs of curves approach each other more and more closely. Similar patterns can be observed in the curves representing yield versus F/V. The most likely source of error in these computations is an overestimation of k_1 at the lower pressures. The assumption of a smaller value of k_1 would lead to a more gradual decline in conversion with flow rate and would improve the fit to the data.

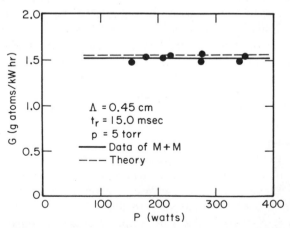

FIGURE 1.33.   Conversion versus power for reactor B.

The effect of power on the conversion and yield obtained from reactor B are shown in Figures 1.32 and 1.33. Both the experimental and the computed conversions illustrate a linear dependence on power. The nearly perfect agreement between theory and experiment shown in Figures 1.32 and 1.33 may be misleading since the effects of increased gas and discharge tube wall temperature have not been taken into account. It should be noted, however, that the introduction of these effects should not produce significant variations in the results because the power densities corresponding to the conditions in Figures 1.32 and 1.33 are quite low, e.g., 0–6 W/cm^3.

REFERENCES

[1] W. P. ALLIS, "Motion of Ions and Electrons," *in Handbuch der Physik*, Vol. 21, Springer-Verlag, Berlin, 1956, p. 383.

[2] H. DREICER, *Phys. Rev.* **117**, 343 (1960).

[3] H. MARGENAU, *Phys. Rev.* **73**, 309 (1948).

[4] A. VON ENGEL, *Ionized Gases*, Oxford Press, New York, 1955.

[5] E. W. MCDANIEL, *Collision Phenomena in Ionized Gases*, Wiley, New York, 1964.

[6] H. S. W. MASSEY, E. H. S. BURHOP, and H. B. GILBODY, *Electronic and Ionic Impact Phenomena*, Oxford Press, New York, 1971.

[7] W. P. ALLIS and D. J. ROSE, *Phys. Rev.* **93**, 84 (1954).

[8] E. W. MCDANIEL, V. CERMAK, A. DALGARNO, E. E. FERGUSON, and L. FRIEDMAN, *Ion-Molecule Reactions*, Wiley-Interscience, New York, 1970.

[9] D. RAPP and P. ENGLANDER-GOLDEN, *J. Chem. Phys.* **43**, 1464 (1965).

[10] E. W. ROTHE, L. L. MARINO, R. H. NEYNABER, and S. M. TRUJILLO, *Phys. Rev.* **125**, 582 (1962).

[11] D. RAPP, P. ENGLANDER-GOLDEN, and D. D. BRIGLIA, *J. Chem. Phys.* **42**, 4081 (1965).

[12] D. RAPP and D. D. BRIGLIA, *J. Chem. Phys.* **43**, 1480 (1965).

[13] H. MYERS, *J. Phys. B* **2**, 393 (1969).

[14] F. C. FEHSENFELD, E. E. FERGUSON, and A. L. SCHMELTEKOPF, *Planet. Space Sci.* **13**, 579 (1965).

[15] R. F. STEBBINGS, A. C. H. SMITH, and H. B. GILBODY, *Phys. Rev.* **38**, 2280 (1963).

[16] J. J. LEVENTHAL and L. FRIEDMAN, *J. Chem. Phys.* **46**, 997 (1967).

[17] D. A. DURDEN, P. KEBARLE, and A. GOOD, *J. Chem. Phys.* **50**, 805 (1969).

[18] R. M. SNUGGS, D. J. VOLZ, J. H. SCHUMMERS, R. D. LASER, I. R. GATLAND, D. W. MARTIN, and E. W. MCDANIEL, Technical Report, School of Physics, Georgia Institute of Technology, Atlanta, Ga. March 29, 1970.

[19] E. E. FERGUSON, F. C. FEHSENFELD, and A. L. SCHMELTEKOPF, *Adv. Chem. Ser.* **80,** 83 (1969).

[20] F. C. FEHSENFELD, E. E. FERGUSON, and A. L. SCHMELTEKOPF, *J. Chem. Phys.* **45,** 1844 (1966).

[21] B. R. TURNER, D. M. J. COMPTON, and J. W. McGOWAN (eds.), General Atomics Report G.A. 7419 (1966).

[22] L. G. McKNIGHT and J. M. SAWINA, DASA Conference, Palo Alto, Ca., (June 1969).

[23] F. C. FEHSENFELD, E. E. FERGUSON, and D. K. BOHME, *Planet. Space Sci.* **17,** 1759 (1970).

[24] T. L. BAILEY and P. MAHADEVAN, *J. Chem. Phys.* **52,** 179 (1969).

[25] F. C. FEHSENFELD, D. L. ALBRITTON, J. A. BURT, and H. I. SCHIFF, *Can. J. Chem.* **47,** 1793 (1969).

[26] S. C. BROWN, *Basic Data of Plasma Physics, 1966*, M.I.T. Press, Cambridge, Mass., 1967, p. 213.

[27] W. H. ABERTH and J. R. PETERSON, *Phys. Rev.* **1,** 158 (1970); R. E. OLSON, J. R. PETERSON, and J. MOSELEY, *J. Chem. Phys.* **53,** 3391 (1970).

[28] K. SCHOFIELD, *Planet. Space Sci.* **15,** 643 (1967).

[29] H. S. JOHNSTON, "Gas Phase Reaction Kinetics of Neutral Oxygen Species," *Nat. Stand. Ref. Data Ser., Nat. Bur. Stand. NSRDS-NBS* 20 (1968).

[30] J. C. GREAVES and J. W. LINNETT, *Trans. Farad. Soc.* **55,** 1355 (1959).

[31] E. GERJUOY and S. STEIN, *Phys. Rev.* **97,** 1671 (1955).

[32] R. D. HAKE and A. V. PHELPS, *Phys. Rev.* **158,** 70 (1967).

[33] H. MYERS, *J. Phys. B* **2,** 393 (1969).

[34] H. SABADIL, *Beitr. Plasma Phy.* 53 (1971).

[35] S. C. BROWN, *Introduction to Electrical Discharges in Gases*, Wiley, New York, 1966.

[36] D. J. ROSE and S. C. BROWN, *Phys. Rev.* **85,** 310 (1955).

[37] A. T. BELL, *I&EC Fund. Quart.* **9,** 160, 679 (1970).

[38] A. T. BELL, *I&EC Fund. Quart.* **11,** 209 (1972).

[39] E. E. PETERSEN, *Chemical Reaction Analysis*, Prentice–Hall, Englewood Cliffs, N.J., 1965.

[40] A. M. MEARNS and A. J. MORRIS, *CEP Symp. Series* **112,** 37 (1971).

Chapter 2

Applications of Nonequilibrium Plasmas to Organic Chemistry

Harald Suhr

2.1. INTRODUCTION

Chemists have always been fascinated by the various types of electrical discharges which they have observed in nature as well as physical studies. As soon as techniques became available for producing discharges in the laboratory, chemists attempted to use them for chemical syntheses. The first experiments of this kind were reported as early as 1796, when four Dutch chemists [1] subjected ethylene to spark discharges and obtained an oily substance. More extensive investigations were not undertaken, however, until the middle of the last century when "silent" discharges became available [2]. The successful preparation of ozone in these discharges stimulated a number of experiments with organic compounds [3–6]. With time other types of discharges became available and were explored as well. Particular attention was focused on high-voltage glow discharges after the discovery that they could be used to prepare atomic hydrogen [7,8], oxygen [9–12], and nitrogen [13]. Most recently the availability of radio-frequency and microwave generators has focused attention on the use of electrodeless discharges.

The plasmas produced by electrical discharges can be divided into two types. The first is the "hot" or "equilibrium" plasma, which is characterized by a high gas temperature and an approximate equality between the gas and electron temperatures. Typical examples of such plasmas are those produced in arcs and plasma torches. The second type of plasma is termed the "cold" or "nonequilibrium" plasma and is characterized by a low gas temperature and a high electron temperature. The plasmas produced in silent discharges and various types of glow discharges are examples of this type of plasma.

Hot plasmas have been used for a number of inorganic reactions. Their use in organic chemistry is limited to special cases due to the low thermal stability of most organic substances. Those applications which are feasible have been known for a long time and recent studies have concentrated on technological improvements. Since this work has been reviewed recently [14–18], it will not be covered in this article. Cold plasmas have also been used predominantly in inorganic chemistry and metallurgy [19], the most important industrial application being the generation of ozone. Experiments with organic substances have shown that numerous complicated reactions can occur, but not until very recently has it been possible to carry out such plasma reactions selectively and with high yields. In the following sections a review is given of these recent developments, and their possible significance to preparative organic chemistry is discussed.

2.2.   EXPERIMENTAL TECHNIQUES

From the viewpoint of a chemist, a plasma provides a new way of transfering energy to molecules. For many reactions it is of little importance whether the energy necessary for a certain process is transferred through collision, light absorption, or electron impact. Of even lesser importance in plasma chemistry

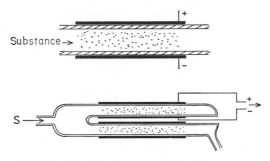

FIGURE 2.1. Low-current discharges.

is the type of discharge equipment used, provided nearly identical electron energies and gas temperatures can be obtained. At present, there is no indication that certain discharge types or frequencies favor specific reactions. For this reason, the criteria for the selection of discharge equipment are based on practical considerations such as stable performance, easy handling, simplicity, size, and cost.

Four different techniques have been used to generate plasmas suitable for organic reactions: silent discharges, direct-current and low-frequency glow discharges, high-frequency, and microwave discharges. Each of these devices shall be treated separately in the following sections.

Silent Discharges

The silent discharge, typified by that observed in an ozonizer, is simple to produce and can operate at pressures up to 1 atmosphere. An example of the apparatus required to sustain such a discharge is shown in Figure 2.1. The discharge tube is usually made of glass or quartz. The inner and outer surfaces of the coaxial sections are connected to a high-voltage power supply (>1000 kV) through metal electrodes or conducting liquids. Since it is possible to sustain a silent discharge at the line frequency (50 or 60 Hz), the power supply can be as simple as a step-up transformer. Higher frequencies can also be used and offer the advantage of higher power dissipation at lower voltages. The principal disadvantages of silent discharges are the small gaps and large surface areas required. Organic reactants frequently form polymers which can bridge the gap or accumulate on the vessel walls. Such occurrences can cause problems with the cleaning of the apparatus.

Direct-Current and Low-Frequency Glow Discharges

Another simple device for carrying out plasma reactions is the high-voltage glow discharge. In typical arrangements, two metal electrodes are placed inside a reactor and connected to a variable high voltage source (see Fig. 2.2). The required voltage depends on the pressure, the nature of the gas, and the distance between the electrodes. At pressures of about 1 torr, voltages of 10–100 V/cm are required.

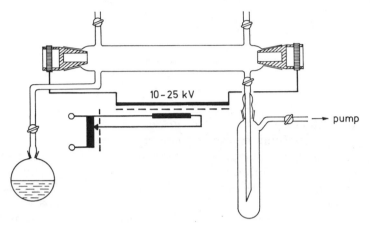

FIGURE 2.2.   High-voltage discharge.

The current is limited by the resistance in the primary circuit and by the surface area of the electrodes. At low currents only part of the surface is covered by a luminous glow. If the current is raised until the electrodes are covered, only a drastic increase in the voltage can achieve a further increase in current. Frequently hollow electrodes are used which permit the passage of higher currents [20]. A particularly convenient means of constructing electrodes is to fashion them to fit into the ground joints of standard glassware. For power levels below 50 W, no special cooling of the electrodes is necessary.

When such equipment is used for organic reactions, the electrodes are quickly covered with polymeric material. To prevent this contamination, it is necessary to isolate the electrodes from the organic gases by a shroud of rare gas. This can be accomplished by passing the rare gas through the electrodes or by introducing it in their vicinity [21,22].

Radio-Frequency Discharges

At frequencies above 1 MHz, direct contact between the electrodes and the plasma is no longer necessary. The energy can be fed to the plasma indirectly by capacitive or inductive coupling (see Fig. 2.3). Normally, an inductive coupling includes a small percentage of capacitive coupling and vice versa. In the case of capacitive coupling, the electrodes enclose the plasma tube.

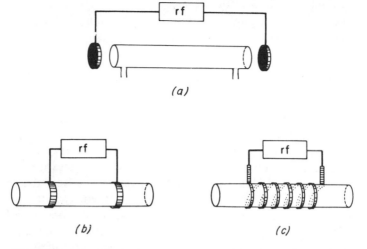

FIGURE 2.3. Typical arrangements for electrodeless high-frequency discharges; (*a*) plates outside the reactor, (*b*) rings, and (*c*) a coil around the reactor.

For inductive coupling, the tube lies on the axis of the coil. The coupling components could, in principle, also be placed inside the reactor, but this would cause side effects due to the interaction of the vapors with the metal surfaces.

If higher and higher working frequencies are chosen, the rf circuit elements become very small in both dimension and value. Since it is not convenient to work with fractions of a coil winding or coupling capacitances on the order of the parasitic capacitance, set-ups such as that shown in Figure 2.3 are limited to frequencies below 200 MHz. A wealth of information on rf transmitters in the range of $1-10^4$ MHz can be found in the literature [23–26]. Since there is no frequency dependence of plasma reactions, the choice of operating frequency is dictated chiefly by aspects of handling or shielding. If the experiments are carried out in a Faraday cage, any frequency may be used. For work in normal unshielded laboratories, care must be taken not to interfere with radio or other equipment. Only certain frequency bands, e.g., the industrial frequencies, or radio-control frequencies, are permitted for scientific experiments, provided the user takes care to maintain low-radiation levels of the fundamental frequency and its harmonics.

For plasma experiments, generators with fixed frequency and adjustable power are preferred. In addition to the generator, it is necessary to have a matching network which transforms the impedance of the generator to that of the load. The proper matching of impedances is necessary to obtain an efficient coupling of power to the discharge. For laboratory purposes, transmitters with power outputs up to 300 W are usually sufficient. Both single- and multiple-stage generators are suitable for this purpose.

Microwave Discharges

Microwave generators have frequently been used in plasma experiments. Since equipment suitable for this purpose has been reviewed recently [27], it will be discussed here only briefly.

Microwave generators with outputs between a few watts and a kilowatt are commercially available and can be used for plasma experiments. The microwave power is led by a coaxial cable or wave guide from the generator (magnetron or klystron) to a resonant cavity which encloses the reactor. A number of resonant cavities have been described in the literature [28–30]. In the simplest case, the cavity is a rectangular section of wave guide terminated at one end by an adjustable short. The reaction tube is positioned at a point about one-quarter wavelength from the shorted end of the cavity. By adjusting the short, the maximum of the electrical field strength is placed at the center of the discharge tube. If the cavity and the generator are not properly matched, a large fraction of the incident power is reflected back towards the generator where it may cause damage. To protect the magnetron or klystron, a directional coupler is used or an arrangement in which the reaction tube is placed between the generator and a water load [31]. In this case all the power not taken up by the plasma is absorbed in the water load. In all experiments with microwaves, the reactors must be made of silica to minimize dielectric losses.

While microwave discharges have been used successfully with inorganic compounds [32], it has been found that organic substances are normally almost completely destroyed. These results show an inherent problem in the use of microwaves. Since the dimensions of the reactor are on the order of the field wavelength, the magnitude of the electrical field can vary significantly within the reaction vessel. In some regions the starting material may be completely destroyed while in other regions the discharge can hardly be sustained. The decomposition of organic material often leads to carbon deposits on the walls which then absorb a large fraction of the incoming energy. At sufficiently high temperatures such deposits will participate in further pyrolytic reactions.

It should be noted that, although microwave equipment is easily set up, microwave discharges are more difficult to initiate and to sustain at low pressures ($\leqslant 1$ torr) than dc or rf discharges. Most experiments reported in the literature have therefore been carried out at pressures of several torr. As a result of the higher pressure, the gas temperature can rise to several hundred degrees, which often leads to pyrolytic decomposition of the reactants.

Selection of Discharge Type

The various kinds of electrical discharges lead to nearly identical results. Therefore, the choice of equipment is not determined by the chemical problem but by questions of flexibility, ease of operation, and cost. High-voltage discharges require the least equipment and may offer the best solution for carrying out reactions on a large scale. For laboratory purposes, radio-frequency equipment is best suited because of its great versatility. The frequency range from about 2–60 MHz is particularly convenient because the dimensions of the rf-coupling elements allow easy handling. Microwave discharges will only become useful for organic reactions when arrangements become available which offer better coupling and more uniform plasma zones.

Plasma Reactors

In addition to the power source for the discharge, equipment is needed to generate organic vapors, to adjust them to the desired pressure, to bring them in contact with the plasma, and to collect products and unused starting material.

For organic plasma experiments, vapor pressures of a few torrs are required. To obtain this pressure, low-boiling material can be injected into the vacuum system directly through an adjustable valve or through a capillary. Substances with higher boiling points can be heated in a stirred distillation flask until the desired vapor pressure is reached. For high-melting compounds it is advisable to heat the material to a point above the melting point and then to bleed it into the reactor through a valve.

The simplest plasma reactor is a tube of glass or silica. The walls should be heated in order to prevent condensation of starting materials or the crystallization of high-melting reaction products. This can often be accomplished by use of double-walled reactors in which the desired wall temperature is maintained by a circulating thermostated liquid, which for radio or microwave experiments must have a low dielectric constant. The reaction tubes should be placed in a vertical position to allow liquid products to flow out of the tube.

The details of discharge reactors depend on the reaction to be studied. A very versatile arrangement suitable for exploratory work is shown in Figure 2.4. The organic material is vaporized in a double-walled distillation flask fitted with a plastic stopper (PTFE) which serves as a cut-off or control valve. An additional opening in the stopper allows the introduction of a

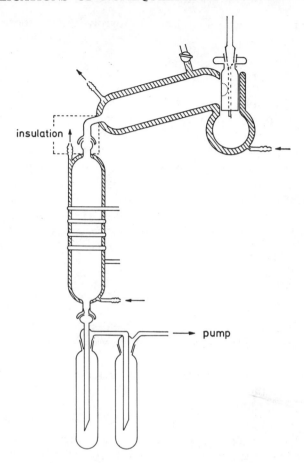

insulation

pump

FIGURE 2.4.   Discharge equipment for laboratory purposes.

second reactant. Connected to the distillation flask is a double-walled vertical reaction tube which leads into a set of cold traps. The distillation flask and the reaction tube can be thermostated together or separately.

Figure 2.5 shows a set-up developed for reactions which yield products with lower boiling points than the starting material [33]. The starting material is refluxed with the temperature of the condensor carefully adjusted to hold back all material except the low-boiling products. Another arrangement which is suited for low-boiling starting materials [34] is shown in Figure 2.6. The vapors are circulated by a membrane or piston pump and only the higher boiling products are condensed.

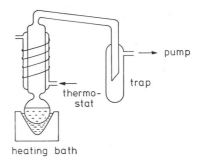

FIGURE 2.5. Discharge equipment for reactions leading to products with a boiling point which is lower than that of the starting material.

Experimental Conditions

The results of plasma experiments strongly depend on the experimental parameters such as pressure, electrical field strength, gas velocity, and tube diameter [35].

To initiate a plasma, a certain field strength is necessary. Its magnitude depends on the pressure and the nature of the gas. After initiation the field strength drops to a lower value. In a plasma the electrical field accelerates the electrons and these electrons transfer their energy to the molecules by collisions. The higher the field strength is, the more energy the electrons can pick up between collisions. An increase in field strength thus raises the average electron energy. This leads to a greater number of inelastic collisions and thus increases the chemical yield† (see Fig. 2.7).

An increase in pressure reduces the average mean free path of the electrons. As a consequence they pick up less energy between collisions. An increase

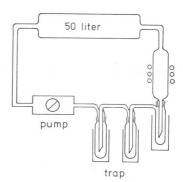

FIGURE 2.6. Discharge equipment with circulation of low-boiling starting material.

† An increase in power after ignition increases the degree of ionization and thus reduces the resistance. Therefore in glow discharges the field strength changes only slightly over a large range of power settings.

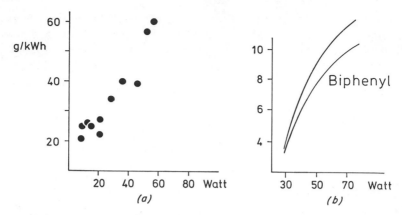

FIGURE 2.7. Influence of field strength on the yield; (*a*) rearrangement of *N,N*-dimethylaniline (monomolecular) [94], (*b*) dimerization of benzene (bimolecular) [95].

in pressure thus shifts the electron-energy distribution to lower values, and reduces the rate constant for inelastic collisions.

In bimolecular processes there is an additional effect since the rate is proportional to pressure. As a typical feature of a bimolecular process, the yield first rises with increasing pressure due to the higher reaction rate and then drops because of the lower electron energy (see Fig. 2.8).

The energy of the electrons also depends on the dimensions of the reactor. A decrease in tube diameter or a constriction in the plasma increases the average energy of the electrons (see Chapter 1).

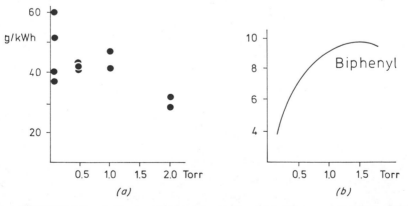

FIGURE 2.8. Influence of pressure on the yield; (*a*) rearrangement of *N,N*-dimethylaniline (monomolecular) [94], (*b*) dimerization of benzene (bimolecular) [95].

The gas flow rate is a very important parameter for plasma experiments with organic compounds. Since organic substances are destroyed if they remain in a discharge for long periods, it is necessary to remove them rapidly from the reaction zone. If the products are very sensitive to electron collisions (e.g., hydrazine), extreme flow rates are necessary to get reasonable yields. For fairly stable compounds, the flow rates are less critical.

Since pressure, field strength, tube diameter, and flow rate influence the average electron energy, the degree of ionization, the gas temperature, and the rate of conversion, it is not possible to give accurate values of optimal working conditions valid for a variety of reactions. As a general rule, pressures of 1–5 torr, gas velocities of 1–5 m/sec and power density leading to conversions of no more than 30% should be used.

Since no simple methods of plasma diagnostics are available at present, it is not possible to give values of electron densities and electron-energy distributions necessary for a specific reaction. As a consequence, for every reaction a series of optimizing experiments is necessary. Within the limits indicated above, the experimental parameters are varied and the influence of this procedure on the yield is determined.

Several techniques which have found less frequent use in plasma chemistry should also be mentioned. Organic plasma reactions normally do not require a carrier gas [21,36]. However, in some instances it has been possible to improve the results by use of a carrier gas which can act as a diluent, reducing the number of bimolecular collisions between organic molecules. The carrier can also deactivate reactive species by collision, or, if it has low-lying excited states, it can effectively transfer energy from the electrons to the organic molecules. Since the carrier must be inert, only a few substances such as the rare gases and to some extent water, carbon dioxide, and carbon monoxide are suitable for this purpose.

For sensitive products, pulsing of the discharge can improve the yields. In the period between two pulses, the product can be flushed out of the reaction zone, thereby minimizing its decomposition. This concept has been applied to the synthesis of hydrazine from ammonia [37–39].

A phenomenon which appears to have great promise but has only rarely been applied in plasma chemistry is the SESER† effect [40,41]. This effect is manifested when one or more constrictions are introduced into the plasma. On the cathode side of the constriction, a small "plasma sac" or double sheath is formed in which the average electron energy has considerably higher (e.g., 10 eV) values than in the surrounding plasma regions. When several constrictions are used in parallel, the plasma oscillates between the various holes and thus generates a pulsed plasma.

---

† SESER: Source of electrons of selected energy range.

A behavior somewhat similar to the SESER effect is observed when the reactor is filled with small glass rings. It has been demonstrated that this can change the product distribution and yields [42]. At present too little is known to decide whether the glass rings act because of their large surface, their small tube diameter, or because they produce a SESER effect.

2.3. PREPARATION OF ORGANIC COMPOUNDS IN GLOW DISCHARGES

Since its inception, attempts have been made to apply plasma chemistry to synthesize new compounds and to prepare known substances in an easier way. Both goals have been reached to some extent. Plasma chemistry occasionally creates new compounds and more often leads in a single step to substances which by classical methods can only be synthesized in a number of reaction steps. However, until very recently the results of plasma reactions were rather unspecific and the yields were much too small for preparative applications. Now a number of plasma reactions are known which are highly selective and give high yields.

Our present knowledge of the plasma chemistry of organic substances is very incomplete. A few hundred substances at the most have been subjected to discharges and only a small number of them has been studied in more detail. It is therefore not possible to present general rules for the behavior of organic substances in discharges. Only a rough pattern has emerged indicating that several types of reactions are very common in glow discharges. These have been investigated more thoroughly using simple model substances. In some cases the studies of the model systems have been extended to more complicated molecules. Typical reactions which occur in glow discharges are:

1. Generation of atoms or radicals

2. Isomerization

3. Elimination of atoms or small groups

4. Dimerization and polymerization

5. Reactions involving a complete scrambling or destruction of the starting material

All but the last type have shown results which are of value to preparative chemistry and will be discussed below.

Reactions of Atomic Hydrogen, Oxygen, and Nitrogen with Organic Compounds

One of the areas of plasma chemistry which has been studied extensively is the generation and reaction of atomic hydrogen, oxygen, and nitrogen [32,43–47]. The reactions of atoms with organic substances can be carried out in different ways. Either the atoms are generated in the discharge and the organic material is injected behind the plasma zone, or the organic substance is fed into the discharge together with molecular hydrogen, oxygen, or nitrogen.

Atomic hydrogen can be prepared easily by hot or cold plasmas [43,48,49]. It is a powerful reducing agent for inorganic materials and has been used for the preparation of several metals and hydrides [50–54]. The reactions of atomic hydrogen with organic compounds are limited. With saturated compounds, atomic hydrogen abstracts other atoms [55,56] (commonly hydrogen), while with unsaturated compounds it is added [56–62]. Thus both saturated and unsaturated compounds are converted into radicals by atomic hydrogen.

$$-\overset{|}{\underset{|}{C}}-\overset{|}{\underset{|}{C}}-H \quad + \quad H \quad \longrightarrow \quad -\overset{|}{\underset{|}{C}}-\overset{|}{\underset{|}{C}}\cdot \quad + \quad H_2$$

$$\overset{\backslash}{\underset{/}{C}}=\overset{/}{\underset{\backslash}{C}} \quad + \quad H \quad \longrightarrow \quad H-\overset{|}{\underset{|}{C}}-\overset{|}{\underset{|}{C}}\cdot$$

Atomic oxygen reacts with saturated compounds also by abstraction of atoms and the formation of radicals [56,63–68]. With olefins atomic oxygen forms three-membered rings and opens up interesting synthetic possibilities

$$\overset{\backslash}{\underset{/}{C}}=\overset{/}{\underset{\backslash}{C}} \quad + \quad O \quad \longrightarrow \quad \overset{\backslash}{\underset{/}{C}}\overset{O}{\diagup\diagdown}\overset{/}{\underset{\backslash}{C}} \quad \longrightarrow \quad -\overset{O}{\overset{\|}{C}}-\overset{|}{\underset{|}{C}}-$$

[56,69–74]. Under the reaction conditions the epoxides partly isomerize to carbonyl compounds. The addition of atomic oxygen to olefins has been studied extensively but mostly with atomic oxygen prepared by photochemical methods. Thorough plasma chemical studies have been carried out for the system oxygen–propylene [74], since the product of this reaction (propyleneoxide) is of great interest in polymer chemistry. Various techniques have been tried. The best results were obtained when oxygen and propylene passed through the discharge together. Since most oxygen–propylene mixtures are explosive, only those with very high or very low oxygen content are suitable for these experiments.

Atomic nitrogen has fascinated many investigators. It is easily prepared in a plasma, is fairly stable, and can be carried away from the plasma zone for use in further reactions. Most inorganic and organic substances react vigorously with atomic nitrogen. Since the results are covered extensively in several recent reviews [75–78], only the field of organic synthesis will be briefly discussed here.

On contact with atomic nitrogen, all organic compounds are destroyed with formation of hydrocyanic acid, ammonia, and low molecular weight gases. Sometimes these reactions are nearly quantitative and allow the determination of the amount of atomic nitrogen by the hydrocyanic acid formed [79].

Practically all attempts to incorporate nitrogen into organic molecules have been unsuccessful. Notable exceptions are the reactions of some dienes [80–82]. When atomic nitrogen reacts with butadiene, the main product is hydrocyanic acid, but in addition some 20% of nitrogen-containing products are formed. These include the products of addition of hydrocyanic acid and acetonitrile to the starting material and also some pyrrole (and croton nitrile which is formed from pyrrole, see p. 79). This is of special interest since it has evidently been formed in a 1,4 addition of a nitrogen atom to the diene.

Another field of plasma chemistry which has attracted special interest is the synthesis of complicated molecules from very simple starting materials [83–92]. Such reactions present simple models for the chemistry in the earth's primitive atmosphere [93]. It is postulated that the primitive atmosphere contained methane, water and ammonia, carbon monoxide and nitrogen. These compounds reacted under the influence of heat, radiation, or electrical discharges to form somewhat larger molecules which in turn, by photochemical, thermal, or plasma reactions, led to more complicated species [93].

Experiments to simulate the primitive atmosphere have been quite successful. Mixtures of simple hydrides or oxides when subjected to spark or glow discharges led to various kinds of oxygen- and nitrogen-containing compounds, including the formation of amino acids. However, the preparatory significance of these experiments is limited since the reactions are non-specific and lead to mixtures of simple amino acids and other products.

Cis-trans Isomerizations

The cis-trans isomerization of stilbene [94] has been studied as a simple model of a rearrangement reaction. When the trans isomer is distilled through a glow discharge, *cis*-stilbene is the predominant reaction product. With

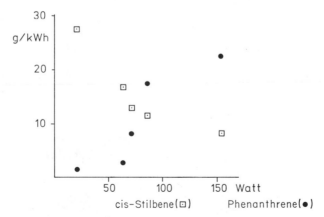

low-power levels and correspondingly smaller conversion rates, *cis*-stilbene makes up for more than 90% of the reaction products. At higher power levels the percentage of *cis*-stilbene drops while that of phenanthrene increases (see Fig. 2.9). Dihydrophenanthrene which is an intermediate in the photocyclization of stilbene [95,96] has not been detected in plasma reactions. This result is not surprising since dehydrogenations and aromatizations both are highly favored in glow discharges (see p. 80).

Since *cis*- and *trans*-stilbene are easily separated by various methods [97], the plasma isomerization is a simple way to prepare *cis*-stilbene. Yields of

FIGURE 2.9. Plasma reaction of *trans*-stilbene. Dependence of yields of *cis*-stilbene and phenanthrene on power.

about 30 g/kWh based on the power consumption of the rf generator are reported [94]. Because of the low efficiency of the generator and incomplete coupling, only a fraction of the power is transferred to the discharge. Based on the amount of rf energy absorbed by the plasma the yield would be about 100–150 g/kWh.

Similar cis-trans isomerizations have been tried with several other olefins [98, 99]. The results obtained with stilbazol are almost identical to those of stilbene. *cis*-Stilbazol is the main product and the predominant side reaction is the cyclization to azaphenanthrene. Crotonnitrile and cinnamonitrile have also been converted to the cis isomers in fairly good yields. With other olefins the cis-trans isomerization is suppressed by various side reactions. For compounds like fumaric acid, the fragmentation to acetylene dominates the cis-trans isomerization. Evidently, such plasma isomerizations are limited to molecules without highly reactive groups.

Migration of Substituents

Another type of isomerization frequently observed in plasmas is the migration of substituents, especially in aromatic compounds. Sometimes such iso-merizations are negligible side reactions, in other cases, the predominant reaction path.

When, for example, xylenes are subjected to a glow discharge, the main reaction is the formation of dixylyl (see p. 90). The recovered starting material, however, shows a small degree of isomerization (at total conversions of 50–70%, about 1–2%). This isomerization could proceed via a tropyllium ion as postulated in mass spectroscopy. More likely, however, is a reaction through a benzvalene type of intermediate as observed in some photo-reactions [100–104].

Examples where the migration of substituents is the main reaction are found for aromatic ethers and amines [105–108]. Alkylaryl ethers like anisole isomerize in glow discharges to the corresponding *ortho*- and *para*-alkyl-phenols. The only significant side reaction is the fragmentation to a phenol.

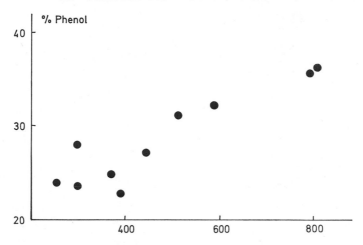

FIGURE 2.10. Plasma reaction of anisol. Percent product of phenol dependence on power (arbitrary units).

With low-power levels and accordingly low rates of conversion, the rearrangement accounts for about 80% of the reaction products. An increase in power raises the percentage of phenol among the reaction products (see Fig. 2.10). The ortho/para ratio of isomeric alkylphenols however remains constant for all experimental conditions. The meta isomer is formed only in very small amounts (about one-tenth or less of the para isomer).

The results obtained for various alkylphenyl ethers follow the same general pattern [107,108] (Table 2.1). When phenetole passes a glow discharge, the fragmentation to phenol is slightly more important than in the case of anisole. *Ortho-* and *para*-ethylphenols are formed in an almost statistical ratio. To a small extent, the ethyl group fragments, thus generating cresols. Such a fragmentation is of greater importance in the plasma reaction of *n*-propylphenyl ether where cresols and ethylphenols are found among the products. The yields of propylphenols are satisfactory only at low-power levels. The isomerization of isopropylphenyl ether leads almost without side reactions to *ortho-* and *para*-isopropylphenol (>50%) and phenol.

Substitution on the aromatic system has little influence on the isomerization [108]. Methylanisoles rearrange to dimethylphenols under migration of the methyl group to the ortho and para positions. The *meta*-methylanisole rearranges with especially high yields and shows less fragmentation (Table 2.2). This probably is due to substitution effects. In electrophilic and homolytic substitutions alkoxy and methyl groups activate the ortho and para positions. For *m*-methylanisole, methyl and methoxy groups favor the same

TABLE 2.1. Rearrangement of Phenylalkyl ethers [107]

Starting material	Power (W)	Conversion (%)	Total yield g/kWh	Phenol	Products (mole %)					
					Cresol		Ethylphenol		Propylphenol	
					o	p	o	p	o	p
Anisole	94	38	39	21	41	23				
Phenetole	69	13	16	36	3	2	37	22		
	97	18	15	39	4	3	33	21		
n-Propyl-phenyl ether	75	7	15	30	7	4	1	—	38	19
	96	13	10	37	14	8	4	3	22	13
Isopropyl-phenyl ether	69	6	16	42	—	—	—	—	36	22
	100	12	14	45	4	2	—	—	27	22

TABLE 2.2. Rearrangement of Various Ethers in Glow Discharges [108]

Starting material	Conversion (%)	Total yield (g/kWh)	Products (weight %)
o-Cresylmethyl ether	17	113	o-Cresol, 32; 2,6-Dimethylphenol, 20; 2,4-Dimethylphenol, 44
m-Cresylmethyl ether	25	187	m-Cresol, 19; 2,5-Dimethylphenol, 28; 2,3-Dimethylphenol, 25; 3,4-Dimethylphenol, 26
p-Cresylmethyl ether	15	107	p-Cresol, 54; 2,4-Dimethylphenol, 31
2,6-Dimethylanisole	22	8	2,6-Dimethylphenol, 65; 2,4,6-Trimethylphenol, 34
α-Naphthylmethyl ether	13	50	α-Naphthol, 5; 2-Methyl-α-naphthol, 48; 4-Methyl-α-naphthol, 35
	50	129	α-Naphthol, 47; 2-Methyl-α-naphthol, 24; 4-Methyl-α-naphthol, 19
β-Naphthylmethyl ether	21	119	β-Naphthol, 30; 1-Methyl-β-naphthol, 7
Diphenyl ether	44	320	Phenol, 18; 2-Hydroxybiphenyl, 44; 4-Hydroxybiphenyl, 13

positions while for the other isomers the inductive effects in part cancel out. With increasing substitution, the yields of rearrangement products drop. For example, only 30–40% of 2,6-dimethylanisole isomerizes to 2,4,6-trimethylphenol the main reaction being the decomposition to 2,6-dimethyl phenol.

Naphthylmethyl ethers rearrange very smoothly with only little fragmentation to naphthols. The α-naphthylmethyl ether isomerizes to 2- and 4-methyl-α-naphthol, while the β isomer forms only 1-methyl-β-naphthol. Particularly high yields have been found for isomerizations involving the migration of aryl groups. Diphenyl ether forms ortho- and paraphenyl-phenols in a ratio of about 3/1. Under the reaction conditions part of the orthoisomer cyclizes to dibenzofuran.

The reaction of 2,6-dimethyldiphenyl ether evidently takes a different course. Only small quantities of the isomerized product (3,5-dimethyl-4-hydroxy-biphenyl) are formed. The predominant reaction product is 4-methyl-xanthene formed either by migration of the phenyl group to a methyl group followed by cyclization or by direct ring closure of the starting material.

A rearrangement similar to that of alkylaryl ethers was found for vinyl ethers. Vinylethyl ether in a glow discharge rearranges or fragments, thus forming butanal and acetaldehyde.

Isomerization similar to those of aromatic ethers have been observed for the analogous nitrogen compounds (Table 2.3). For example, N,N-di-

TABLE 2.3. Rearrangement of *N*-Alkylanilines [109]

Starting material	Conversion (%)	Total yield (g/kWh)	Products (weight %)	
N,N-Dimethylaniline	16	57	*N*-Methyl-*o*-toluidine, 22 *N*-Methyl-*p*-toluidine, 12.5 *o*-Toluidine, 9.5 *p*-Toluidine, 2.6	*N*-Methylaniline, 41 *N*-Methyleneaniline, 19
N-Methylaniline	20	84		Aniline, 43 *N*-Methyleneaniline, 29
N,N-Dimethyl-*o*-toluidine	8	39	*N*-Methyl-2,6-dimethylaniline, 16 *N*-Methyl-2,4-dimethylaniline, 12	*N*-Methyl-*o*-toluidine, 38
N,N-Dimethyl-*m*-toluidine	5	25	*N*-Methyl-2,5-dimethylaniline, 10 *N*-Methyl-2,3-dimethylaniline, 13	*N*-Methyl-*m*-toluidine, 52 3-Methyl-*N*-methylene-aniline, 8
N,N-Dimethyl-*p*-toluidine	3	68	*N*-Methyl-3,4-dimethylaniline, 17 *N*-Methyl-2,4-dimethylaniline, 16	*N*-Methyl-*p*-toluidine, 71 4-Methyl-*N*-methylene-aniline, 8

methylaniline rearranges to *ortho-* and *para-N*-methyl toluidine or eliminates a methyl group under formation of *N*-methylaniline.

When *N*-methylaniline is the starting material the products are *ortho-* and *para*-toluidine as well as aniline. In side reactions, *N,N*-dimethylaniline and *N*-methylaniline fragment to *N*-methyleneaniline.

In plasmas, *N,N*-dimethyltoluidines isomerize in the same manner as methylcresol ethers. Attempts to rearrange arylacetates and acetanilines in glow discharges were unsuccessful since other reactions predominated. Phenylacetate decomposes to phenol and ketene or cyclizes to cumerane-2-one. Cresylacetates and 2,6-dimethylphenylacetate also react under fragmentation and cyclization. Experiments with acetanilides led to numerous products which have not been fully analyzed.

Rearrangements of Cyclic Compounds

A number of isomerizations have been found for cycloolefins and hetero-cyclics. There is, however, not enough experimental material available to decide how common such isomerizations are.

Certain cycloolefins readily isomerize to aromatic compounds [42,100]. For example, cycloheptatriene forms toluene, and cyclooctatetraene re-arranges to styrene almost quantitatively. Isomerizations like that of benzyl-cations to tropyllium systems which are postulated in mass spectroscopy have not been observed in glow discharges, possibly because the electron energy is insufficient for such a process. Examples of ring enlargements have been observed for α- and β-methylindole [98,110]. These substances isomerize to dihydroquinoline which under the reaction conditions dehydrogenates to quinoline.

It was also possible to isomerize styrene to cyclooctatetraene in small amounts ($\sim$2%) when the product was trapped quickly at the cold ($-40°C$) walls of the reactor [42].

Ring-chain isomerizations which have been observed for several nitrogen-containing compounds [98,110] offer a number of interesting preparative approaches (Table 2.4). Pyrrole, for example, rearranges in glow discharges

$$\text{pyrrole} \longrightarrow CH_3-CH=CH-CN$$

$$\text{indole} \longrightarrow C_6H_5-CH_2-CN$$

TABLE 2.4. Ring-Chain Isomerizations of Nitrogen Compounds [98,110]

Starting material	Product	Yield (%)
pyrrole	$H_3C-CH=CH-CN$ cis, trans	85
indole	$C_6H_5-CH_2-CN$	22
quinoline	$C_6H_5-CH=CH-CN$ cis,trans	15
isoquinoline	$C_6H_5-CH=CH-CN$ cis,trans	15
aniline ($C_6H_5-NH_2$)	$H_3C-CH=CH-CH=CH-CN$	25
benzotriazole	$H_3C-CH=CH-CH=CH-CN$	10

easily to crotonnitrile. In a similar way indole isomerizes to benzylcyanide. Six-membered heterocyclics show similar rearrangements [98]. When pyridine is subjected to a glow discharge, it is almost completely converted to new products. Besides large quantities of polymers, considerable amounts of unsaturated nitriles are also isolated, probably formed by opening of the pyridine ring. Similarly quinoline and isoquinoline (!) form cinnamo nitrile. A related ring chain isomerization is observed for aniline. In plasmas this forms unsaturated nitriles to a large extent.

$$\langle\!\!\!\bigcirc\!\!\!\rangle\!-NH_2 \longrightarrow \quad CH_3-CH=CH-CH=CH-CN$$

During the plasma reactions of nitrogen heterocycles and of aniline, a considerable fraction of the starting material is always converted to polymers, possibly through unsaturated acyclic intermediates.

Eliminations

A very interesting field of plasma chemistry with numerous preparative applications is that of elimination reactions [111]. Collisions with high-energy electrons can destroy molecules completely. At low energies, however, such collisions may eliminate small groups or atoms without altering the rest of the molecule. The remaining species often are reactive intermediates (radicals, carbenes) which stabilize through the formation of multiple bonds, through cyclization, ring contraction or dimerization.

Dehydrogenation is a common elimination reaction occurring in plasmas. The driving force for this reaction is the large concentration of atoms and free radicals found in organic plasma systems which can react to abstract hydrogen. Even in mixtures of organic compounds and molecular hydrogen, dehydrogenation takes place more readily than hydrogenation.

The yields of dehydrogenated products depend on the nature of the starting material [110]. Compounds such as paraffins normally form a variety of products, which, depending on reaction conditions, may contain considerable amounts of olefins. These dehydrogenations are however highly unselective. Since aliphatic molecules contain a number of carbon–carbon and carbon–hydrogen bonds of comparable strengths, elimination reactions may start at various locations with equal probability. Thus different radical intermediates are generated which stabilize to a number of saturated and unsaturated products. The dehydrogenation of aliphatic hydrocarbons, though occasionally capable of producing up to 60% olefins is of limited preparative value because of the low selectivity.

If, however, molecules with only one reactive site or one bond which is considerably weaker than all the others are subjected to glow discharges, the dehydrogenation can be highly selective. These possibilities have been studied with a number of simple model compounds. Substances such as ethylbenzene, acenaphthene, tetraline, or indane eliminate hydrogen easily and with high yields (Table 2.5).

TABLE 2.5. Preparation of Unsaturated Compounds by Plasma Eliminations [42,98,114,118]

Starting material	Product	Yield (%)
Ethylbenzene	Styrene	7
β-Bromoethylbenzene	Styrene	80
α-Phenylethanol	Styrene	70
1,2-Diphenylethane	Stilbene	28
	Diphenylacetylene	9
Styrene	Phenylacetylene	40
Indane	Indene	80
Dihydroindol	Indol	30
Acenaphthene	Acenaphtylene	90
Tetraline	Naphthalene	60
Tetrahydroisochinoline	Isoquinoline	47

The dehydrogenation of olefins to acetylenes can also be achieved in plasmas as demonstrated with stilbene and styrene. The dehydrogenation of stilbene leads mainly to phenanthrene, but in addition small amounts of diphenylacetylene are formed. When styrene is subjected to a glow discharge, it is frequently converted completely into polymeric material. Under certain reaction conditions, the polymer formation can be suppressed and phenylacetylene is the main reaction product [42].

Eliminations of hydrogen halides or water also can lead to unsaturated compounds. Only two examples of this type have been studied. Both β-

phenylethanol and β-bromoethylbenzene form styrene in high yields. A number of eliminations of the type

$$R-X-R' \xrightarrow{-X} R \ + \ R' \longrightarrow R-R'$$

lead to fragments which stabilize by formation of new single bonds. Among these eliminations, the decarbonylation and decarboxylation reactions are of particular interest for preparative chemistry.

The formation of hydrocarbons by decarbonylation of aldehydes involves at least two intermediate steps. This becomes evident from the by-products, which are observed when the decarbonylation is carried out under mild conditions. Thus benzaldehyde in most plasma experiments forms benzene almost exclusively. With very low electron energies and fast gas flow, considerable amounts of biphenyl, benzophenone, and benzil are isolated [110] (Table 2.6).

The presence of benzil and benzophenone demonstrate the intermediate existence of benzoyl radicals. Since both benzil and benzophenone decarbonylate easily, they normally are not observed. The phenyl radicals are stabilized by abstraction of hydrogen or by dimerization. At low conversion rates, benzene may be the sole reaction product. At higher conversion rates, biphenyl is formed in quantities up to 20%. In an analogous reaction, pyridine-2-aldehyde forms pyridine and α,α-dipyridyl [98]. It has also been possible to decarbonylate thiophene aldehyde without any desulfuration [112].

The decarbonylation of ketones was studied first with simple model substances [112]. Acetophenone in glow discharges decompose to toluene; benzophenone forms diphenyl (Table 2.7) but also some fluorenone and biphenylene. The decarbonylation of benzil to biphenyl is almost quantitative.

The decarbonylation of an isocyanate has been tried with phenylisocyanate [112]. This compound easily fragments to a reactive intermediate (probably a nitrene) which then dimerizes to azobenzene. Since azobenzene also fragments in plasmas (see below), there is always some biphenyl and diphenylamine among the products obtained from the decarbonylation of phenylisocyanate. Fragmentations similar to those of ketones are found for azo compounds. These substances under plasma conditions eliminate nitrogen and form two radicals which stabilize by dimerization.

Because the elimination reactions of benzophenone and azobenzene proceed smoothly, a similar decomposition of benzalanilide and stilbene was anticipated:

TABLE 2.6. Plasma Decarbonylation of Aldehydes [98,110,112]

Starting material	Power (W)	Conversion (%)		Products (mole %)
Benzaldehyde	14	4	Benzene, 95	Biphenyl, 2.5
	50	63	Benzene, 81	Biphenyl, 16
		20[a]	Benzene, 68	Biphenyl, 8; Benzophenone, 17; Benzil, 9
Thiophenaldehyde	7	4	Thiophene, 100	
	40	40	Thiophene, 86	
Pyridine-2-aldehyde	69	24	Pyridine, 67	2,2-Dipyridyl, 23

[a] High-Voltage discharge.

83

TABLE 2.7. Plasma Decarbonylations of Ketones [112, 114]

Starting material	Conversion (%)	Products (weight %)	
Cyclohexanone	8	Cyclopentane, 7	n-Pentane, 79
Camphor	29 (76)	Trimethylbicyclo-2,1,1-hexane, 57 (42)	
p-Benzoquinone	4	8,9-Dihydroindenone (1), 60	
Acetophenone	32 (54)	Toluene, 57 (34)	Benzene, 19 (24)
Benzophenone	3 (68)	Diphenyl, 70 (27)	Fluorenone, 10 (36) Biphenylene, 3 (36)
Benzil	18	Diphenyl, 98	
Fluorenone	20	Biphenylene 99	
	50	Biphenylene,[a] 99	
Phenol	20	Cyclopentadiene, 70	Benzene, 33
α-Naphthol	50	Indene, 76	
β-Naphthol	40	Indene, 92	
9-Anthrol	77	Fluorene, 77	
2,3-Dihydroxynaphthol	30	Indanone-2, 26	
6-Methylnaphthol-2	14	5- and 6-methylindene, 43	

[a] Refluxing.

These compounds however react differently. Benzalanilide splits predominantly into benzene and benzonitrile and eliminates hydrogen cyanide only to a minor extent. The principal products from stilbene are *cis*-stilbene and phenanthrene (see p. 71) and only very small amounts of diphenyl are found [96].

High yields are found in the decarboxylation of carboxylic acids and their esters [113]. Benzoic acid, for example, decarboxylates to benzene almost quantitatively. The methyl ester of benzoic acid forms toluene and benzene as major products.

For most of the decarboxylation reactions of acids and esters studied so far there was a pronounced similarity between the results obtained in a plasma and those by pyrolysis.

Elimination reactions of cyclic compounds often lead to intermediates which stabilize by ring contraction. This form of stabilization is quite common for cycloketones. The yields of such reactions differ widely, depending on the nature of the starting material. For cyclohexanone, the predominant reaction is the dehydrogenation to phenol while the decarbonylation to cyclopentane has only minor importance. Decarbonylations of unsaturated cycloketones give better results. Two molecules of benzoquinone lose three molecules of carbon monoxide to form dehydroindenone. The overall yield of this reaction sequence is about 40%. The reaction probably involves cyclopentadienone which dimerizes and loses a second molecule of carbon monoxide to an intermediate which rearranges to dehydroindenone. An alternative mechanism via cyclobutadiene is less likely.

The best cases are decarbonylations of bicyclic or tricyclic ketones. A very impressive example is the decarbonylation of fluorenone [113]. When this compound is subjected to a discharge, the decarbonylation product biphenylene is obtained in yields of 99%.

The plasma decarbonylation of fluorenone has also been carried out by a different technique. In these experiments fluorenone was boiled under reflux and the temperature of the condenser was adjusted to hold back the starting material but not the biphenylene (see Fig. 2.5). This technique allowed higher conversion rates and gave better yields.

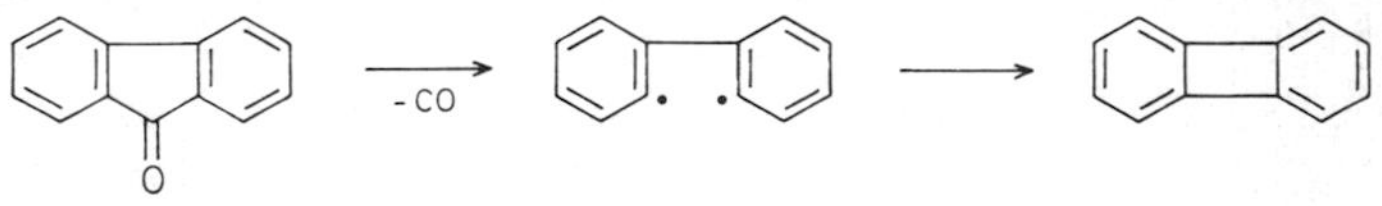

Bicyclic aliphatic ketones decarbonylate with considerably higher yields than monocyclics. It was possible to transfer camphor and norcamphor to trimethylbicyclohexane in yields up to 60%:

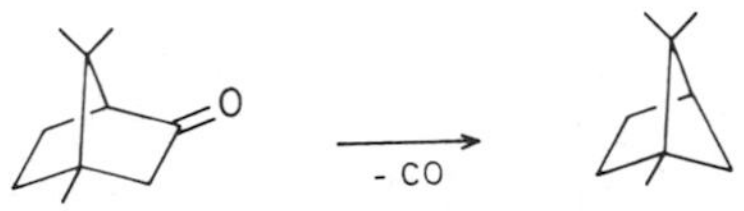

Elimination of carbon monoxide followed by ring contraction is also observed for aromatic hydroxy compounds [107,114]. The yields of decarbonylation products roughly parallel the tendency to form the tautomeric ketones. When phenol is subjected to a discharge, the main reaction is the elimination of the hydroxy group. The competing decarbonylation to cyclopentadiene can reach 40–70% in favorable cases. In the naphthalene series decarbonylation is the predominant reaction; β-naphthol forms indene almost quantitatively and the α-isomer to an extent of 70–80%; 9-anthrol easily eliminates carbon monoxide forming fluorene. The decarbonylation of substituted phenols is also possible but has only been studied in a few cases.

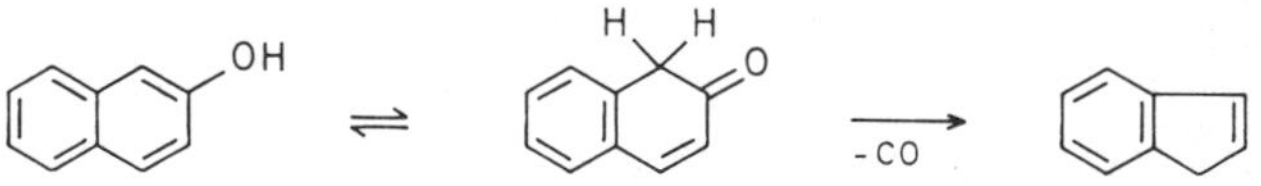

A reaction similar to the decarbonylation of naphthols has been observed for α-nitronaphthalene [42]. From mass spectroscopy it is known that the isomerization of nitro compounds to nitrites is followed by a loss of nitric oxide [115]. In glow discharges α-nitronaphthalene eliminates nitric oxide in a similar way to form a naphthalene oxide radical. This radical may decarbonylate and stabilize to indene by hydrogen abstraction The decarbonylation of nitro compounds is not a very smooth reaction. Frequently, the starting material decomposes completely to undefined black products. In some cases, however, indene was obtained in yields up to 40% [42].

Another reaction which seems promising for preparative chemistry is the formation of unsaturated compounds from cyclic anhydrides [114]. Maleic anhydride, for example, fragments to carbon monoxide, carbon dioxide, and acetylene. Bicyclic anhydrides are suitable starting materials for the preparation of cycloolefins. Hexahydrophthalic anhydrides fragment easily in plasmas to cyclohexenes:

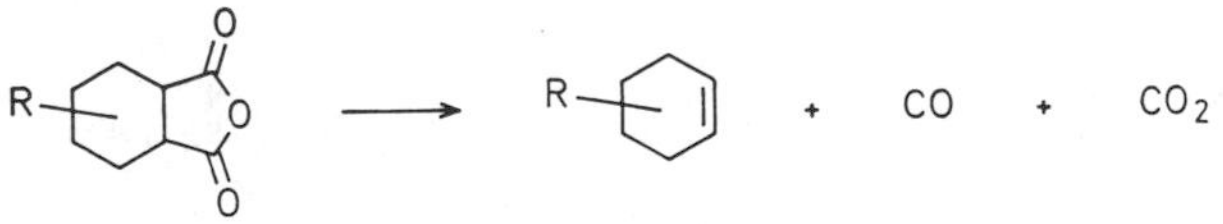

Unsaturated bicyclic anhydrides like tetrahydrophthalic anhydrides are suitable starting materials for the preparation of cyclodienes.

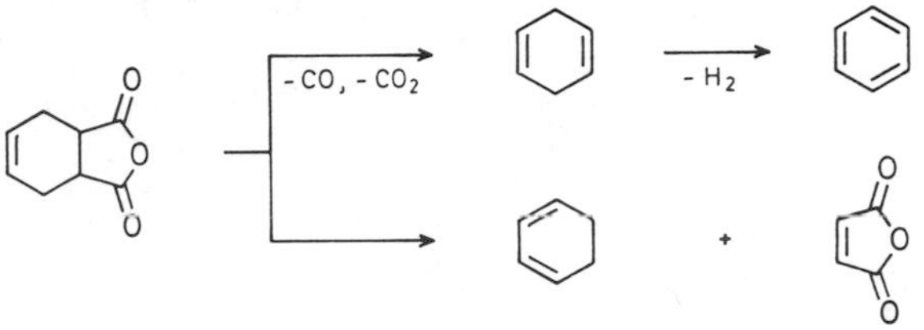

Tricyclic anhydrides, like the addition products of maleic anhydride to cyclodienes, also fragment easily. The cyclodienes sometimes stabilize to aromatic systems by elimination of hydrogen. In the reaction of unsaturated anhydrides a retro Diels–Alder reaction aways competes with the elimination reaction. For this reaction the yields are often unsatisfactory.

Extensive studies have been made on the fragmentation of phthalic anhydride [116,117]. This compound eliminates carbon monoxide and carbon dioxide with formation of dehydrobenzene. This reactive species quickly dimerizes, trimerizes or polymerizes. In glow discharges at low conversion rates, the products are about 60% biphenylene, 20% triphenylene, and 10% polymer [117]. The dehydrobenzene may be trapped by addition of other products (as will be discussed on p. 94).

Cyclizations

In certain cases the intermediates formed in elimination reactions stabilize by cyclization. This type of reaction provides a simple method for synthesizing carbocyclics and heterocyclics. The success of these reactions depends on the proper choice of the starting material. Simple paraffins, olefins or unsaturated nitriles do not form cyclic products to an appreciable extent. The best results have been obtained with aromatic substances. Five-membered rings are easily formed by dehydrogenation of compounds with two aromatic rings linked by a carbon or hetero atom.

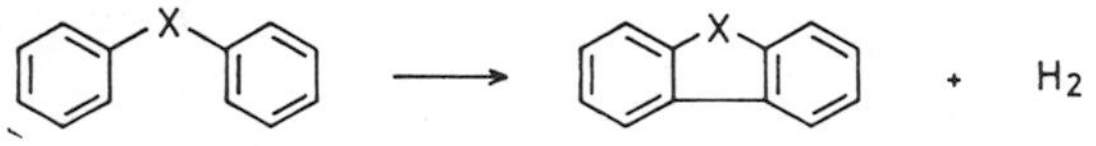

$$X = CH_2, O, NH$$

TABLE 2.8. Syntheses of Five Membered Rings by Plasma Reactions [110, 119–121]

	Conversion (%)		Yield (%)
	30		26
	68		27
	39		9
	40		30
	22		62
	24		82
	25		40–50
	13		57
	50		15–20
	52		15

Other suitable starting materials are 2-substituted biphenyls:

$$X = CH_2, O, NH$$

The possibilities for synthesizing five membered carbocyclics and hetero-cyclics on a preparative scale have been studied with simple model substances (Table 2.8). Though most of the reactions have not been optimized for the cyclization, the results look very promising.

Diphenylmethane cyclizes in glow discharges to fluorene in amounts of up to 25% [118]. Diphenyl ether [109] and diphenylamine [119] form five membered rings to a lesser extent since part of the starting material rearranges to para substituted biphenyls. The cyclization of 2-substituted biphenyls is much more efficient.† 2-Hydroxybiphenyl cyclizes to dibenzofurane in yields of about 60% [109], 2-aminobiphenyl yields carbazol to more than 80% [119], and 2-nitrobiphenyl acts similar to hydroxybiphenyl. The nitro compound probably isomerizes to a nitrite. With loss of nitric oxide this product forms a radical which cyclizes to dibenzofurane.

Other suitable starting materials for the synthesis of furanes are phenol-acetates [33].

The unsubstituted phenylacetate yields about 20% of cumerane-2-one, *meta-* and *para*-cresylacetates lead to methylcumeranones in somewhat smaller yields. The *ortho*-cresylacetate in a different reaction cyclizes to dehydrocumerene [33].

A number of six-membered carbo- and heterocyclic rings have been pre-pared via plasma eliminations [110]. Suitable starting materials are molecules containing two phenyl rings separated by three bonds.

---

† Sometimes both synthetic routes are related. Substances like diphenyl ether and diphenyl-amine are known to rearrange to substituted biphenyls (see pp. 72–78). These compounds may cyclize directly or after rearrangement to structures equal or similar to 2-substituted biphenyls.

X—Y = CH=CH , CH=N , N=N

Other routes start with molecules containing phenyl rings separated by two bonds. A reactive group in the ortho position initiates the cyclization.

X = NH_2 , OH , CH_2

All the suitable links and reactive groups present a great number of possibilities for preparing six-membered carbocyclics and heterocyclics, but only a few have been tested so far (Table 2.9). Six-membered carbocyclics may be formed from dibenzyl or stilbene. Dibenzyl yields about 20% stilbene and up to 50% of cyclic products. Stilbazol cyclizes like stilbene. Benzalanilide and azobenzene form cyclic products only in small amounts.

Experiments with nitrodiphenylamine, nitrodiphenyl ether and amino-diphenylamine showed very promising results [110]. Nitrodiphenylamine cyclizes to phenoxacine in yields up to 45%. Cyclizations through methyl groups also are smooth reactions. 2,6-Dimethyldiphenyl ether led to 4-methylxanthene in yields up to 64%.

Bimolecular Reactions

Most organic plasma reactions investigated recently have been mono-molecular processes such as rearrangements and eliminations. Bimolecular reactions are also possible and some of them may be useful for preparative work. They have, however, been studied only in a few cases since they normally require more complex equipment, and a more careful optimization of pressure, field strength, and rate of distillation.

Certain types of bimolecular processes such as dimerizations, oligomerizations and polymerizations are quite common in glow discharges. Frequently, the fragments generated in elimination reactions stabilize by dimerization. When, for example, methane passes a glow discharge, up to 50% of the starting material may be converted to ethane. The mechanism is probably complex and involves a number of steps but certainly the dimerization of methyl radicals plays an important role. A similar dimerization of fragments is found for carbon tetrachloride which in glow discharges produces mainly hexachloroethane and tetrachloroethylene. Both the formation of ethane and of hexachlrooethane are too unspecific for preparative work.

With aromatic compounds dimerization occurs as a result of the loss of hydrogen or substituent groups from the reactants. These reactions proceed

TABLE 2.9. Syntheses of Six Membered Rings by Plasma Reactions [110, 120, 121]

	Conversion (%)		Yield (%)
$\bigcirc$—CH_2—CH_2—$\bigcirc$	30	(phenanthrene, dihydro)	19
$\bigcirc$—CH=CH—$\bigcirc$	77	(phenanthrene)	48
$\bigcirc$—CH=N—$\bigcirc$	29	(phenanthridine)	3
$\bigcirc$—$\overset{CH_3}{\underset{\|}{C}}$=$N$—$\bigcirc$	18	(methyl phenanthridine)	4
$\bigcirc$—N=N—$\bigcirc$	80	(N=H ring system)	3
$\bigcirc$—CH_2=CH_2—$\bigcirc_N$	22	(benzoquinoline)	12
$\bigcirc\overset{NH}{\underset{NO_2}{}}\bigcirc$	90	(phenoxazine type)	45
$\bigcirc\overset{O}{\underset{NO_2}{}}\bigcirc$	~25	(dibenzodioxin)	~10
$\bigcirc\overset{NH}{\underset{NH_2}{}}\bigcirc$	93	(dibenzodiazine)	14

91

TABLE 2.9 (*continued*)

	Conversion rate		Yield (%)
(structure: 2,6-dimethylphenyl acetate)	41	(structure: dihydrocoumarin derivative)	17
(structure: 2,6-dimethylphenyl phenyl ether)	22	(structure: methyl-dimethyl xanthene)	40

smoothly and can be applied for preparative work [122,123]. For example, benzene forms diphenyl in a fairly high yield. At higher conversion rates terphenyl is also isolated [123–130].

$$2\ Ar-X \longrightarrow Ar-Ar + X_2$$

$$2\ Ar-CH_2X \longrightarrow Ar-CH_2-CH_2-Ar + X_2$$

$$X = H, OH, Hal, Alkyl$$

The bimolecular nature of the process is manifested by the pressure dependence of the yields (see Fig. 2.13). The dimerizations of aromatic methyl compounds also give exceptionally high yields [106,118,124,125, 128,131–137].

The plasma chemistry of toluene has been studied extensively [118,137,138]. Toluene reacts in a bimolecular process to dibenzyl and to benzene and ethylbenzene in a monomolecular process. With increasing pressure, the bimolecular reaction is enhanced and dibenzyl becomes the predominant product. Polymethylbenzenes also dimerize easily to diarylethanes. The yields decrease slightly with an increasing number of methyl groups in the starting material [110]. The opposite effect has been observed when toluene is substituted with electron-attracting groups. Methylnaphthalenes dimerize more easily than toluene and especially high yields have been found for toluol nitriles [110,136] (Table 2.10).

Alkyl aromatics with straight or branched side chains also form diarylethanes or substituted diarylethanes. In the first reaction step, a C—C bond in the α-position is broken and the remaining radicals stabilize by dimerization. The elimination of groups other than hydrogen as the initial steps in a dimerization has been studied with benzyl alcohol and benzyl halides [139].

TABLE 2.10.   Examples of Dimerization by Elimination of Hydrogen [106, 110, 118, 136]

Starting material	Product	Yield (mole %)
Benzene	Biphenyl	40
Toluene	Bibenzyl	90
p-Xylene	Bixylyl	90
Durene	Bisdurene	77
Isopropylbenzene	2,3-Diphenylbutane	90
tert-Butylbenzene	2,3-Dimethyl-2,3-diphenylbutane	90
1-Methylnaphthalene	1,2-Dinaphthylethane	48
p-Methylbenzonitril	1,2-Di(4-cyanophenyl)ethane	90

Both the halogen and the hydroxy compounds form dibenzyl as the main plasma product.

For unsaturated compounds, dimerizations of the fragments of elimination reactions are paralleled by true dimerizations. In this way propene leads to a mixture of hexadiene and hexenes [110]:

$$H_3C-CH=CH_2 \xrightarrow[-H]{} \cdot CH_2-CH=CH_2$$

$$\cdot CH_2-CH=CH_2 \longrightarrow H_2C=CH-CH_2-CH_2-CH=CH_2$$

$$H_3C-CH=CH_2 \longrightarrow H_3C-\overset{\cdot}{C}H-CH_2-CH_2-CH=CH_2$$

$$\downarrow {+H}$$

$$H_3C-CH_2-CH_2-CH_2-CH=CH_2$$
and isomers

In contrast to dimerizations and polymerizations [140–146] which are typical plasma reactions, oligomerizations have only been observed in rare cases. One example of a tetramerization and trimerization has been observed for acetylene. In most plasma experiments, acetylene is polymerized completely to a yellow product (cuprene) [147]. If however the surface of the reactor walls is greatly enlarged, new products, instead of polymers, appear with 6, 8, or 10 carbon atoms like styrene, phenylacetylene, cyclooctatetraene, naphthalene, and benzene [42].

Plasma reactions between two different compounds have been studied only in a few cases [148]. Such reactions can be carried out in two different ways. One of the components is subjected to the discharge and the other is used as a trapping agent, or both compounds are passed through the discharge together. The two variations have been tested in the reaction between propene and oxygen [74]. The highest yields have been obtained when the olefin and oxygen were mixed before they entered the discharge (see p. 69). Similar results have been observed for the reaction between styrene and oxygen [42]. When mixtures of benzene and oxygen are passed through a glow discharge, the main reaction product is phenol [149,150]. Initial experiments with mixtures of benzene and ammonia indicate the possibility of synthesizing aniline in an analogous way [110,151].

New and interesting preparative possibilities arise when one of the components is a reactive species which can react with a second substance. Studies of this type of synthetic route have been made with phthalic anhydride [117]. Phthalic anhyride fragments in a two-step reaction with the loss of carbon dioxide and carbon monoxide to dehydrobenzene.

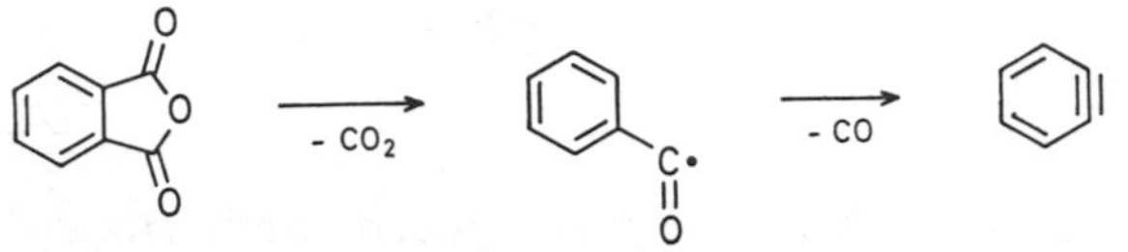

If no other reaction partner is present, dehydrobenzene dimerizes to a radical which can either cyclize to diphenylene or add another molecule of dehydrobenzene to form triphenylene:

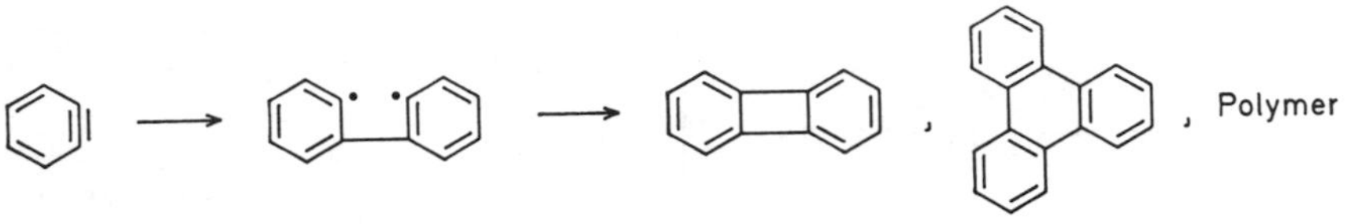

The intermediates in the decomposition of phthalic anhydride or the dimerization of dehydrobenzene may be scavenged by addition of suitable substances [117] (Table 2.11). When a mixture of phthalic anhydride and hydrogen is passed through a discharge, the yield of diphenylene goes down

TABLE 2.11. Trapping of Intermediates in the Decomposition of Phthalic Anhydride by Added Substances [117]

	Products (Mole %)					
Addition	Benzene	Phenyl-acetylene	Bi-phenyl	Bi-phenylene	Tri-phenylene	Others
NO	0.5	2	10	42	19	
H_2	5	3	34	4	8	
C_2H_2	12	31	4	5	—	
CS_2	0.5	1	6	23	11	Dibenzothiophene, 31
NH_3	1	1	1	—	—	2-Aminobenz-aldehyde, 21 Aniline, 5 Carbazol, 3

while that of benzene and diphenyl is increased, probably because hydrogen adds to dehydrobenzene and the biphenyl diradical. Mixtures of phthalic anhydride and acetylene produce considerable amounts of phenylacetylene, probably by the addition of acetylene to dehydrobenzene. Sulfur atoms generated by decomposition of carbon disulfide scavenge the diphenyl diradical during the formation of dibenzylthiophene. This reaction is of particular interest since both reactants are plasma products. When ammonia is added, three different intermediates can be trapped. The benzoyl radical leads to aminobenzaldehyde and benzamid, dehydrobenzene forms aniline, and the diphenyl diradical is trapped as carbazole:

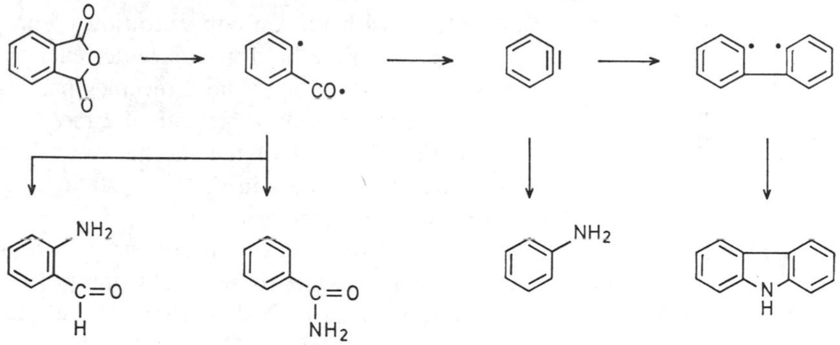

2.4. MECHANISMS OF ORGANIC PLASMA REACTIONS

The study of mechanism of plasma reactions is interesting, but rather difficult. The chemistry involved is very complex and only in rare cases is it possible to obtain information on reaction intermediates. Of the experimental parameters, only pressure, gas velocity (rate of distillation), and power uptake can be measured easily. The pressure dependence of the yields indicates whether a product has been formed by a mono- or bimolecular process (see p. 66). The degree of ionization, the average electron energy, the field strength, the concentration, and the nature of intermediates are normally not known and can only be determined by complicated measurements.

Elementary Processes

In a plasma various species are present. The concentration of charged particles in organic plasmas has not been determined but probably is on the order of $10^9 - 10^{12}$ cm^{-3}. Since the concentration of neutrals is 10^{16} cm^{-3} (at 1 torr), the degree of ionization is low ($10^{-4} - 10^{-7}$). The function of the electrons is visualized easily. The electrons are accelerated by the electric field and transfer energy to the molecules. Elastic collisions are of negligible importance for the chemistry in glow discharges since they transfer only very little energy. Inelastic collisions are very important and can lead to various species. Electrons of low energy may add to the molecules and form negative ions. Even if the electrons are attached to the molecules for very short times, they transfer part of their energy into the vibrational energy of the molecules.

When molecules collide with electrons of higher energy, they may become excited or ionized. Excited molecules can undergo various chemical processes as is known from photochemistry and high-temperature chemistry. The role of positive ions in the chemistry of glow discharges is still not clear. It is difficult to imagine that a small volume with an ion concentration of only 10^{11} cm^{-3} could account for yields up to 1 mole/hr. Ion–molecule reactions are known to be very fast. A fairly large fraction of the molecules passing the discharge could therefore be ionized for a short period of time. This however does not necessarily mean that the part of the mechanism which involves bond breaking and bond formation occurs in the ionic state, since the ions may loose their charge by charge-transfer reaction.

When an ion captures an electron, a considerable amount of energy (about 10 eV) is liberated. This amount is high enough to seriously damage the molecule. Normally recombinations occur in three-body collisions, the third body being another molecule, an electron, or the wall.

For organic molecules which tend to form negative ions or ion–molecule complexes, other recombination processes may be important:

$$M^+ + M^- = M^* + M^* \qquad \text{or} \quad M + M^*$$

$$MM^+ + e^- = M^* + M^* \qquad \text{or} \quad M + M^*$$

$$MM^+ + M^- = M^* + M^* + M \quad \text{or} \quad 2M + M^*$$

These processes, being two-body collisions, have a much higher probability. Organic molecules by means of the numerous bonds can take up a lot of energy in the form of vibrations and may therefore withstand such recombination processes.

The various species generated by electron impact may form the products through monomolecular processes (rearrangements, fragmentations) or by interactions with neighboring molecules. It is also conceivable that they are first transformed into reactive intermediates which then lead to the final products. The formation of the intermediates as well as their transfer to the reaction products could proceed through monomolecular or bimolecular processes (see Fig. 2.11). Evidently, such a system is extremely complex and only in the most simple cases one can hope for a complete analysis.

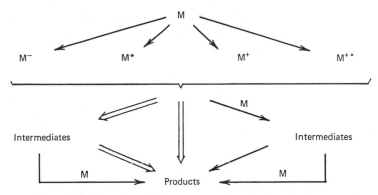

FIGURE 2.11. Possible elementary processes in plasma reactions.

Comparison of Plasma Reactions with Reactions Activated by Related Methods

A rather simple but not very accurate method to determine the mechanism of plasma reactions is to compare the products obtained from a plasma with those obtained when other energy sources are used. If the products are identical, it is reasonable to assume the existence of similar mechanisms. Such comparisons have been carried out and sometimes have revealed similarities between plasma reactions and pyrolysis, photolysis, and mass spectroscopy.

Certain substances form the same products in plasma reactions as in pyrolysis [152, 153]. Aliphatic hydrocarbons, for example, are transformed into mixtures of paraffins, olefins, and acetylenes. Similar mixtures are produced by high-temperature reactions [154,155]. In both methods, the predominant reactions probably are the generation of radicals, various types of radical transfers, and the stabilization of radicals. Azo compounds, carboxylic acids and esters also exhibit pronounced similarities in their behavior when pyrolyzed or passed through a plasma.

The relationship between plasma chemistry and pyrolysis is not surprising. Both techniques transfer energy to the molecules and thus cause the rupture of bonds. There is, however, also an important difference between the two methods. In pyrolysis, most of the molecules are at high-energy levels. In a plasma experiment, only few molecules gain a large amount of energy, and after reaction are immediately deactivated by collision with other molecules. If a fast deactivation is not important, both methods may lead to the same results. If, however, the reaction products are sensitive compounds, the yields of the plasma reaction are expected to be considerably higher than those obtained by pyrolysis.

In a number of cases (e.g., the isomerization of olefins), similarities are observed between plasma reactions and photochemical reactions. It is very unlikely that the light emitted by a glow discharge initiates photo reactions to a considerable extent. It is more reasonable to assume that molecular excitation is brought about by electron collisions.

Frequently, similarities are found between the behavior of a substance in plasma reactions and in mass spectroscopy. In both cases, an electron collision is the initial step, and plasma chemistry may therefore be considered as a kind of preparative mass spectroscopy. To a certain extent, one can predict the results of a plasma experiment on the basis of the mass spectrum of the starting material. It is, however, necessary to keep in mind three important differences between plasma chemistry and mass spectroscopy. A mass spectrometer monitors only ions while in plasma chemistry all ions are neutralized before identification. Bimolecular processes are rarely observed in mass spectroscopy but play an important role in plasmas. Finally, the electron energy differs, being about 70 eV in mass spectroscopy and only a few electron volts in glow discharges.

Sometimes there is no similarity to known mechanisms. An example is the plasma reaction of nitrobenzene [156]. In glow discharges nitrobenzene is converted to a number of compounds, all of which are products of the three intermediates—namely, the phenoxy radical, the phenyl radical and dehydro-benzene (see Fig. 2.12). The fragmentation of nitrobenzene to phenyl and phenoxy radicals is known from mass spectroscopy [115], photochemistry [157], and pyrolysis [158,160]. These fragmentations require 57 kcal/mole and about 40 kcal/mole [156]. The third reaction, the fragmentation of nitrobenzene to dehydrobenzene, has not been observed by other methods. The energy requirement of this reaction can be estimated to be about 250 kcal/mole. Surprisingly, this last reaction contributes as much to the fragmentation of nitrobenzene as the other two reactions. A process which could account for the occurrence of a reaction with such a high-energy

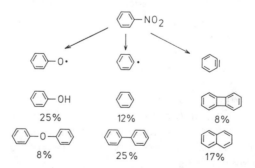

FIGURE 2.12. Decomposition of nitrobenzene in plasmas; only major products shown.

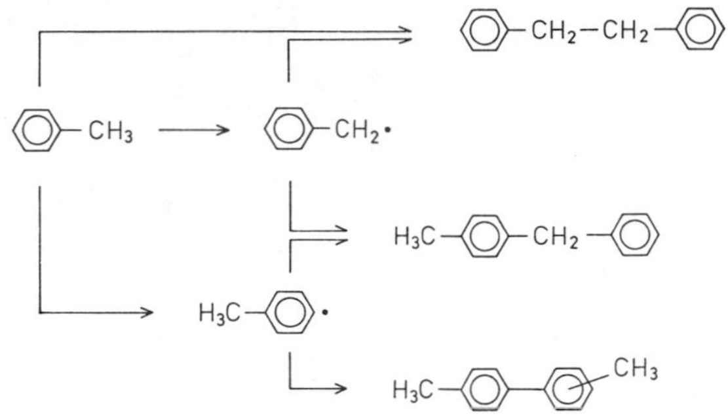

FIGURE 2.13. Possible mechanisms for the formation of dibenzyl from toluene.

demand is the recombination of nitrobenzene ions which liberates about 10 eV $\sim$ 230 kcal/mole).

Detailed Reaction Mechanisms

A few studies of the mechanisms of organic plasma reactions have been carried out in detail. The results of the investigations on toluene, acetylene, and anisole shall be discussed briefly.

When toluene is subjected to a glow discharge, the main products are dibenzyl, benzene, and ethylbenzene. Two different routes could lead to dibenzyl [118] (see Fig. 2.13). Since the yields of dibenzyl increase with pressure, a reaction path involving the monomolecular formation of benzyl radicals and their dimerization is unlikely. There is a higher probability of a bimolecular process involving an excited molecule. Such an interpretation is supported by a comparison of the yields of dibenzyl and methyldiphenyl methane which is formed as a by-product. If benzyl radicals were the intermediate, both compounds should always appear in a constant ratio. This, however, is in contrast to the experimental results.

Benzene and ethylbenzene show a different pressure dependence and are probably formed by different mechanisms [118]. The slight decrease in yield with increasing pressure indicates a monomolecular process, probably the elimination of the methyl group. The resulting methyl and phenyl radicals then lead to ethylbenzene and benzene.

Some interesting results have been obtained from studies of the mechanism of the reactions of acetylene in a plasma [42]. In various kinds of discharges, acetylene has been converted into a yellow polymer [147] of unknown structure [161] (the ir spectra resembles polystyrene). This process is interpreted in various ways. There is, however, general agreement that the polymerization follows a second-order rate [162–164] with a first intermediate

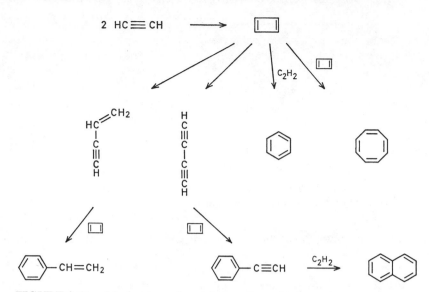

FIGURE 2.14. Mechanism of acetylene polymerization.

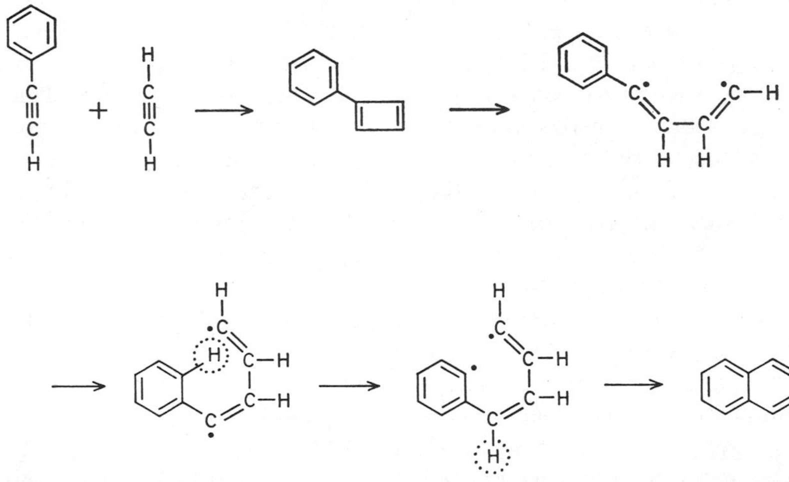

FIGURE 2.15. Mechanism of the formation of naphthalene from acetylene and phenylacetylene.

100

which is not vinylacetylene [162]. In light of the new experiments (which led to liquid and gaseous hydrocarbons, see p. 93), it seems highly likely that the first step in the acetylene polymerization is a dimerization to cyclobutadiene [42]. This intermediate, by ring opening and hydrogen shift, may lead to vinylacetylene, by further elimination of hydrogen to diacetylene (see Fig. 2.14). Addition of cyclobutadiene to acetylene, diacetylene and vinylacetylene would result in the Dewar forms of benzene, phenylacetylene, and styrene [165] which rapidly isomerize under the reaction conditions to aromatic systems [166]. Finally, the reaction of acetylene with phenylacetylene may lead in an analogous way via phenylcyclobutadiene to naphthalene (see Fig. 2.15). The proposed mechanism predicts enhanced formation of styrene and phenylacetylene if vinylacetylene or diacetylene is added to the acetylene. Similarly, mixtures of acetylene and phenylacetylene should give higher yields of naphthalene. The experimental results are totally in accord with these predictions.

Another mechanistic study has been carried out on the rearrangement of aromatic ethers (see p. 72). In this reaction the group attached to oxygen migrates to an ortho or para position or is eliminated [108]:

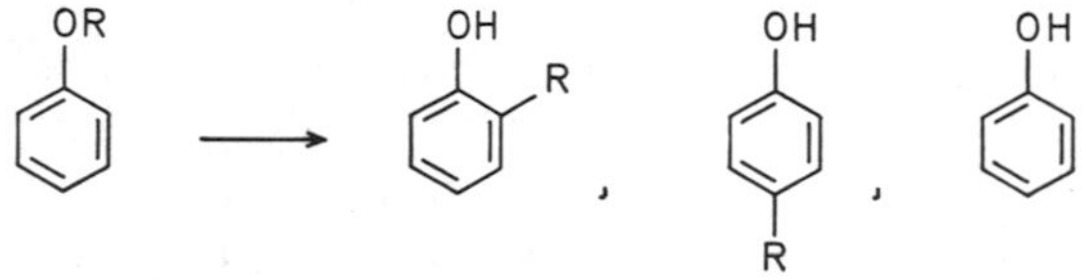

A result of great importance for the elucidation of the mechanism is the absence of compounds with an increased number of carbon atoms (e.g., $C_8H_{10}O$) among the reaction products. This rules out a route involving free methyl radicals. A concerted intramolecular migration is difficult to visualize in view of the fact that the isomer distribution of the rearranged products is nearly statistical and almost independent of the reaction conditions and the size of the migrating groups.

Both the inter- and the intramolecular reaction cannot account for the experimental results. Another possibility is a migration within a complex of two molecules. Only ion–molecule complexes have enough stability [167–170] for such a process.

The migration of an alkyl group within an ion–molecule complex could not account for the change of product distribution with increasing power (see Fig. 2.10). This result is best explained by a chain mechanism with an ion as chain propagator (see Fig. 2.16).

The initial steps probably are the ionization of the molecule and the homolysis of the weakest bond. The resulting ion forms a fairly stable complex with another ether molecule. Within this complex the methyl group migrates. The product of the migration is a complex of the ion with a cyclohexadienone.

$ArO^+ + ArOR \longrightarrow [ArO, ArOR]^+$ Start

$Ar{-}O{-}R \xrightarrow[-2e]{+e} [Ar{-}O{-}R]^+ \longrightarrow ArO^+ + R\cdot$

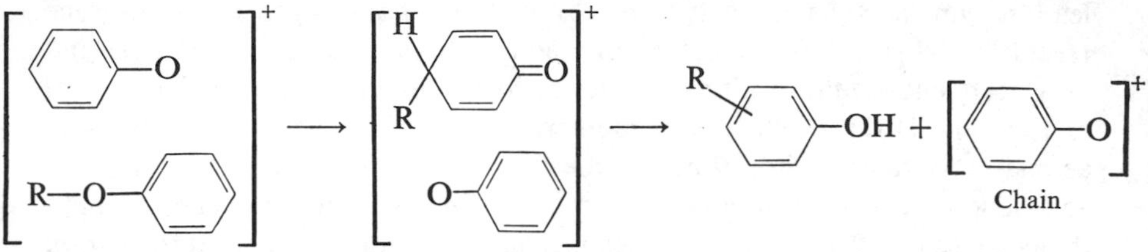

$[ArO, ArOR]^+ + e \longrightarrow ArOR + ArO\cdot$ Termination

FIGURE 2.16.   Mechanism of the arylalkyl ether rearrangement.

Since the cyclohexadienone is less basic than the ether, the complex is less stable and may decompose. The hexadienone tautomerizes to a cresol while the ion enters the chain again.

The chain is finally terminated when the complex is neutralized by electron capture. On neutralization the complex decomposes, leading to a phenoxy radical which may form phenol by hydrogen abstraction. An increased electron density (power) thus would shorten the chains and enhance the yields of phenol.

A final check on this mechanism can be made when two ethers are subjected to the plasma in common. Such experiments have been performed with mixtures of m-methylcresyl ether and phenetol [108]. These experiments yielded five crossover products in addition to the substances which are obtained when the ethers react separately and thus gave a final proof to the intermediate existence of an ion–molecule complex.

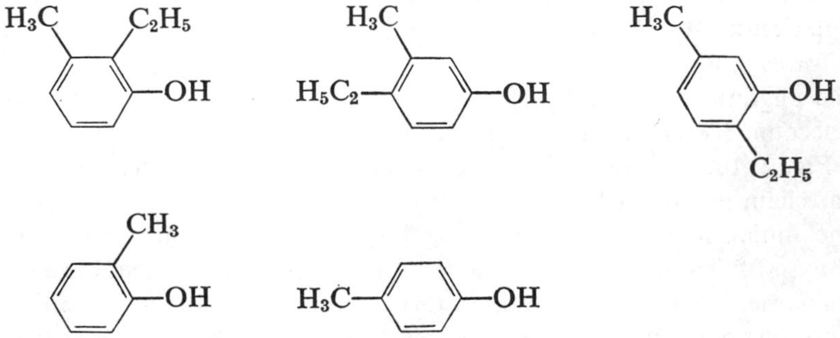

2.5.   SUMMARY AND CONCLUSIONS

The field of organic plasma chemistry is rather new. Only a limited number of reactions has been studied and much remains to be done. Some reactions which are typical for the behavior of organic substances in cold plasmas have been studied in more detail by using simple model systems. Aromatic compounds have mainly been used as models. Due to the stability of the aromatic ring, relatively simple results have been obtained. The model reactions themselves are normally not of preparative value, but they do supply mechanistic information which can be applied to problems of more preparative interest.

Many of the plasma reactions discussed in the preceding sections easily fulfill the requirements for a useful preparative reaction because of good yields and high selectivity. In other cases, mixtures of varying complexity are obtained in which the compound of interest may form a major or a minor constituent. The usefulness of such reactions depends on the ease of separation and purification.

The possibility of transferring the results of model systems to other molecules depends on the reactivity of all the groups in the molecule. Compounds with fairly unreactive substituents will probably give the same results as the model systems. For example, benzaldehyde and p-fluorbenzaldehyde will react alike since fluorine is a substituent with little reactivity. Molecules which have highly reactive substituents may give different results. In order to predict the course of a reaction, the reactivity of all substituents in the molecule must be considered. Present experience indicates that there is no completely unreactive substituent but that there are great differences in reactivity.

The most inert substituent for plasma reactions is the cyano group which either reacts not at all or, at a much slower rate than all other groups. Due to their great bond strength, fluorine, the phenyl group, and condensed aromatic systems are fairly inert. All other substituents are far more reactive. A sequence of reactivity may be estimated from bond energies, since normally the weakest bond of the molecule is broken most easily. Exceptions are known, however, and predictions on the basis of bond energies are only reliable if the reaction mechanism is known.

The selectivity of a plasma reaction depends on the number of reactive sites within the molecule. In an ideal case, one reaction path requires considerably less energy than all others. Under these conditions, the reactions can be optimized to lead to a single product. If the starting material can react in two or more different ways and all routes have comparable energy requirements, the result will always be a mixture of products. The energy difference between the two routes with the least energy requirement mainly determine the product distribution.

The influence of bond strength on the product distribution may be illustrated by results obtained with isopropylbenzene and toluene.

In isopropylbenzene, homolysis of the C—C bond in the side chain requires only 63 kcal/mole. Since this bond is weaker than all other bonds in the molecule, it is broken almost exclusively. The final product of this reaction, 2,3-diphenylbutane, can be obtained in an almost pure state.

In toluene the fragmentations of the C—C single bond and the benzene C—H bond require similar energies and therefore compete effectively. The predominant reaction products are dibenzyl, formed by breakage of a C—H bond, as well as benzene and ethylbenzene formed by homolysis of the C—C bond. An increase in the electron energy with decreasing pressure shifts the product distribution towards more C—C fragmentation (Table 2.12).†

TABLE 2.12. Influence of Pressure on the Product Distribution in Toluene Plasma Reactions [118]

	Dissociation energy, (kcal/mole)	Products (mole %)		
		$p = 0.8$	$p = 1.7$	$p = 3.5$
$\langle\bigcirc\rangle$—CH$_2$·	78	44	55	65
$\langle\bigcirc\rangle$·	85	53	44	34
H$_3$C—$\langle\bigcirc\rangle$·	102	3	1	0.5

† The picture is somewhat more complicated since the dibenzyl formation is a bimolecular process while the benzene and ethylbenzene formation is monomolecular.

The success of a plasma experiment depends very much on experimental conditions. Often only a narrow range of pressure, field strength, and flow rate is suitable for a specific reaction. Since in the present state of our knowledge, it is not possible to predict optimal working conditions for a new system based on known reactions, each new reaction requires a series of optimizing experiments.

Since no simple and fast techniques of plasma diagnostics are available, it is difficult to characterize and to reproduce a plasma experiment. It has also not yet been possible to study relationships between the various plasma properties and reaction mechanisms or yields. With improved diagnostics and better understanding of plasma mechanisms, it may become possible to predict the experimental parameters best suited for a certain reaction.

Another problem in plasma chemistry is the quantity of products. For organic plasma reactions, a low pressure is desirable to keep the gas temperature at low values. This, however, results in a small mass flow. The pressure of 1–5 torr used in preparative plasma reactions is a compromise between the requirements of reasonable mass throughputs and low gas temperatures. The situation may be improved by pulsing of the plasma. Calculations indicate that pulsed plasmas maintain low gas temperatures even at higher pressures [171].

All the recent plasma experiments have been carried out in a laboratory scale with mass flows of 10–100 g/hr. The scale-up of these experiments to a size suitable for industrial production has still to be solved.

A final limitation is that all plasma experiments are bound to the gas phase. Thus only compounds which can be vaporized without decomposition are suitable for plasma experiments.

REFERENCES

[1] BONDT, DEIMANN, P. VAN TROOSTWYK, and LAUWERENBURG, *Ann. Chim.* **21**(48), 58 (1796).

[2] W. V. SIEMENS, *Pogg. Ann.* **102**(66), 120 (1857).

[3] M. BERTHELOT, *Compt. Rend.* **67**, 1141 (1869); **82**, 1283 (1876).

[4] P. THENARD and A. THENARD, *Compt. Rend.* **78**, 219 (1874).

[5] P. DE WILDE, *Ber.* **7**, 352 (1874).

[6] S. M. LOSANITSCH, *Ber.* **42**, 4394 (1909).

[7] R. W. WOOD, *Proc. Roy. Soc. A* **97**, 455 (1920).

[8] R. W. WOOD, *Proc. Roy. Soc. A* **102**, 1 (1922).

[9] P. HARTEK, *Z. Phys. Chem.* **54**, 881 (1929).

[10] P. HARTEK and U. KOPSCH, *Z. Phys. Chem. B* **12**, 327 (1931).

[11] C. COPELAND, *Phys. Rev.* **36,** 1221 (1930).

[12] C. COPELAND, *J. Am. Chem. Soc.* **52,** 2580 (1930).

[13] R. J. STRUTT, *Proc. Roy. Soc. A* **85,** 219 (1911).

[14] C. P. BEGUIN, J. B. EZELL, A. SALVEMINI, J. C. THOMPSON, D. G. VICKROY, and J. L. MARGRAVE, *in* R. F. BADDOUR and R. S. TIMMINS (eds.), *The Application of Plasmas to Chemical Processing*, Pergamon Press, New York, 1967, Chap. 4.

[15] R. S. TIMMINS and P. R. AMMANN, *in* R. F. BADDOUR and R. S. TIMMINS (eds.), *The Application of Plasmas to Chemical Processing*, Pergamon Press, New York, 1967, Chap. 7.

[16] J. T. CLARKE, *in* R. F. BADDOUR and R. S. TIMMINS (eds.), *The Application of Plasmas to Chemical Processing*, Pergamon Press, New York, 1967, Chap. 8.

[17] C. S. STOKES, *in* M. VENUGOPALAN (ed.), *Reactions under Plasma Conditions*, Vol. 2, Wiley-Interscience, New York, 1971, Chap. 15.

[18] F. VURSEL and L. POLAK, *in* M. VENUGOPALAN (ed.), *Reactions under Plasma Conditions*, Vol. 2, Wiley-Interscience, New York, 1971, Chap. 16.

[19] P. L. SPEDDING, *Chem. Eng.* **17** (1969).

[20] G. FRANCIS, *in* S. FLÜGGE, *Handbuch der Physik, Encyclopedia Physics*, Vol. 22, 97 (1956).

[21] H. SCHÜLER and E. LUTZ, *Z. Naturf.* **12a,** 334 (1957).

[22] H. SCHÜLER and M. STOCKBURGER, *Z. Naturf.* **14a,** 229 (1959).

[23] H. SOBOTKA, *HF-Industriegeneratoren für induktive Erwärmung*, Philips Techn. Bibliothek, N. V. Philips Gloeilamperfabr. Eindhoven, Holland 1962.

[24] RCA Transistor Manual (1967), Harrison, N. J., Technical Series SC-13.

[25] H. MEINKE and F. GUNDLACH, *Taschenbuch der HF-Technik*, Springer-Verlag, Berlin, 1968.

[26] H. KOCH, *Transistorsender*, Vol. 2, Verlag, Munich, 1970.

[27] F. K. MCTAGGART, *Plasma Chemistry in Electrical Discharges*, Elsevier, New York, 1967.

[28] Ref. 25 pp. 457-483.

[29] F. C. FEHSENFELD, K. M. EVENSON, and H. P. BROIDA, *Nat. Bur. Std. U.S. Ann. Rep. No. 8701* and *Rev. Sci. Instr.*, **36,** 294 (1965).

[30] Ref. 27, Chap. 4.

[31] R. G. BOSISIO, M. R. WERTHEIMER, and C. F. WEISSFLOCH, *J. Phys. E* **6,** 628 (1973).

[32] R. L. MCCARTHEY, *J. Chem. Phys.* **22,** 1360 (1954).

[33] R. I. WEISS, Thesis, University of Tübingen, Tübingen, West Germany, 1971.

[34] U. Schöch, Diplomarbeit, University of Tübingen, Tübingen. West Germany, 1969.

[35] K. G. MÜLLER, *Chem. Ing. Tech.* **3,** 122 (1973).

[36] H. SCHÜLER and I. REINEBECK, *Z. Naturf.* **5a,** 657 (1950).

[37] J. C. DEVINS and M. BURTON, *J. Am. Chem. Soc.* **76,** 2618 (1954).

[38] D. SAVAGE, *J. Dep. Chem. Eng. Univ. Newcastle Upon Tyne* (5), 33 (1969).

[39] J. D. THORNTON, W. D. CHARLTON, and P. L. SPEDDING, *in* Advances in Chemistry Series 80, *Chemical Reactions in Electrical Discharges*, Am. Chem. Soc., Washington, D.C., 1969, Chap. 13.

[40] J. G. ANDREWS and J. E. ALLEN, *Proc. Roy. Soc.* (*London*) *A* **320,** 459 (1971).

[41] J. G. ANDREWS and R. H. VAREY, *Nature* **225,** 270 (1970).

[42] G. ROSSKAMP, Thesis, University of Tübingen, Tübingen, West Germany, 1971.

[43] Ref. 27, Chaps. 6–9.

[44] F. KAUFMAN, *in* Advances in Chemistry Series 50, *Chemical Reactions in Electrical Discharges*, Am. Chem. Soc., Washington, D.C., 1969, Chap. 3.

[45] F. KAUFMAN, *J. Chem. Phys.* **28,** 352 (1958).

[46] F. KAUFMAN and J. R. KELSO, *J. Chem. Phys.* **32,** 301 (1960).

[47] P. HARTECK and E. POLDER, *Z. Phys. Chem.* **178,** 389 (1937).

[48] R. W. WOOD, *Proc. Roy. Soc.* (*London*) *A* **102,** 1 (1922).

[49] H. G. POOLE, *Proc. Roy. Soc.* (*London*) *A* **163,** 415,424 (1937).

[50] K. F. BONHOEFFER, *Z. Phys. Chem.* **113,** 199 (1924).

[51] H. F. SMYSER and H. M. SMALLWOOD, *J. Am. Chem. Soc.* **55,** 3499 (1933).

[52] I. E. NEWNHAM and J. A. WATTS, *J. Am. Chem. Soc.* **82,** 2113 (1960).

[53] T. G. PEARSON, P. L. ROBINSON, and E. M. STODDART, *Proc. Roy. Soc.* (*London*) *A* **142,** 275 (1933).

[54] H. I. SCHLESINGER and A. B. BUNG, *J. Am. Chem. Soc.* **53,** 4321 (1931).

[55] B. A. THRUSH, *Progress in Reaction Kinetics*, Vol. 3, Pergamon Press, Oxford, 1965, p. 63.

[56] H. G. WAGNER and J. WOLFRUM, *Angew. Chem.* **83,** 561 (1971).

[57] P. E. M. ALLEN, H. W. MELVILLE, and J. C. ROBB, *Proc. Roy. Soc.* (*London*) *A* **218,** 311 (1953).

[58] K. H. GEIB and P. HARTECK, *Ber. dtsch. chem. Ges.* **66,** 1815 (1933).

[59] B. S. RABINOWITCH, S. G. DAVIS, and C. A. WINKLER, *Can. J. Res. B* **21,** 251 (1943).

[60] R. KLEIN and M. D. SCHEER, *J. Am. Chem. Soc.* **80,** 1007 (1958).

[61] R. J. CVETANOVIC, *Advances in Photochemistry*, Vol. 1, Wiley-Interscience, New York, 1963, p. 115.

[62] K. HOYERMANN, H. G. WAGNER, J. WOLFRUM, and R. ZELLNER, *Chem. Ber.* **75,** 22 (1971).

[63] P. HARTECK and U. KOPSCH, *Z. Phys. Chem. B* **12,** 327 (1931).

[64] E. W. STEACIE and N. A. D. PARLEE, *Can. J. Res. B* **16,** 203 (1938).

[65] R. J. CVETANOVIC, *J. Chem. Phys.* **23,** 1375 (1955).

[66] R. J. CVETANOVIC, *J. Chem. Phys.* **30,** 19 (1959).

[67] G. BOOCOCK and R. J. CVETANOVIC, *Can. J. Chem.* **39,** 2436 (1961).

[68] W. K. STUCKEY and J. HEICKLEN, *J. Chem. Phys.* **46,** 4843 (1967).

[69] R. J. CVETANOVIC, *Can. J. Chem.* **36,** 623 (1958).

[70] S. SATO and R. J. CVETANOVIC, *Can. J. Chem.* **36,** 1668 (1958).

[71] S. SATO and R. J. CVETANOVIC, *Can. J. Chem.* **37,** 953 (1959).

[72] R. J. CVETANOVIC, *J. Chem. Phys.* **30,** 19 (1959).

[73] J. M. S. JARVIE and R. J. CVETANOVIC, *Can. J. Chem.* **37,** 529 (1959).

[74] R. WEISBECK and R. HÜLLSTRUNG, *Chem. Ing. Tech.* **42,** 1302 (1970).

[75] G. G. MANNELLA, *Chem. Rev.* **63,** 1 (1963).

[76] O. K. FOMIN, *Russ. Chem. Rev.* (Engl. transl.) **36,** 725 (1967).

[77] A. M. WRIGHT and C. A. WINKLER, *Active Nitrogen*, Academic Press, New York, 1968.

[78] G. R. BROWN and C. A. WINKLER, *Angew. Chem.* **82,** 187 (1970).

[79] A. N. WRIGHT, R. L. NELSON, and C. A. WINKLER, *Can. J. Chem.* **40,** 1082 (1962).

[80] A. TSUKAMOTO and N. N. LICHTIN, *J. Am. Chem. Soc.* **82,** 3798 (1960).

[81] A. FUJINO, S. LUNDSTED, and N. N. LICHTIN, *J. Am. Chem. Soc.* **88,** 775 (1966).

[82] T. HANAFUSA and N. N. LICHTIN, *Can. J. Chem.* **44,** 1230 (1966).

[83] S. L. MILLER, *Science* **117,** 528 (1953).

[84] S. L. MILLER, *J. Am. Chem. Soc.* **77,** 2351 (1955).

[85] S. L. MILLER, *Biochem. Biophys. Acta* **23,** 488 (1957).

[86] S. L. MILLER and H. C. UREY, *Science* **130,** 245 (1959).

[87] K. HEYNS, W. WALTER, and E. MEYER, *Naturwiss.* **44,** 385 (1957).

[88] C. PALM and M. CALVIN, *J. Am. Chem. Soc.* **84,** 2115 (1962).

[89] C. PONNAMPERUMA, R. M. LEMMON, R. MARINES, and M. CALVIN, *Proc. N. A. S.* **49,** 737 (1963).

[90] J. ORO, *Nature* **197,** 862 (1963).

[91] J. ORO, *Nature* **197,** 971 (1963).

[92] C. PONNAMPERUMA, F. WOELLER, J. FLORES, M. ROMIEZ, and W. ALLEN, *in* Advances in Chemistry Series 80, *Chemical Reactions in Electrical Discharges*, Am. Chem. Soc., Washington, D.C., 1969, Chap. 23.

[93] M. CALVIN, *Chemical Evolution*, Clarendon Press, Oxford, 1969, Chap. 6.

[94] H. SUHR and U. SCHÜCKER, *Synthesis* **1970,** 431.

[95] E. V. BLACKBURN and C. J. TIMMONS, *Quart. Rev.* **23,** 482 (1969).

[96] J. SALTIEL and E. D. MEGARITY, *J. Am. Chem. Soc.* **91,** 1265 (1969).

[97] L. ZECHMEISTER and W. H. MCNEELY, *J. Am. Chem. Soc.* **64,** 1919 (1942).

[98] U. SCHÖCH, Thesis, University of Tübingen, Tübingen, West Germany, 1971.

[99] A. Szabo and H. Suhr, unpublished results.

[100] L. Kaplan, K. E. Wilzzach, W. G. Brown, and S. S. Yang, *J. Am. Chem. Soc.* **87**, 675 (1965).

[101] K. E. Wilzbach, A. L. Harlmess, and L. Kaplan, *J. Am. Chem. Soc.* **90**, 116 (1968).

[102] L. Kaplan and K. E. Wilzbach, *J. Am. Chem. Soc.* **90**, 3291 (1968).

[103] H. G. Viehe, *Angew. Chem.* **77**, 768 (1965).

[104] K. E. Wilzbach and L. Kaplan, *J. Am. Chem. Soc.* **37**, 4004 (1965).

[105] R. I. Weiss, Thesis, University of Tübingen, Tübingen, West Germany, 1971.

[106] H. Suhr, G. Rolle, and B. Schrader, *Naturwiss.* **55**, 168 (1968).

[107] H. Suhr and R. I. Weiss, *Z. Naturf.* **25b**, 41 (1970).

[108] H. Suhr and R. I. Weiss, *Liebigs Ann. Chem.* **760**, 127 (1972).

[109] R. I. Weiss and H. Suhr, *Liebigs Ann. Chem.* 301 (1973).

[110] H. Suhr, W. Hohnefeld, G. Kruppa, W. Reck, G. Rosskamp, U. Schöch, A. Szabo, and R. I. Weiss, publication in preparation (1974).

[111] H. Suhr, *Angew. Chem.* **84**, 876 (1972); *Int. Ed.* **11**, 781 (1972).

[112] H. Suhr and G. Kruppa, *Liebigs Ann. Chem.* **744**, 1 (1971).

[113] H. Suhr and R. I. Weiss, *Angew. Chem.* **82**, 295 (1970); *Int. Ed.* **9**, 312 (1970).

[114] A. Szabo and H. Suhr, *Liebigs Ann. Chem.* in press (1974).

[115] H. Budzikiewicz, C. Djerassi, and P. H. Williams, *Mass Spectroscopy of Organic Compounds*, Holden-Day, San Francisco, 1967, p. 515.

[116] E. K. Fields and S. Meyerson, *Chem. Comm.* **1965**, 474.

[117] H. Suhr and A. Szabo, *Liebigs Ann. Chem.* **752**, 37 (1971).

[118] H. Suhr, *Z. Naturf.* **23b**, 1559 (1968).

[119] H. Suhr, U. Schöch, and G. Rosskamp, *Chem. Ber.* **104**, 674 (1971).

[120] H. Suhr, *Angew. Chem.* **84**, 876 (1972); *Int. Ed.* **11**, 781 (1972).

[121] H. Suhr, *Fortschr. Chem. Forschg.* **36**, 39 (1974).

[122] J. K. Stille and C. E. Rix, *J. Org. Chem.* **31**, 1591 (1966).

[123] H. Schüler, K. Prehaf, and E. Kloppenhey, *Z. Naturf.* **15a**, 308 (1960).

[124] H. Schüler and V. Degenhart, *Z. Naturf.* **7a**, 753 (1952).

[125] C. Bolhouwer and H. I. Waterman, *Chem. Abstr.* **50**, 13617 (1956).

[126] N. Nishi, *Chem. Abstr.* **50**, 12660 (1956).

[127] H. Schüler, K. Pochal, and E. Kloppenburg, *Z. Naturf.* **15a**, 308 (1960).

[128] H. Schüler and E. Lutz, *Spectrochim. Acta.* **10**, 61 (1957).

[129] F. Swift, *J. Org. Chem.* **30**, 3114 (1965).

[130] R. Weisbeck, *Chem. Ing. Tech.* **43**, 721 (1971).

[131] C. Boehouwer and H. I. Waterman, *Res. London* **9**, 11 (1956).

[132] H. Kokad, T. Mori, and I. Tanake, *Chem. Abstr.* **51**, 16111 (1957).

[133] A. STREITWIESER and H. R. WARD, *J. Am. Chem. Soc.* **84**, 1065 (1962).

[134] A. STREITWIESER and H. R. WARD, *J. Am. Chem. Soc.* **85**, 539 (1963).

[135] M. VACHER and Y. CORTIE, *J. Chem. Phys.* **56**, 732 (1959).

[136] H. SUHR, U. SCHÖCH, and U. SCHÜCKER, *Synthesis* **1971**, 426.

[137] K. TAKI, *Bull. Chem. Soc. (Japan)* **43**, 1574 (1970).

[138] F. J. DINAN, S. FRIDMAN, and P. J. SCHIRMANN, *in* Advances in Chemistry Series 80, *Chemical Reactions in Electrical Discharges*, Am. Chem. Soc., Washington, D.C., 1969, Chap. 24.

[139] H. SCHÜLER and M. STOCKBURGER, *Spectrochim. Acta* **841** (1959).

[140] R. VANIER DE SAINT-AUNAY, *Chim. Ind.* **29**, 1011 (1933).

[141] W. STROHMEIER and H. GRÜBER, *Z. Naturf.* **20b**, 11 (1965).

[142] A. BRADLEY, *Ind. Eng. Chem. Prod. Res. Dev.* **9**, 101 (1970).

[143] D. R. SECRIST, *in* Advances in Chemistry Series 80, *Chemical Reactions in Electrical Discharges*, Am. Chem. Soc., Washington, D.C., 1969, Chap. 19.

[144] J. R. HOLLAHAN and R. P. MCKEEVER, *in* Advances in Chemistry Series 80, *Chemical Reactions in Electrical Discharges*, Am. Chem. Soc., Washington, D.C., 1969, Chap. 22.

[145] D. D. NEISWENDER, in Advances in Chemistry Series 80, *Chemical Reactions in Electrical Discharges*, Am. Chem. Soc., Washington, D.C., 1969, Chap. 29.

[146] P. M. HAY, *in* Advances in Chemistry Series 80, *Chemical Reactions in Electrical Discharges*, Am. Chem. Soc., Washington, D.C., 1969, Chap. 30.

[147] Ref. 27, p. 198.

[148] J. H. WAGENKNECHT, *Ind. Eng. Chem. Prod. Res. Dev.* **10**, 184 (1971).

[149] K. SUGINO and E. INOUE, *J. Chem. Soc. (Japan)* **71**, 518 (1950).

[150] J. CHIN CHU, H. C. AI, and D. F. OTHMER, *Ind. Eng. Chem.* **45**, 1266 (1953).

[151] K. SUGINO, E. INOUE, and K. ODO, *J. Chem. Soc. (Japan)* **71**, 343 (1950).

[152] A. G. LOUDON, A. MACCOLL, and S. K. WONG, *J. Chem. Soc. B* 1733 (1970).

[153] Y. KAWAHARA, *J. Phys. Chem.* **73**, 1648 (1969).

[154] A. S. GORDON, *J. Am. Chem. Soc.* **70**, 395 (1948).

[155] N. FRIEDMANN, H. H. BOVEE, S. L. MILLER, *J. Org. Chem.* **35** (1970).

[156] H. SUHR and G. ROSSKAMP, *Liebigs Ann. Chem.* **742**, 43 (1970).

[157] R. HURLEY and A. C. TESTA, *J. Am. Chem. Soc.* **88**, 4330 (1966).

[158] E. K. FIELDS and S. MEYERSON, *J. Am. Chem. Soc.* **89**, 3224 (1967).

[159] E. K. FIELDS, *J. Am. Chem. Soc.* **89**, 724 (1967).

[160] E. K. FIELDS and S. MEYERSON, *Acc. Chem. Res.* **2**, 273 (1969).

[161] S. C. LIND, *Radiation Chemistry of Gases*, Reinhold, New York, 1961, p. 172.

[162] G. B. SKINNER and E. SOKOLOSKI, *J. Phys. Chem.* **64**, 1952 (1960).

[163] C. F. CULLIS, G. J. MINKOFF, and M. A. NETTLETON, *Trans. Farad. Soc.* **58**, 1117 (1962).

[164] E. K. FIELDS and S. MEYERSON, *Tetrahedron Letters* **6,** 571 (1967).

[165] J. FONT, S. C. BARTON, and O. P. STRAUSZ, *Chem. Comm.* **1970,** 499.

[166] J. F. M. OTH, *Angew. Chem.* **80,** 633 (1968).

[167] F. H. FIELD, P. HAMLET, and W. F. LIBBY, *J. Am. Chem. Soc.* **89,** 6035 (1967).

[168] S. WEXLER and R. P. CLOW, *J. Am. Chem. Soc.* **90,** 3940 (1968).

[169] F. H. FIELD, P. HAMLET, and W. F. LIBBY, *J. Am. Chem. Soc.* **91,** 2839 (1969).

[170] C. LIFSHITZ and B. G. REUBEN, *J. Chem. Phys.* **50,** 951 (1969).

[171] K. G. MÜLLER, *Chem. Ing. Tech.* **3,** 122 (1973).

Chapter 3

Plasma Treatment of Solid Materials

Martin Hudis

3.1. INTRODUCTION

A growing interest has arisen over the past 15 years in the use of plasmas to modify the surface structure and composition of solid materials. This interest has been motivated in part by the ability of plasmas to produce unique surface modifications and by the ease with which the extent of modification can be controlled. This latter property can afford plasma treatment processes an edge over more conventional processes.

The unique surface modifications which can be achieved in a plasma are due to the effects of the ultraviolet radiation and chemically active species produced by the plasma (see Chapter 1). By contrast, thermal surface treatment processes require temperatures which can either damage or distort the solid being treated, and ultraviolet radiation processes are limited by the spectrum available from uv lamps. A significant advantage of plasma processes is that they can take advantage of both the uv radiation and the active species simultaneously. Thus in the case of a polymer, the uv radiation can produce polymer free radicals which react with free radicals produced in the plasma. A second important advantage is that plasma processes can be controlled easily through the large number of independent parameters influencing the properties of the plasma. Thermal, chemical, and high-energy radiation processes usually do not possess equivalent degrees of freedom in their control.

There are three general means by which a plasma and a solid can interact. In the first case, the plasma and the solid are physically in contact, are electrically isolated, and have a steady-state interaction. Physical contact is produced by placing the solid directly in the volume where the plasma is generated or by placing it downstream from this position and allowing the plasma to expand over the solid. Electrical isolation means that there is no net current arriving at the solid boundary and that the potential of the solid floats relative to the plasma. An example of this type of interaction is a sheet of polyethylene placed inside of a rf electrodeless discharge. In the second case, the plasma and solid are both physically and electrically in contact and again have a steady-state interaction. As a result of the electrical contact, current is drawn from the plasma through the solid material. The solid material may be an electrode used in the production of the plasma or it may be independent of the plasma and electrically biased relative to the plasma. A current flow to the solid material can alter the reactions at the surface by producing a change in the plasma and gas composition around the solid. A typical example is the metal cathode of a dc glow discharge. The metal cathode is used to produce the plasma and at the same time undergoes surface chemical modifications by interacting with the plasma. The third case involves a nonsteady-state interaction in which the solid has physical but no electrical contact. The nonsteady-state condition can be achieved by dropping solid particles through the core of a plasma. A typical example is zircon ($ZrSiO_4$) particles dropped through a 1 atmosphere, 10 kW argon plasma.

The interactions of a plasma with a solid covers a broad spectrum of topics bordering on many disciplines. An attempt to cover all of the topics is beyond the scope of this chapter. This chapter will be confined to a discussion of steady-state interactions of plasmas with polymers and metals. Specific attention will be given to the mechanisms of energy transfer from the plasma, the dissipation of energy at the solid, and the resulting surface modifications.

3.2. PLASMA–SOLID ENERGY-TRANSFER AND DISSIPATION PROCESSES

Energy can be transferred from a plasma to a solid through optical radiation, through neutral particle fluxes, and through ionic particle fluxes. The energy transferred from the plasma is dissipated within the solid by a variety of chemical and physical processes. These dissipation processes are the origin of the desired surface property modifications.

The optical radiation emitted by a plasma contains components in the infrared (ir), the visible, and the ultraviolet (uv). For polymers, the visible radiation is weakly absorbed and does not produce interesting chemistry. The ir radiation can be strongly absorbed but is dissipated through thermal reactions which produce heat and thus also fails to produce any interesting chemistry. The uv radiation is strongly absorbed by polymers producing polymer free radicals. The polymer free radicals are active sites which can then react with gas components within the plasma.

The neutral particles in the plasma continually bombard the solid transferring energy from the plasma to the solid. The neutral particle flux contains energy in four forms: Kinetic, vibrational, dissociation (free radicals), and excitation (metastables). The kinetic and vibrational forms of energy heat the solid. They provide an oxygen-free heat source and are of no direct interest to the type of chemistry being discussed here. For polymers, the free-radical dissociation energy is dissipated through surface chemical reactions such as abstraction, addition, and oxidation as well as through thermal heating caused by free-radical recombination on the surface of the solid. For metals, the dissociation energy is dissipated through absorbed surface reactions and recombination. Metastable atoms and molecules are the principal carriers of the energy stored in electronic excitation not released by radiation. These metastable particles cannot radiate and hence must dissipate their energy by collisions with the surface. For polymers, the metastable energy in general is larger than the polymer dissociation energy and tends to produce polymer free radicals. For metals, the metastable energy is dissipated by thermal heating and does not induce any chemical reactions.

The ionic flux to a surface is the third means for energy transfer. The ions which impinge on the surface carry kinetic, vibrational, and electronic energy. The kinetic energy can be large and is a function of the electrical contact. For a surface which floats relative to the plasma potential, the ions arrive at the surface with up to 5 V of energy. This energy is dissipated by physical effects such as ablation and electrostatic charging. For a biased metallic surface, the ions are accelerated by the cathode-fall voltage and can hit the surface with hundreds of volts of energy. The kinetic energy is dissipated through heating and sputtering. Ions produced by dissociative ionization can also bombard the surface leading to reactions similar to those caused by dissociated neutral particles. The ionization energy is released by neutralization at the surface and is dissipated by heating for metals and by the production of free radicals for polymers.

3.3. APPLICATIONS OF PLASMA–SOLID REACTIONS TO POLYMERS

Survey of Applications

The principal changes brought about by exposure of a polymer to a plasma are in the surface wettability, the molecular weight of a surface layer, and the chemical composition of the surface. For the most part, the effects of the plasma treatment are confined to a layer 1–10 μ in depth. As a result, the bulk properties of the treated polymer remain unchanged.

The wettability of a polymer surface, defined as the contact angle between a drop of liquid and the surface [1], is an important characteristic which relates to the adherence of dyes, inks, and adhesives to the polymer. Plasma processes which lead to an improved wetting have found application in the packaging, electronics, construction, and clothing industries. Some typical examples are discussed below.

Labeling the surface of polymers and polymeric coatings is a major problem. Polymers tend to be inert and cannot be printed, painted, or dyed. Wire and cables sealed in polymer jackets are a typical example. Plasma processes are available which increase the surface wettability and allow the polymer jacket to be labeled. The plasma process is valuable because it produces no changes in the bulk properties which led to the use of the polymer originally.

In general, it is difficult to bond, graft, or glue to polymer surfaces. Polymeric-potting materials are a typical example. High voltage and small compact electronic equipment are commonly potted in various materials to prevent dielectric breakdown, to provide structural support, and to prevent environmental attack from moisture, sunlight, and dust. The superior electrical and thermal properties of silicone (high dielectric strength, high thermal stability, the ability to be molded) have been recognized for some time, but its weak environmental resistance has prevented its use as a potting agent. The environmental resistance can be improved through the use of a plasma process which allows epoxy resins to be bonded to the silicone. A number of pieces of electronic equipment are now being potted in silicone, exposed to an oxygen plasma, and sealed in epoxy. This new technique provides superior potting and decreases the weight and size of the total package. Capacitors which use polymers as their dielectric material are a second example. Good contact between the electrode and the polymer is a major problem. Improved bonding between the electrode and the polymer has recently been achieved by a vapor depositing aluminum directly to the polymer material. Increased surface wetting is a prerequisite for good results and is achieved using a corona discharge. Again, the advantage of the plasma is its ability to confine its effect to the outer surface. Chemical or thermal processes could affect the bulk properties of the polymer which could lead to a deterioration in the dielectric properties.

Similar techniques have been used to bond polymers to dissimilar materials while producing no change in their bulk properties. Such processes are having a large impact on the general use of polymers. Typical examples are silicone and polytetrafluoroethylene (Teflon), bonded to polyethylene, and polymers in general bonded to epoxies which in turn can be bonded to a variety of materials.

Molecular weight changes are a second property modification which can be produced with a plasma. Variations in this characteristic affects a number of physical properties of the polymer such as permeability, solubility, melt temperature, and cohesive strength. The cohesive strength is important for adhesion and again is a surface effect.

The principal processes by which the molecular weight of the surface layer is changed are scissioning, branching, and cross-linking [2]. Scissioning is a polymer dissociation process which reduces the length of the polymer molecule. Branching refers to the creation of side chains which are attached to the main polymer molecule. Cross-linking produces a three-dimensional structure by forming covalent bonds at random between adjacent polymer chains.

Increased adhesion due to increased wettability results from improved bonding between the epoxy and the polymer. Whereas increasing the surface wettability helps to form a strong adhesive bond, it is not always sufficient due to poor surface (0.1 μ) cohesive properties. Adhesion to some of these polymers can be improved by cross-linking the outer surface. The plasma process has the advantage that it acts within seconds and produces no noticeable effects on the bulk properties of the polymer.

Exposure of a polymer to a plasma can also be used to create reactive sites on the polymer surface. These changes in the surface composition can be produced by bond rearrangement reactions which lead to unsaturation, or by attachment of amine and halogens to the polymer surface. The groups are attached through covalent bonds and can act as "hooks" for the addition of new compounds which can further change the surface chemical and electrical properties. A typical example is the grafting of heparin. Capillary tubes made of inert flexible polymer material are ideal for catheters but, unfortunately, cause blood clotting. By attaching amine groups to polyethylene and then coating them with an anticoagulant like heparin, the clotting time can be increased.

A summary of those polymers which have been studied is shown in Table 3.1. Common polymer trade names are shown in Table 3.2. Table 3.1 is

TABLE 3.1. Summary of Polymer Plasma Results

SUMMARY OF POLYMER PLASMA RESULTS

POLYMER MATERIAL	WETTABILITY						MOLECULAR WEIGHT CHANGES						CHEMICAL REACTIVITY
	OXYGEN	HYDROGEN	NITROGEN	HELIUM	ARGON	INERT GASES	OXYGEN	HYDROGEN	NITROGEN	HELIUM	ARGON	INERT GASES	
FLUOROCARBON POLYMERS													
POLYTETRAFLUOROETHYLENE	d	f[o]	f[o]	f[o]	f[o]	f[o]	d[o]	f	f	f	f	f	NH_3[k]
POLY(VINYL FLUORIDE)	d[o]						d[-]	g		g			
POLYPERFLUOROPROPYLENE	t						e[o]				t		
POLY(VINYLIDENE FLUORIDE)	t										t		
POLYTRIFLUOROCHLOROETHYLENE	d						d[o]						
HYDROCARBON POLYMERS													
POLYETHYLENE	a	f[o]	f	f[o]	f[o]	f[o]	d[t]	f	b	f	f	f	CO_2[r]
POLYPROPYLENE	d	w[o]	p[o]f	p[o]	p[o]	p[o]	d[t]	w[o]	c	h[o]	p[o]	p[o]	NH_3, N_2O[p], H_2S[p]
POLYSTYRENE							e[-]		h				
POLYVINYLCYCLOHEXANE							e[-]						
POLYISOBUTYLENE										c[o]			
POLYBUTENE-I							c[-]						
POLYBUTADIENE	n												
HYDROCARBONS (OXYGEN)													
POLY(ETHYLENE TEREPHTHALATE)	d[-]						d[-]		h				
POLYCARBONATE	e						e[-]		h				NH_3[k]
POLY(OXYMETHYLENE)	t						e[-]		h				
CELLULOSE ACETATE							e[-]						
POLY(METHY METHACRYLATE)							e[-]						NH_3[k]
HYDROCARBONS (OXYGEN, NITROGEN)													
POLYHEXAMETHYLAMINE	e						e[-]		h				
POLYURETHANE													NH_3[k]
HYDROCARBON (CHLORINE)													
POLY(VINYL CHLORIDE)													NH_3[k]
HYDROCARBON (SILICONE, OXYGEN)													
POLYDIMETHYSILOXANE	m	w	m[d]	s									NH_3[k]

a Kurt Rossmann, 1956 [5].
b J. L. Weininger, 1960 [15].
c J. L. Weininger, 1961 [16].
d R. M. Mantell and W. L. Ormand, 1964 [3].
e R. M. Hansen, et al., 1965 [9].
f R. H. Hansen and H. Schonhorn, 1966 [23].
g H. Schonhorn and R. H. Hansen, 1967 [25].
h J. R. Hall, et al., 1969 [10].
i P. Blais, et al., 1971 [19].
j C. Y. Kim, et al., 1971 [7].
k J. R. Hollahan, et al., 1969 [14].
l Neon, krypton, and zenon.
m J. R. Hollahan and G. L. Carlson, 1970 [11].
n M. L. Kaplan and P. G. Kelleher, 1970 [45].
p H. Schonhorn, et al., 1970 [13].
q P. Blais, et al., 1971 [19].
r C. Y. Kim, et al., 1971 [7].
s N. J. Delollis, 1972 [12].
t J. R. Hall, et al., 1972 [21].
w M. Hudis, 1972 [24].

TABLE 3.2

COMMON TRADE NAME	POLYMER NAME
MYLAR[a]	POLY (ETHYLENE TEREPHTHALATE)
NYLON[g]	POLYHEXAMETHYLAMINE
DELRIN[b], CELCON[b]	POLY (OXYMETHYLENE)
PLEXIGLAS[c] LUCITE[a]	POLY (METHY METHACRYLATE)
LEXAN[d]	POLYCARBONATE
TFE TEFLON[a]	TETRAFLUOROETHYLENE
FEP TEFLON[a]	POLYPERFLUOROPROPYLENE
KEL-F[e]	POLYTRIFLUOROCHLOROETHYLENE
TEDLAR[a]	POLY (VINYL FLUORIDE)
KYNAR[f]	POLY (VINYLIDENE FLUORIDE)

[a] E. I. du Pont de Nemours and Co.
[b] Celanese Plastic Co.
[c] Rohm and Haas Co.
[d] General Electric Co.
[e] Minnesota Mining and Manufacturing Co.
[f] Pennwalt Corp.
[g] Resin name.

subdivided according to the chemical composition of the polymers, the property changes induced in the polymers, and the gases used for the plasma. The lowercase letters in the table reference the original work. The 0 and − superscripts in the wettability column indicate no change in wettability and a decrease in wettability, respectively. The unscripted letters indicate an increase in wettability. The +, 0, and − superscript in the molecular weight change column indicate an increase in molecular weight, no change in molecular weight, and a decrease in molecular weight, respectively. Polymers can scission and cross-link simultaneously in which case the molecular weight change notation is somethat ambiguous. The − superscript indicates ablation at the polymer surface due to scissioning but does not indicate that cross-linking cannot or does not occur simultaneously. The i and j superscript

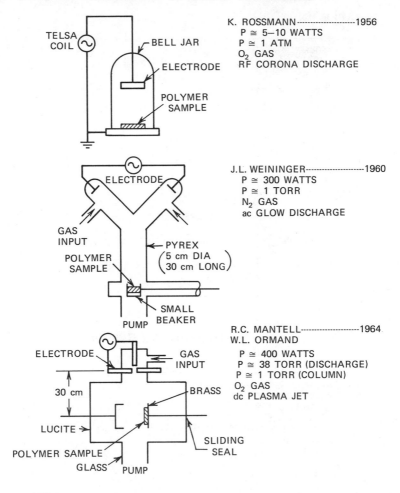

FIGURE 3.1. Typical reactors used for the plasma treatment of polymers.

for polyethylene and polypropylene indicate cases where independent experiments gave different results. Polyethylene, polypropylene, and polytetrafluoroethylene have received the majority of the attention. This is due to the simple structure of these polymers and their widespread use. Oxygen is the most commonly used diatomic gas and helium is the most commonly used noble gas.

In general, an oxygen plasma simultaneously produces wettability and molecular weight changes. Improved wettability is produced in all of the polymers with the exception of poly(ethylene terephalate) and poly(vinyl

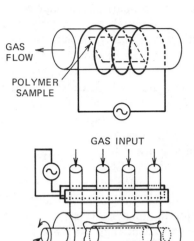

GAS
FLOW

POLYMER
SAMPLE

R.H. HANSEN----------------------1965
H. SCHONHORN
P ≅ 500 WATTS
P ≅ 1 TORR
"INERT GASES"
RF ELECTRODELESS GLOW
DISCHARGE
(f = 15 MHz)

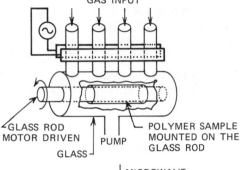

GAS INPUT

J.R. HOLLAHAN----------------------1968
P ≅ 500 WATTS
P ≅ 1 TORR
"INERT GASES"
RF ELECTRODELESS GLOW
DISCHARGE
(f = 15 MHz)

GLASS ROD
MOTOR DRIVEN
PUMP
GLASS
POLYMER SAMPLE
MOUNTED ON THE
GLASS ROD

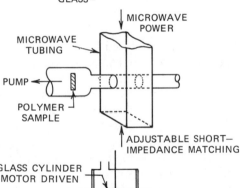

MICROWAVE
POWER

MICROWAVE
TUBING

PUMP

POLYMER
SAMPLE

ADJUSTABLE SHORT—
IMPEDANCE MATCHING

J.R. HALL---------------------1969
C.A. WESTERDAHL
A.J. DEVINE
M.J. BODNAR
P ≅ 100 WATTS
P ≅ 1

GLASS CYLINDER
MOTOR DRIVEN

P. BLAIS---------------------1971
D.J. CARLSON
D.M. WILES
P ≅ 10 WATTS
P ≅ 1 ATM
N_2 GAS
RF CORONA DISCHARGE

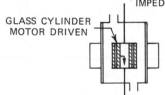

ELECTRODES

POLYMER SAMPLE
MOUNTED ON A
GLASS CYLINDER

121

fluoride) [3]. Noble gas plasmas tend to produce primarily molecular weight changes with little or no effect on wettability. Although most of the data is for helium, all of the noble gases should react similarly. When oxygen is part of the polymer structure, wettability changes can be produced although this is not clear from the data in Table 3.1. Polypropylene is a unique case. Noble gases do not cross-link polypropylene and do not cause increased wetting. The results for polypropylene, unlike the other hydrocarbons, are pressure-sensitive. For corona systems (1 atmosphere gas pressure), argon causes increased wetting while at lower pressures (glow discharge), it appears to have no effect. Insufficient data is available to generalize the chemical changes caused by N_2, H_2, NH_3, N_2O, and CO_2.

An illustration of the types of plasma reactors used in various studies is shown in Figure 3.1. In general, gas pressure, electrode configuration, and power-source frequency are the three major parameters. Gas pressure has been varied from 100 μ to 1 atmosphere, electrode configurations have been varied from electrode to electrodeless, and frequency has been varied from dc to microwave (GHz).

Surface Property Modifications

Wettability

Changes in the wettability of a polymer are produced by adding oxygen- and nonoxygen-containing functional groups to the polymer surface. Oxygen-containing functional groups produce larger wettability changes than those produced by nonoxygen-containing functional groups. Unsaturation effects, electrostatic charging, and surface morphology changes can also cause wettability changes, but the wettability changes produced by these modifications are small and, in general, are unimportant.

Oxidation reactions produce oxygen-containing functional groups (carbonyl —C=O, hydroperoxide —OOH, hydroxyl —OH) which are attached to the polymer surface. These functional groups have surface densities which are of the order of the monomer density (10^{14}–10^{15} sites/cm^2) and cause large increases in the wettability [4]. Plasma oxidation is a relatively fast process and changes in wettability are produced in seconds [5]. By contrast, thermal oxidation (thermal pyrolysis) requires minutes to hours to produce the same change in wettability [6]. Typical time dependent oxidation data are shown in Figures 3.2 and 3.3. Figure 3.2 shows the growth of ir bands for oxygen-containing functional groups and Figure 3.3 shows wettability changes. The changes in wettability caused by plasma treatment saturate in minutes and, in general, are unaffected by longer exposure times, although cases have been reported where longer plasma exposures cause a decrease in wettability [3]. These effects have been observed by a number of investigators and hence

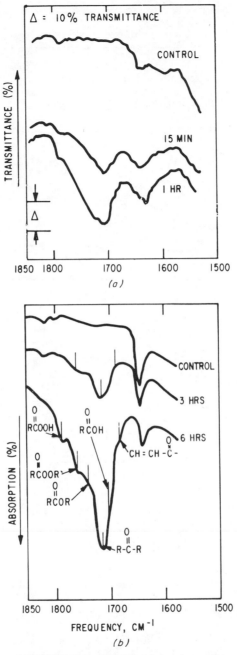

FIGURE 3.2. Carbonyl ir absorption measurements for polyethylene; (*a*) Plasma data taken by Kim et al. [7] and (*b*) thermal data taken by Luongo [6].

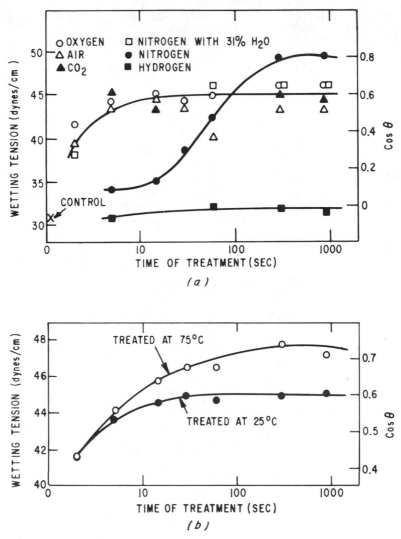

FIGURE 3.3. Wetting angle measurements for polyethylene measured by Kim et al. [7]: (*a*) Effects of gas composition and (*b*) effects of polymer temperature for treatment in an air plasma.

may be assumed to be independent of the plasma reactor. Polymer temperature is another important parameter which affects the oxidation process [7] (see for example Figure 3.3*b*). In general, increasing the polymer temperature with no change in the plasma, causes an increase in the wettability, an increase in the ablation rate, and for some polymers, an increase in the average molecular weight of the polymer present near the surface.

The initial step in the oxidation process is the rate-limiting reaction and is different for the plasma and the thermal processes. The initial steps for thermal oxidation are shown by (3.1a) and (3.1b) [6], where RH and RF

$$
\left.\begin{array}{c} RH \\ RF \end{array}\right\} \xrightarrow[\text{activation}]{\text{Thermal}} \left\{\begin{array}{c} R\cdot + H\cdot \\ R\cdot + F\cdot \end{array}\right. \tag{3.1a}
$$

$$
\left.\begin{array}{c} RH \\ RF \end{array}\right\} \xrightarrow[\text{activation}]{\text{Thermal}} R\cdot + R'\cdot \tag{3.1b}
$$

are the polymers and R· is the polymer free radical. Reactions (3.1a) and (3.1b) are valid for simple hydrocarbon and fluorocarbon polymers. For more complex polymers, gas emission and monomer decomposition can also occur [8]. In any case, the reactions are driven by the thermal energy and proceed on a time scale of hours. The initial steps for the plasma oxidation of simple hydrocarbon and fluorocarbon polymers are shown by (3.2a)–(3.2c):

$$
RH + O\cdot \longrightarrow \left\{\begin{array}{c} R'\cdot + R''O\cdot \\ R\cdot + OH\cdot \end{array}\right. \tag{3.2a}
$$

$$
\left.\begin{array}{c} RH \\ RF \end{array}\right\} \longrightarrow + 2\,O\cdot \longrightarrow \left\{\begin{array}{c} R\cdot + H\cdot + O_2 \\ R\cdot + F\cdot + O_2 \end{array}\right. \tag{3.2b}
$$

$$
\left.\begin{array}{c} RH \\ RF \end{array}\right\} \longrightarrow + uv \longrightarrow \left\{\begin{array}{c} R\cdot + H\cdot \\ R\cdot + F\cdot \end{array}\right. \tag{3.2c}
$$

Reaction (3.2a) is an oxygen-abstraction reaction whereas (3.2b) and (3.2c) are bond-dissociation reactions driven by the absorption of the dissociation energy of O_2 (5.08 eV) and uv radiation, respectively. Both energy sources have ample energy to break the RH (3.2 eV) or the RF (4.2 eV) bond.

Additional reactions can occur in the presence of a small concentration of admixed gases. These might be either impurities present in the oxygen, gases added intentionally (i.e., noble gases), or impurities released by gas emission from the polymer. The admixed gases can contribute reactions such as

$$RH + H\cdot \rightarrow R\cdot + H_2 \tag{3.3a}$$

$$RH + M^* \rightarrow R\cdot + H\cdot + M \tag{3.3b}$$

$$RF + H\cdot \rightarrow R\cdot + HF \tag{3.3c}$$

$$RF + M^* \rightarrow R\cdot + F\cdot + M \tag{3.3d}$$

in which M^* is an excited metastable state. Reactions (3.3a) and (3.3c) are abstraction reactions and are not limited to hydrogen. Reactions (3.3b) and (3.3d) are dissociation reactions driven by the metastable-excitation energy. Noble gases and hydrogen all have metastable states with sufficient energy to dissociate the polymer (see Table 3.3).

TABLE 3.3. Noble Gas and Diatomic Gas Properties

GAS	DISSOCIATION ENERGY (VOLTS)	METASTABLE ENERGY (VOLTS)	IONIZATION ENERGY (VOLTS)
He		19.8	24.58
Ne		16.6	21.56
A		11.5	15.76
Kr		9.9	14.0
Xe		8.32	12.13
H_2	4.5		15.6
H		10.1	13.6
N_2	9.8		15.5
N		2.38	14.5
O_2	5.1		12.5
O		1.97	13.6

Once oxidation is initiated, the ensuing reactions are similar for both plasma and thermal oxidation of hydrocarbon polymers. Plasma and thermal oxidation of fluorocarbon polymers and more complex polymers may also have the same reactions but detailed studies on these polymers have not been conducted. The dominant oxidation reactions for hydrocarbon polymers are illustrated in (3.4a)–(3.4e),

$$R \cdot + O_2 \rightarrow ROO \cdot \tag{3.4a}$$

$$ROO \cdot + R'H \rightarrow ROOH + R' \cdot \tag{3.4b}$$

$$2\,ROOH \rightarrow R{=}O + H_2O + ROO \cdot \tag{3.4c}$$

$$ROOH \rightarrow RO \cdot + OH \cdot \tag{3.4d}$$

$$RO \cdot + R'H \rightarrow R \cdot + R'OH \tag{3.4e}$$

$$RH + OH \cdot \rightarrow R \cdot + H_2O \tag{3.4f}$$

$$R \cdot + O \cdot \rightarrow RO \cdot \tag{3.4e}$$

Reaction (3.4d) is a slow decomposition and is not very important. The hydroxyl functional groups are produced by (3.4e). The carbonyl, hydroperoxide, and hydroxyl groups are easily detected using ir measurements. These measurements demonstrate that oxidation is confined to a thin layer typically of dimensions 1 μ or less and that the bulk properties are completely unaffected.

Molecular weight changes and unsaturation can occur simultaneously with oxidation. The presence of unsaturated sites does not change the oxidation reactions, but since these sites are chemically active they are easily oxidized and can speed up the rate of oxidation. There is some evidence that molecular weight changes, caused by cross-linking, increases the oxidation rate. Oxidation rate measurements for cross-linked and uncross-linked polyethylene indicates cross-linking may increase the oxidation rate by 20–30% [9]. Unfortunately, the cross-linking for the example just cited was produced by high-energy electron irradiation which also produces unsaturation and induces reactions with the impurities (antioxidants, catalytic agents, lubricants, sensitizers, etc.) present in the polymer. As a result, the measurements are not a clear demonstration that cross-linking is responsible for the change in the oxidation rate although it certainly is possible. Decreases in the molecular weight caused by scissioning appear to have no affect on wettability.

Antioxidants are commonly placed in commercial polymers to prevent oxidation. Plasma oxidation unlike thermal oxidation is unaffected by the antioxidants [9]. Thermal oxidation is a chain process involving reactions (3.4a) and (3.4b). The production of polymer free radicals by thermal activation is a very slow process. Once oxidation has begun, the polymer free radicals are generated through the oxidation process itself. The antioxidants terminate the chain reaction through the series of steps shown by (3.5a)–(3.5c):

$$ROO\cdot + AH \rightarrow ROOH + A\cdot \qquad (3.5a)$$

$$RO\cdot + AH \rightarrow ROH + A\cdot \qquad (3.5b)$$

$$R\cdot + A\cdot \rightarrow RA \qquad (3.5c)$$

in which AH is the antioxidant. In plasma oxidation, the antioxidants still terminate the oxidation chain reaction but they do not decrease the oxidation rate. These observations are independent verification that polymer free radicals, in plasma oxidation, are primarily generated by (3.2) and (3.3) and that these reactions differentiate plasma from thermal oxidation.

Plasma oxidation studies of fluorocarbon polymers have demonstrated the existence of oxygen functional groups attached to the polymer and changes in the surface wettability [3,10]. Detailed information about the oxygen groups and the reaction mechanisms is not available.

The largest wettability changes are produced in polymers containing nitrogen and/or oxygen [10]. In general, detailed information about the plasma oxidation of such polymers is not available. A few plasma-oxidation studies of polydimethylsiloxane are available [10,11]. In this case, the oxygen-containing functional groups are primarily hydroxyl and are attached to the methyl groups ($SiCH_2OH$). A reaction mechanism has been suggested and is shown in (3.6a) and (3.6b):

$$ (3.6a) $$

$$ (3.6b) $$

The oxidation mechanism and products appear to be the same for both a corona and a glow discharge. The ir data suggest that most of the hydroxyls are bound up in hydrogen-bonded structures and that there is relatively little free —OH concentration.

Plasma oxidation of polymers containing nitrogen and/or oxygen is not confined to an oxygen plasma. Oxidation can be produced by hydrogen and noble gas plasmas as well. For this class of polymers, the oxygen is supplied by the polymer and the polymer free radicals are generated through (3.2b), (3.2c), (3.3a), and (3.3b). Poly(oxymethylene) [10] and polydimethylsiloxane [12] are two examples.

Oxidation is not restricted to a pure oxygen plasma. Similar results are produced with oxygen-containing gases like air, CO_2 and N_2O [11,13].

The attachment of groups which do not contain oxygen can also be used as a means to alter wettability. Two procedures have been developed for attaching such functional groups to polymer surfaces. The first is a single-step process in which the polymer is exposed to a nonoxygen-containing plasma. The plasma produces polymer free radicals and gas-free radicals. The gas-free radicals then bond covalently to the polymer free radicals. The second procedure is a multistep process. The polymer is first exposed to a nonoxygen-containing plasma like argon and then immersed either in a warm monomer gas or dipped into a solution containing a component capable of reacting with the polymer free radicals.

An example of a single-step process is the attachment of amine groups to the surface of a polymer when the polymer is placed in an ammonia or nitrogen and hydrogen plasma [14]. These groups are confined to a thin layer with dimensions which are less than 1 μ. The gas pressure and discharge power are not critical parameters and do not have a large effect on the amine group surface density. However, the polymer temperature is an important parameter. Increasing the temperature in general causes an increase in the thickness of the amine layer. There is some evidence that a nitrogen–hydrogen gas mixture is more effective than ammonia but the data is not conclusive [14]. Detailed kinetics studies do not exist but a mechanism has been suggested [14]:

$$RH \xrightarrow[\text{activation}]{\text{Plasma}} R\cdot + H\cdot \qquad (3.7)$$

$$R\cdot + NH_2\cdot \rightarrow RNH_2 \qquad (3.8)$$

Infrared absorption measurements do not reveal the amine groups but they can be detected by measuring the "effective thickness" of the heparin layer produced by heparinization to the amine sites. The process is not confined to a single class of polymers but appears to work on all polymers which have been studied in a plasma—i.e., hydrocarbon polymers, fluorocarbon polymers, etc.

There is some evidence that a nitrogen plasma alone can change the surface wettability by forming nitrogen compounds on the polymer surface [15,16]. These nitrogen compounds have not been directly detected but indirect evidence through gas emission does exist which suggests that nitrogen atoms can react with polymer free radicals or unsaturation groups to form nitrogen compounds. Some possible reactions are exemplified by (15, 16):

$$H_2C{=}CH(CH_2)_m{:} + N{:} \longrightarrow$$

$$:N{-}CH_2{-}\overset{\cdot}{C}H(CH_2)_m{:} \longrightarrow \left[\begin{array}{l} \longrightarrow H_2\underset{\underset{N}{\diagdown\diagup}}{C}{-}CH(CH_2)_m{\cdot} \\[12pt] \longrightarrow \cdot CH_2(CH_2)_m{:} + HCN \end{array}\right. \qquad (3.9)$$

$$R{\cdot} + N{:} \longrightarrow RN{:} \qquad (3.10)$$

In the multistep process, a graft to the surface is initiated by exposing the polymer to an inert gas discharge. The polymer is then contacted with a vinyl monomer or it is first exposed to air and then treated with the monomer [17]. The air bath converts the polymer free radicals to peroxide sites which act as graft initiators. The reaction mechanism is shown in (3.11),

$$ROOH + H_2C{=}CH{-}COOH \rightarrow RO{-}CH_2{-}\overset{\cdot}{C}H{-}COOH + OH{\cdot} \qquad (3.11)$$

In the second procedure, the polymer free radical initiates the graft directly. The reaction mechanism is shown in (3.12),

$$R{\cdot} + H_2C{=}CH{-}COOH \rightarrow R{-}CH_2{-}\overset{\cdot}{C}H{-}COOH \qquad (3.12)$$

In both cases the results are the same. The peroxide is a stable functional group and can be used at any time as a graft initiator. The process should be a general technique since peroxide sites can be produced on all the polymers studied to date. In the second procedure, the plasma treated polymer must be exposed immediately to the monomer gas. Polymer free radicals decay with time and the process will only be effective for polymers which have a slow decay time [18]. The multistep grafting process using a liquid dip has only been used to graft nylon to polypropylene but appears to be a less general procedure than the peroxide process [19].

The improvements in bondability due to increased wetting falls within the domain of adhesion chemistry and will not be discussed here. Adhesion results are summarized in the literature [20]. In general, improved bondability due to various plasma processes saturates in time and for a large group of polymers the time dependence has been least square fit to the function in (3.13) [21]:

$$S = \frac{1}{b} \log (a + bt) \tag{3.13}$$

where S is the bond strength and a and b are the constants.

Molecular Weight

Molecular weight changes occur when a polymer is placed in a plasma. In general, for noble gas, nitrogen, and hydrogen gas plasmas, the molecular weight increases due to cross-linking while for an oxygen plasma, the molecular weight decreases due to scissioning. The molecular weight changes can be detected using thermogravimetric analysis which indicates a change in the melt index, gel permeation chromatography which indicates cross-linking beyond the incipient gel point, and swelling which indicates an increase in molecular weight [22].

Molecular weight changes have been measured in polyethylene, polypropylene, and perfluorokerosene [23]. The cross-linking density can be large and is equivalent to the density produced with a 100 MR dose of high-energy radiation [24]. The degree of cross-linking in polyethylene is larger than the degree in polypropylene. Molecular weight changes have been detected indirectly in other polymers as indicated by the data in Table 3.1. The indirect tests are based on improved adhesion which is related to molecular weight changes through the weak boundary layer concept [25]. Typical data are shown in Figure 3.4. The increase in bondability found after treatment in an oxygen plasma is ascribed to a change in wettability. The increase in bondability found after treatment in a helium plasma is ascribed to a change in molecular weight.

For nonoxygen-containing plasmas, the molecular weight changes penetrate below the surface to a depth of 1–10 μ depending upon the time of exposure. The polymers exhibit less than a 1% mass loss and surface properties such as wettability are unchanged when noble gas plasmas are used. These observations are general and have led to the suggested cross-linking reaction mechanism [25]

$$RH \xrightarrow[\text{activation}]{\text{Plasma}} R\cdot + H\cdot \tag{3.14a}$$

$$R_1\cdot + R_2\cdot \rightarrow R_1R_2 \tag{3.14b}$$

Reaction (3.14b) represents the cross-linking process which can occur either at the end or middle of a polymer chain. Both of these possibilities are shown

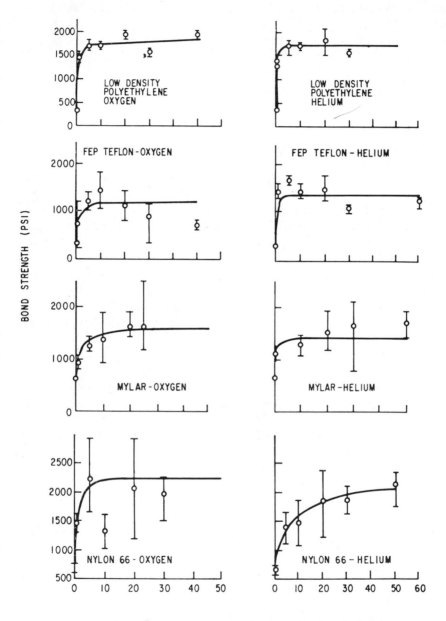

FIGURE 3.4. Bond strength measurements for polymers treated in oxygen and helium plasmas. The data was taken by Hall, et al. [10].

in (3.15a) and (3.15b):

$$R_1-CH_2\cdot \ + \ \cdot\overset{\displaystyle R_2}{\underset{\displaystyle R_3}{CH}} \ \longrightarrow \ R_1CH_2-\overset{\displaystyle R_2}{\underset{\displaystyle R_3}{CH}} \qquad (3.15a)$$

$$\begin{array}{c} R_1\overset{\cdot}{C}HR_2 \\ + \longrightarrow \\ R_3\overset{\cdot}{C}HR_4 \end{array} \quad \begin{array}{c} R_1CHR_2 \\ | \\ R_3CHR_4 \end{array} \qquad (3.15b)$$

The presence of polymer free radicals has been confirmed through electron-spin resonance techniques [23]. This information supports the polymer free radical cross-linking theory although cross-linking by ionic reactions also can occur. Cross-linking with noble gas plasmas have led to the suggestion that excited metastable states are the source of the energy transfer from the plasma to the polymer. This idea is the basis for an adhesion process called "CASING" (cross-linking by activated species of inert gases) [23]. The energy transfer mechanism for casing is shown by (3.16a)–(3.16c)

$$RH + M^* \longrightarrow \begin{cases} RH^* + M & (3.16a) \\ R\cdot + H\cdot + M & (3.16b) \\ R_1\cdot + R_2\cdot + M & (3.16c) \end{cases}$$

The reactions (3.16a)–(3.16c) are not limited to hydrocarbon polymers and also pertain to fluorocarbon polymers. The possibility of excited polymer molecules as indicated by reaction (3.16a) suggest a second type of cross-linking reaction exemplified by (3.17) and (3.18),

$$R_1H^* + R_2H^* \to R_1R_2 + H_2 \qquad (3.17)$$

$$\begin{array}{l} R_1CH{=}CH_2 + R_2H^* \\ (R_1CH{=}CH_2)^* + R_2H \end{array} \longrightarrow \ R_1\underset{\displaystyle R_2}{\overset{\displaystyle |}{C}}HCH_3 \qquad (3.18)$$

The cross-linked surface layer does not dissolve when the polymer is placed in a solvent. The gelation mass (cross-linked surface layer) can be measured and used to estimate the depth of penetration. A study of the cross-linking gelation mass as a function of the exposure time, for polyethylene,

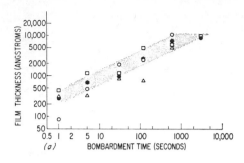

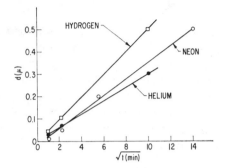

FIGURE 3.5. Gelation mass measurements for polyethylene exposed to noble gas plasmas: (a) Original data taken by Schonhorn and Hanson [25] and (b) same data plotted on a square root of time scale.

indicates that the mass grows with time and that the growth is proportional to the square root of time. This is a general observation for all noble gas plasmas, hydrogen plasmas, and nitrogen plasmas. Typical results are shown in Figure 3.5. Based on this data, diffusion of nitrogen and hydrogen atoms into the polymer has been suggested to explain the penetration of the molecular weight change below the surface [15,25]. The atoms would generate polymer free radicals which could then cross-link. The data in Figure 3.5 seems to confirm the diffusion theory, but more detailed analysis shows the diffusion coefficient as estimated from the data in Figure 3.5 is orders of magnitude smaller than typical measured diffusion coefficients for light-gas atoms in polyethylene. The diffusion coefficient estimated from Figure 3.5 is approximately 2.5×10^{-12} cm^2/sec $[D \approx L^2/t \approx (500 \text{ Å})^2/100$ sec]. Typical measured diffusion coefficients are shown in Table 3.4 [26]. The large

TABLE 3.4. Diffusion Coefficients for Gases through Polyethylene

GAS	TEMPERATURE (°C)	$D(10^6 cm^2/sec)$
He	20 60	8.2 18.2
Ne	20 60	2.16 8.1
Ar	20 60	0.351 2.4
H_2	20 60	3.87 15.6
N_2	20 60	0.317 2.21

difference between the two diffusion coefficients and cross-linking results which are seemingly insensitive to the gas composition has motivated new work which has demonstrated that cross-linking is caused by uv radiation.

The gelation mass time dependence and depth of penetration can be completely accounted for using an attenuated light model. This model assumes that cross-linking is caused by the plasma uv radiation and that the square root of time dependence is produced by the polychromatic light spectrum of the plasma. The possibility that "CASING" may be ascribable to the deactivation of metastable states and free radicals is not precluded. However, their effects are confined to the polymer surface while uv radiation produces the cross-linking below the surface, and is the dominant mechanism.

Cross-linking theory is well defined and has been used for years to explain high-energy radiation cross-linking reactions. The attenuated light model is an extension of those models and is modified to account for the nonhomogeneous free-radical concentration caused by the exponential attenuated light absorption. The number of photons absorbed at position x centimeters per gram of polymer and per unit wavelength, $R_x(\lambda)$, is given by (3.19),

$$R_x(\lambda) = t\bar{v}kI(\lambda)e^{-k(\lambda)x} \tag{3.19}$$

Here t (sec) is the exposure time, $\bar{v}(cm^3/g)$ is the specific volume of the polymer, $I(\lambda)$ (photons/cm²) is the radiation intensity, and k (per cm) is the absorption coefficient. The number of cross-links at position x per primary weight average molecular weight and per unit wavelength, $\delta_x(\lambda)$ [27], is related to $R_x(\lambda)$ through (3.20),

$$\delta_x(\lambda) = \frac{\bar{M}_{w0}\phi(\lambda)}{N}R_x(\lambda) \tag{3.20}$$

Here $\phi(\lambda)$ is the number of cross-links per absorbed photon, $\bar{M}_{w0}$ is the initial weight average molecular weight and N is Avogardo's number. Equation (3.20) neglects scissioning and only considers cross-linking since plasma cross-linking data indicate that scissioning is not an important process. In any event scissioning can be easily included within the theory. The gelation mass is the experimentally measured quantity and is related to the theory [28] through (3.21) and (3.22),

$$g_x(\lambda) = g[\delta_x(\lambda)] \tag{3.21}$$

$$M_G = \rho A \int_0^\infty d\lambda \int_0^{x^*} dx \, g_x(\lambda) \tag{3.22}$$

Here ρ (g/cm) is the polymer density, A (cm²) is the irradiation area, x^* is the incipient gelation point within the polymer, and $g_x(\lambda)$ is the gelation content. Equation (3.21) is a function of the polymer distribution. Equations (3.21) and (3.22) have been evaluated and discussed for a most probable Poisson distribution [24]. The main results of the work are discussed in the next paragraph.

Equation (3.22) is plotted in Figure 3.6a for a monochromatic light beam (1849 Å). The solid curve is the theory. The dashed curve is the measured gelation mass produced by exposing polyethylene to a low-pressure mercury uv lamp (1849 Å). The numbers in the parenthesis are the exposure time; R/R^* is proportional to the exposure time and is a measure of the radiation dose ($R^* = N/\bar{M}_{w0}\phi$). The data demonstrates that 1849 Å uv radiation will cross-link polyethylene and that the cross-linking can be described by the attenuated light model.

Ultraviolet lamps have been used in the past to cross-link polyethylene but generally have not produced much cross-linking [29,30]. Early uv work used high-pressure mercury uv lamps which emit most of their radiation at 2537 Å. Influenced by the high-pressure uv results, radiation effects have been neglected in most plasma polymer work.

Recent plasma work has demonstrated that the plasma uv radiation alone will cross-link polyethylene and that the effective wavelengths are less than 1900 Å [31]. Although uv radiation from both a mercury lamp and a plasma will cross-link polyethylene, the gelation curves are quite different. Plasma and low-pressure mercury uv gelation data are shown in Figure 3.6b. The dashed line is proportional to the square root of time and emphasizes the differences between the two curves. The differences are explained in terms of the polychromatic nature of plasma uv radiation. The curves in Figure 3.6c are a plot of (3.22) for a polychromatic light beam. The spectrum is divided into three components consistent with the polyethylene uv-absorption

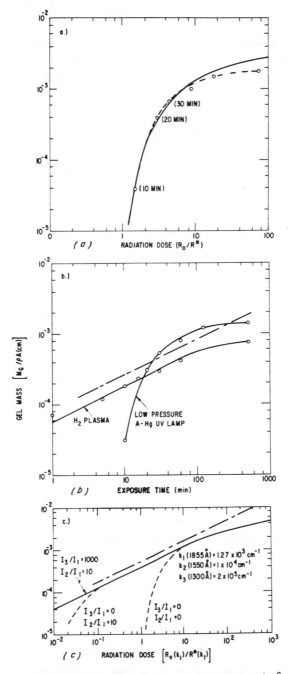

FIGURE 3.6. Gelation mass measurements for polyethylene exposed to a hydrogen plasma and a low-pressure A–Hg uv lamp: (*a*) Low-pressure A–Hg uv lamp data, (*b*) hydrogen plasma data, and (*c*) plot of Eq. (3.22). The measurements were made by Hudis [24].

spectrum [32]. By properly choosing the relative light intensities, the light theory can be matched to the plasma data. The light model also demonstrates that cross-linking is a free radical and not an ionic reaction since the light photons do not have enough energy to ionize the polymer molecules.

The surface molecular weight in general, decreases for polymers exposed to an oxygen plasma. An oxygen plasma produces polymer free radicals which can cross-link but subsequent oxidation prevents rebonding (branching, cross-linking, recombination) and produces a decrease in the molecular weight and volatile compounds which ablate from the polymer surface. Polymers which readily cross-link are an exception. Ablation occurs at the surface but a thin cross-linked layer (typically 500 Å thick) exists below the surface followed by a second layer (typically 1–10 μ thick) which exhibits a decrease in the molecular weight. The thickness of the cross-linked layer saturates and is not a function of the exposure time. Polyethylene is a typical example [13]. The oxygen plasma produces uv radiation which penetrates into the polymer producing free radicals but the concentration decreases exponentially. Oxygen atoms readily diffuse 1–10 μ into the polymer but have a concentration which decreases at a slower than exponential rate. It would thus appear that uv radiation produces cross-linking in the region where the intensity is large, but for penetrations where the intensity is small the oxygen atoms tie up all the free radicals and produce a decrease in the molecular weight.

Ablation produces a continuous and linear (with time) loss of material from the polymer surface. The rate of this process is a function of the polymer composition, and is also a function of the degree of branching, cross-linking, and mechanical stress contained in the polymer. Fluorocarbon polymers have the smallest ablation rate. Hydrocarbon polymers have a larger ablation rate and hydrocarbon polymers containing oxygen have the largest ablation rate. Typical ablation data is shown in Figure 3.7. In general, branching, cross-linking, and mechanical stress increases the ablation rate [9]. Ablation appears to be an exclusive property of oxygen-containing plasmas. Ablation, if it exists at all, is very small for hydrogen plasmas and oxygen-containing polymers. Ablation also causes changes in the surface morphology. Small molecular weight degradation products which are not covalently bonded to the surface migrate around the surface, producing small bumps which have been detected using a scanning electron microscopy [33]. The bumps have a 10 μ scale size and are clear evidence that oxidation causes a decrease in the surface molecular weight.

There are a number of parameters which can affect the molecular weight changes observed upon exposure of a polymer to a plasma. Although detailed studies do not exist, some general observations about the cross-linking reaction can be made.

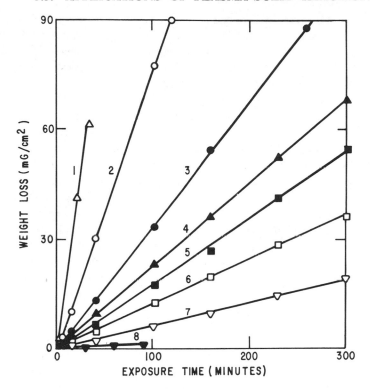

FIGURE 3.7. Ablation rate measurements for various poly-mers exposed to an oxygen plasma: (1) Polysulfide, (2) poly-(oxymethylene), (3) polypropylene, (4) low-density polyethylene, (5) poly(ethylene glycol terephthalate), (6) polystyrene, (7) poly-tetrafluoroethylene) and (8) sulfur-vulcanized natural rubber. The measurements were made by Hanson, et al. [9].

Unsaturation can be produced simultaneously with the molecular weight change. The creation of double bonds changes the uv absorption coefficient which in turn changes the energy absorption within the polymer. For example, ethylene ($-CH_2CH=CHCH_2-$) absorbs in the range between 1700 and 2400 Å while polyenes ($-CH_2[CH=CH]_mCH_2-$) absorb in the range between 2300 and 2900 Å. In addition to shifting the wavelength of light absorbed, the double bonds act as chemically active sites which can affect the reaction rate.

Increasing the polymer temperature tends to increase the degree of cross-linking. There is a large discontinuous change in the degree of cross-linking across the glass transition temperature [16]. This has led to the suggestion that cross linking is confined to the amorphous region of the polymer [25]. Cross-linking data for low and high molecular weight polyethylene, (i.e., low and high degree of crystallinity), do not agree with theory [21]. Here again, detailed studies under clean conditions for pure materials are not available and definite conclusions can not be made.

The weight average molecular weight of the polymer also affects changes in the molecular weight. The number of bond changes per gram of polymer is a function of the polymer temperature, gas composition, and plasma condition. The change in the polymer physical properties is a function of the number of bond changes per polymer molecule which is a function of the weight average molecular weight. This effect is seen in (3.20).

The composition of the gas atmosphere has a large effect on the cross-linking reactions. For example, polypropylene will not cross-link in a noble gas or hydrogen gas plasma, but it does cross-link in a nitrous oxide or oxygen plasma [13]. In both cases the cross-linked layer is confined to 300 Å, but for oxygen the surface is oxidized and has a frosted appearance while for nitrous oxide the surface is not frosted although there is a change in the surface wettability. The complete effect of the gas atmosphere is not clear but it is an important parameter and cannot be neglected.

Finally, it is important to consider the influence of polymer impurities. Impurities can change the uv absorption coefficient and can serve as active sites to initiate molecular weight changes. Typical examples are antioxidants, oxidation products like carbonyl groups ($C=O$), aromatic ring structures, and free radicals such as alkyl ($-CH_2\dot{C}HCH_2-$), allyl ($-CH_2CHCH=CHCH_2-$), and polyenyl ($-CH_2CH[CH=CH]_nCH_2O$). These groups generally have large absorption coefficients over the range 1700–3500 Å, and are very active.

Composition

Nonoxidative changes in the surface composition are produced by either covalently bonding a chemically active group to the polymer surface or by changes in molecular structure such as the creation of unsaturation. A change in the surface chemistry also produces a change in the surface wettability. In general, surface reactivity can not be separated from wettability changes. Beyond surface wettability, changes in the surface chemistry make possible additional surface reactions which would not occur with the initially inert polymer. To the extent possible, covalent bonding has already been discussed in terms of adding amine groups to polyethylene. The possibilities of introducing groups derived from SF_6, H_2S, and CCl_4 have been proposed [14] but have not yet been explored.

An example of a composition change which does not require the addition of new species is the production of *trans*-ethylenic unsaturation in hydrocarbon polymers. Infrared measurements have revealed *trans*-vinylene (RCH=CHR′), vinylidene [R′—C(=CH$_2$)—R], and vinyl (RCH=CH$_2$) unsaturation [23]. Although the mechanisms have not been studied in detail, several reactions have been suggested [25]. Three of these are represented by (3.23), (3.24), and (3.18),

$$\left. \begin{array}{c} \text{RH}^* \\ \text{R· + H·} \end{array} \right\} \longrightarrow \text{RCH=CHR}' + \text{H}_2 \qquad (3.23)$$

$$\text{RH} + \text{UV} \rightarrow \text{RCH=CHR}' + \text{H}_2 \qquad (3.24)$$

Infrared absorption measurements and established molar extinction coefficients have demonstrated that unsaturation occurs to the same depth as cross-linking [24]. As a result, uv radiation appears to be an important participant in the mechanism producing unsaturation.

3.4. APPLICATIONS OF PLASMA–SOLID INTERACTIONS TO METALS

Survey of Applications

Plasmas have also been applied to the modification of the surface properties of metals. Characteristics such as hardness and ductility can be altered through reactions occurring between the metal and active species present in the plasma. Examples of such processes include the nitriding, carbiding, and siliciding of metals. In a somewhat different type of process, silver has been deposited as a coating which acts as a lubricant.

A typical system is shown in Figure 3.8. The metallic sample sits on an insulating pedestal in the center of a vacuum tank. A dc power supply is connected to the sample which serves as the cathode of a dc glow discharge. For ion-nitriding, the system is filled with a 1–10 torr nitrogen–hydrogen gas mixture. For silver coatings, the system is filled with 1–10 torr of argon and silver is introduced by evaporation from a silver-coated tungsten wire. In both cases, a glow discharge is operated between the metallic sample used as the cathode and the vacuum tank used as an anode. Plasma–metal systems have both electrical and physical contact with the plasma. The current is an important part of the interaction and effects the surface chemistry. By adjusting the cathode current and the pressure, the temperature of the cathode and the ionic flux can be regulated. The current at the cathode of

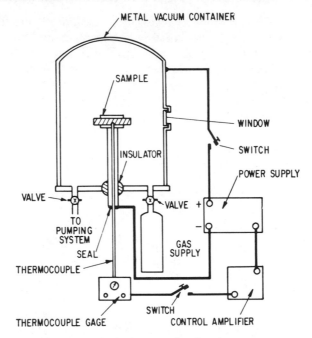

FIGURE 3.8.   Typical system for the plasma treatment of metals.

a glow discharge is almost entirely due to ions as may be seen from (3.25):

$$\frac{I^i}{I} = \frac{1}{1 + \gamma} \tag{3.25}$$

where I^i is the ionic current, I is the discharge current, and γ is the secondary electron coefficient; γ is <1, typically 0.01–0.1. As a result, the electrons present near the cathode do not have a direct influence on the chemistry transpiring there.

Ion-nitriding is used commercially to harden gun barrels, gears, ball bearings, valve stems, and tappets [34,35,36]. Although a thermal-nitriding process has been available for over 40 years [37], the plasma process has a number of advantages over the thermal process and is slowly replacing it. The principal advantages of the plasma process are a rapid case growth [38], low thermal distortion of the treated metal, and superior process control [38].

Silver coatings are used as lubricants for items such as ball bearings which are intended for service in a high temperature vacuum environments where standard lubricants can not be used. Whereas silver coatings can be produced by electroplating or vaporization, the plasma process generally produces a coating with superior adhesion and as a result, longer wear life.

The plasma silver process is a simple example of a plasma–metal interaction in which the silver ions are driven onto the metal sample. The plasma itself is relatively inactive and cannot be used to control the properties of the silver coating. For ion-nitriding, the plasma is an active ingredient and can be used to control the nitriding properties. Since ion-nitriding is a more interesting example of a plasma–metal system, it will be discussed in more detail.

Surface-Nitriding

The nitriding of metal surfaces occurs as a result of the reaction of nitrogen atoms with the alloy elements within the metal sample. Details of these reactions will not be discussed here inasmuch as they have been presented elsewhere [39,40]. Instead, the focus will remain on the production of the nitrogen bearing species required for nitriding.

The introduction of atomic nitrogen into the metal sample is envisioned to occur by two processes: Adsorption of neutral species and ion bombardment. Neutral atomic nitrogen can result from adsorption of atoms produced in the gas phase by dissociation of molecular nitrogen. An indirect path for producing atomic nitrogen also exists. For example, in the nitriding of iron, atoms of iron sputtered from the cathode react with nitrogen in the cathode-fall region to form iron nitrides such as FeN. A fraction of these compounds diffuse back to the cathode where they can be adsorbed. Atomic nitrogen is then produced from the adsorbed iron nitride through (3.26) [41]:

$$FeN(\text{adsorbed}) + FeN(\text{adsorbed}) \rightarrow Fe_2N(\text{adsorbed}) + N(\alpha) \quad (3.26)$$

Here, $N(\alpha)$ represents those nitrogen atoms which diffuse into the α iron phase. Ionic bombardment is also capable of producing atomic nitrogen by direct and indirect processes. In the direct process, nitrogen ions produced in the plasma are driven into the cathode by the cathode electric field [34]. The ions recombine at the surface producing nitrogen atoms. In the indirect process, nitrogen–hydrogen molecular ions (NH^+, NH_2^+, $N_2H_2^+$, etc.) are driven into the cathode by the cathode electric field. The molecular ions dissociate and recombine at the surface again producing nitrogen atoms [42].

The gas composition has a large effect on ion-nitriding. Empirically, a nitrogen–hydrogen gas mixture has been found to produce the best nitriding results. If the hydrogen is replaced with argon or any noble gas, the nitriding case depth and hardness decrease [38]. The data in Figure 3.9 show this effect. These data can be interpreted in terms of the adsorption model by assuming

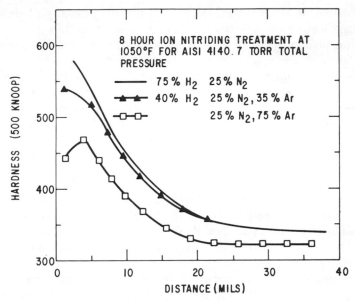

FIGURE 3.9. Effects of gas composition and hardness produced by ion-nitriding. The measurements are by Soccorsy and Ebihara [38].

that adsorption requires a clean surface and that hydrogen is the surface-cleaning ingredient. As hydrogen is replaced by argon, an absorbed impurity layer builds up on the cathode which reduces the nitrogen adsorption and therefore decreases the amount of nitriding. Alternatively, the data can be explained in terms of the molecular ionic bombardment model [42]. As hydrogen is replaced by argon, the nitrogen–hydrogen molecular ion concentration decreases leaving nitrogen and argon ions to bombard the cathode. The nitrogen ions appear to be less effective than the nitrogen–hydrogen molecular ions and produce less nitriding.

Although definitive experiments have not been conducted, the ionic bombardment model appears more consistent with the gas composition effect than the gas adsorption model. If direct gas adsorption is to explain ion-nitriding, then an electrically floated 500°C sample immersed in a nitrogen–hydrogen rf plasma should nitride. An experimental study has demonstrated that for the above conditions, the sample does not nitride [42]. A cathode current is required for nitriding. If indirect gas adsorption is to explain ion-nitriding, then a small change in current should have a large effect on the nitriding results. The sputtering rate is a strong function of the discharge current [43,44] as indicated in (3.27),

$$S \propto J^\mu \tag{3.27}$$

Here, S is the sputtering rate (mG/cm^2), J is the current density (mA/cm^2), and μ is a number which lies between 2.5 and 3.5. A small change in the current will cause a large change in the sputtering rate, a corresponding large change in the iron–nitride concentration and nitriding. This effect has not been observed experimentally. Mass spectrometer data for ions sampled through a hole in the center of the cathode supplies additional data which supports the molecular ionic bombardment theory [42]. For a nitrogen–hydrogen plasma, nitrogen ions (N_1^+, N_2^+, N_3^+, etc.) compose less than 0.1 % of the cathode current while nitrogen–hydrogen molecular ions compose 10–20 %. In a nitrogen–argon plasma, nitrogen ions (N_1^+, N_2^+) compose 80 % of the cathode current but produce very little nitriding. It is possible hydrogen gas is required to maintain a clean surface but this is not consistent with experiments. If hydrogen is required for cleaning, then a critical partial pressure should exist which will maintain a clean surface and provide maximum nitriding. Experimentally, when any fraction of hydrogen is replaced with argon for any total pressure [38,42], the effective nitriding decreases.

The question of ionic bombardment versus gas adsorption is still unresolved. But, the work which has been conducted to date has demonstrated that the plasma parameters (gas partial pressures, plasma temperature, cathode ionic current density) can have a large effect on the nitriding properties (peak hardness and case depth).

REFERENCES

[1] A. W. ADAMSON, *Physical Chemistry of Surfaces*, Interscience, New York, 1960, p. 345.

[2] P. J. FLORY, *Principles of Polymer Chemistry*, Cornell University Press, Ithaca, N.Y., 1967.

[3] R. M. MANTELL and W. L. ORMAND, *I&EC Prod. Res. Dev.*, **3**, 300 (1964).

[4] A. BRADLEY and T. R. HEAGNEY, *Anal. Chem.* **42**, 894 (1970).

[5] K. ROSSMANN, *J. Polymer Sci.* **19**, 141 (1956).

[6] J. P. LUONGO, *J. Polymer Sci.* **42**, 139 (1960).

[7] C. Y. KIM, J. EVANS, and P. A. I. GORING, *J. Appl. Polymer Sci.* **15**, 1365 (1971).

[8] G. J. KNIGHT and W. W. WRIGHT, *J. Appl. Polymer Sci.* **16**, 739 (1972).

[9] R. H. HANSEN, J. V. PASCALE, T. D. E. BENEDICTIS, and P. M. RENTZEPIS, *J. Polymer Sci. A* **3**, 2205 (1965).

[10] J. R. HALL, C. A. L. WESTERDAHL, A. T. DEVINE, and M. J. BODNAR, *J. Appl. Polymer Sci.* **13**, 2085 (1969).

[11] J. R. HOLLAHAN and G. L. CARLSON, *J. Appl. Polymer Sci.* **14,** 2499 (1970).

[12] N. J. DELOLLIS, "The Use of RF Activated Gas Treatment to Improve Bondability," Sandia Laboratory Research Report No. SC-RR-71-0920 (1972).

[13] H. SCHONHORN, F. W. RYAN, and R. H. HANSEN, *J. Adhes.* **2,** 93 (1970).

[14] J. R. HOLLAHAN, B. B. STAFFORD, R. D. FALB, and S. T. PAYNE, *J. Appl. Polymer Sci.* **13,** 807 (1969).

[15] J. L. WEININGER, *Nature* **186,** 546 (1960).

[16] J. L. WEININGER, *J. Phys. Chem.* **65,** 941 (1961).

[17] A. BRADLEY and J. D. FALES, *Chem. Tech.* **1,** 232 (1971).

[18] H. L. BROWNING, JR., H. D. ACKERMANN, and H. W. PATTON, *J. Polymer Sci. A-1* **4,** 1433 (1966).

[19] P. BLAIS, D. J. CARLSSON and D. M. WILES, *J. Appl. Polymer Sci.* **15,** 129 (1971).

[20] S. S. VOYUTSKII, *Autohesion and Adhesion of High Polymers*, Interscience, New York, 1962.

[21] J. R. HALL, C. A. L. WESTERDAHL, M. J. BODNAR, and D. W. LEVI, *J. Appl. Polymer Sci.* **16,** 1465 (1972).

[22] K. E. BACON, *Newer Methods of Polymer Characterization*, Interscience, New York, 1964.

[23] R. H. HANSEN and H. SCHONHORN, *J. Polymer Sci. B* **4,** 203 (1966).

[24] M. HUDIS, *J. Appl. Polymer Sci.* **16,** 2397 (1972).

[25] H. SCHONHORN and R. H. HANSEN, *J. Appl. Polymer Sci.* **11,** 1461 (1967).

[26] R. ASH, R. M. BARRER, and D. G. PALMER, *Polymer* **11,** 421 (1970).

[27] A. CHARLESBY, *Atomic Radiation and Polymers*, Pergamon Press, New York, 1960.

[28] A. R. SCHULTZ, *J. Chem. Phys.* **29,** 200 (1958).

[29] G. OSTER, G. K. OSTER, and H. MOROSON, *J. Polymer Sci.* **34,** 671 (1959).

[30] G. KUJIRAI, S. HASHIYA, H. FURUNO, and N. TARADO, *J. Polymer Sci. A-1* **6,** 589 (1968).

[31] M. HUDIS and L. E. PRESCOTT, *J. Polymer Sci. B* **10,** 179 (1972).

[32] R. H. PARTRIDGE, *J. Chem. Phys.* **45,** 1695 (1966).

[33] C. Y. KIM and D. A. I. GORING, *J. Appl. Polymer Sci.* **15,** 1357 (1971).

[34] C. K. JONES and S. W. MARTIN, *Metal Progress* **85,** 95 (1964).

[35] Y. M. LAKHTIN, Y. M. KRYMSKII, and R. A. SEMENOV, *Metal. Termich. Obr. Metal.* (3), 37 (1964) Russian.

[36] M. L. THORP, *Chem. Week* 119 (June 17, 1970).

[37] C. F. FLOE, *Metal Progress* **50,** 1212 (1946).

[38] W. D. SOCCORSY and W. T. EBIHARA, U.S. Army Weapons Command Science and Technology Laboratory Report No. RD70-156, Rock Island, Ill., 1970.

[39] V. A. PHILLIPS and A. V. SEYBOLT, *Trans. Metal. Soc. AIME* **242,** 2415 (1968).

[40] A. V. SEYBOLT, *Trans. Metal. Soc. AIME* **245,** 769 (1969).

[41] K. KELLER, "Build Up of Glow Discharge Nitrided Layer on Iron Materials," *Haerterei-Tech. Mitt.* **26**(2), 120 (1971).

[42] M. HUDIS, *J. Appl. Phys.* **44,** 1489 (1973).

[43] V. ORLINOV, G. MLADENOV, and B. GORANCHEV, *Int. J. Elec.* **30,** 233 (1971).

[44] G. WEHNER, "Physical Sputtering," *Advances in Electronics and Electron Physics*, Vol. 7, Academic Press, New York, 1955, p. 239.

[45] M. L. KAPLAN and P. G. KELLEHER, *Science* **169,** 1206 (1970).

Chapter 4

Plasma Treatment of Natural Materials

Attila E. Pavlath

4.1. INTRODUCTION

Many natural polymers such as wool, mohair, leather, and cotton, possess properties which have long made them useful for various purposes. In recent years the natural polymers have met increasing competition from synthetic materials which offer greater stability and strength. A portion of this competition has been offset, though, as a consequence of research done on the modification of natural polymers. For example, through chemical modification cotton garments are now made more wrinkle resistant, wool is made machine washable, and leather is tanned by new processes which give softer, more durable products. Research continues to seek new, better, and more economical treatments, including those which provide multipurpose benefits.

This chapter will describe the prospects for improving natural materials through the modification of their surface properties by plasma treatment. The surface properties of the natural materials are important determinants of their usefulness and many of the chemical treatments now in use are aimed at modifying these properties. Although the potential for plasma modification of surfaces has been known for some time, relatively little application has been made of this technique. Below, a preliminary examination is made of the expected benefits of plasma-treated natural materials.

Electric discharges have been explored extensively as a means for altering the surface properties of synthetic polymers [1]. A major observation of these investigations has been that the effects of the plasma do not penetrate below about 100 Å from the surface. In view of this, one can safely expect that the plasma treatment of natural fibers would only affect a small portion of the fiber. Thus, if one considers even the finest wool fiber (diameter 10 μ), 99.8 % of its bulk should remain unchanged after treatment. With such minor changes the treated materials should retain their strength and their highly desired natural appearance.

One of the primary properties of natural fibers which might be altered by plasma treatment is the surface roughness of the fibers. Mohair, for example, has a smooth and slippery surface which creates difficulties in the carding and spinning of this material into yarn. Since glow discharges have been used in the past to roughen the surfaces of metals and polymers [1], one could anticipate a similar effect on the surface of mohair fibers.

Surface modifications could provide other advantages, such as an increase in the compatibility of natural materials with various synthetic ones. In this way paper could be primed to make the adhesion of polymer coatings stronger and more permanent. Another application might be found in the alteration of the surface properties of natural fibers. Most natural fibers have an outer shell (primary wall or cuticle) different in structure from the inside core. In the case of wool, the modification or even removal of this outside hydrophobic layer could cause significant changes in the properties of the fiber. The rate of dyeing should also improve. If the discharge treatment is carried out under oxidative conditions, a hydrophilic surface could form with increased water absorption capability and improved soil repellancy.

A phenomenon of particular significance for the use of wool is felting shrinkage. Although the mechanism of shrinkage is not completely understood, it is believed to be related to either the frictional or hydrophilic characteristics of the fibers. Procedures which utilize a resin are available for reducing shrinkage. However, these procedures tend to alter the appearance and hand of the final product. Consequently, an incentive exists for developing an alternative treatment technique. Since it is known that discharges can be used to alter both the frictional and hydrophilic properties of polymers [1], it is possible to project that the shrinkproofing of wool could be achieved by a similar type of treatment.

A well-established means for altering the surface properties of natural materials is the deposition or grafting of a polymer film onto the surface of the material. Since it is known that a wide variety of monomers can be polymerized in a discharge [2], one can expect that it should be possible to deposit or graft such polymers onto natural materials. Two major difficulties are anticipated for such a process. If the substrate material is porous (for instance, yarn), then gas escaping from the pores could interfere with the penetration of monomer vapor into the interior of the material. Such circumstances could give rise to a nonuniform coating which would be poorly adherent to the substrate. A second potential problem is excessive heating of the substrate which would cause its degradation.

The applications for plasma treatment of natural materials discussed above have all been anticipated on the basis of the known effects of such treatment on polymers and metals. The extent to which these applications are practical is surveyed in the balance of this chapter.

4.2. EFFECTS OF PLASMA TREATMENT ON NATURAL MATERIALS

During the last decades the applications of plasma chemistry have increased, but only a limited number of publications have appeared on the treatment of natural products. Some very promising improvements in the surface properties of natural materials have been observed, but the results are still sporadic and much more information is needed. Our knowledge is especially limited with regard to the mechanism of the interaction of plasmas with natural materials. In many cases, improvements in the desired properties have been obtained but these cannot be explained because the change in the surface is so negligible that conventional diagnostic techniques cannot reveal what has occurred. The reports which have been presented deal with increased adhesiveness of discharge-treated surfaces, increased strength of yarns and fabrics, shrinkproofing of wool and mohair and a few other physical properties. These results are summarized below.

Adhesive Properties

With reference to natural materials, the term *adhesive properties* is used in a broad sense. In one instance, the term refers to those surface properties which affect the bonding of fibers and films to other materials by either heat or adhesives. A second usage of the term is in reference to those properties of a fiber which allow it to be spun into a yarn and subsequently give the yarn its cohesive strength.

Corona discharges have been observed to cause large improvements in the wet-bonding strength of cellulose strips. Goring [3] reports an almost eightfold increase in bond strength: from 11 kg/cm² to 80 kg/cm². The effect of the treatment on adhesion is initially in direct proportion to the exposure time, up to about 5 min, after which the curve flattens and reaches its plateau at 10 min. At the same time the tensile strength does not change substantially since only a small part of the bulk is being modified. The presence of a small amount of moisture is found to be essential. A vacuum-dried (at 105°C) sample gives a bond strength of 45 kg/cm² after corona treatment. If the vacuum-drying is done at 50°C, the resulting bond strength is 75 kg/cm². A summary of Goring's results is shown in Table 4.1.

TABLE 4.1. Effect of Corona Treatment on Bond and Tensile Strength

Sample	Bond strength (kg/cm²)	Tensile strength (kg/cm²)
Cellulose		
Treated	75	1400
Control	11	1500
Handsheet		
Treated	51	710
Control	2	750
Birch wood		
Treated	16	850
Control	1	940

The explanation of this corona effect is still largely unknown. Since Goring found that no improvement of the bond strength occurred in an atmosphere of pure nitrogen, while oxygen and air resulted in high values, it could be inferred that this is an oxidative effect. The occurrence of oxidation was detected by infrared analysis. However, when cellulose strips were treated outside of the corona chamber with corona-generated ozone the bond strength was not enhanced even though ir analysis again showed the presence of oxidation. It was supposed, therefore, that oxidation alone was not sufficient. The active species present in the corona were theorized to have a scouring effect on the surface of the cellulose, removing debris and adsorbed impurities.

The heat-bonding strength of cellulose to synthetic polymers can also be improved by corona treatment [4]. An example of such an effect is shown in Table 4.2. The results, however, cannot be explained by the hypothesis outlined in the previous paragraph. While the bond strength is 5–7 times higher for samples treated in an air or oxygen corona than the bond strength for the untreated ones, treatment in a nitrogen corona gives an unexpected result. The bond is at least 15 times stronger after corona treatment in nitrogen. The exact value could not be determined because the cellulose strip failed cohesively before the bond line released. Infrared analysis failed to show any oxidation. Even surface roughening which is sometimes theorized

TABLE 4.2. Bond Strengths (kg/cm^2) of Cellulose to Polymer Sheets after Treatment in an Oxygen Corona for 15 min

Polymer:[a] Temperature of pressing:	PE 90°C		PS 110°C		PVC 110°C		PVDC 110°C
Source:	Dow	Comm	Dow	Comm	Dow	Comm	Dow
Untreated control	1.3	0.2	0.0	3.3	0.0	3.6	0.5
Cellulose-treated	1.3	0.9	1.3	7.3	0.5	5.4	2.0
Polymer-treated	4.4	12.2	6.7	10.2	3.3	5.8	2.2
Both treated	6.5	13.8	15.2	21.8	5.3	6.9	3.8

[a] PE = polyethylene; PS = polystyrene; PVC = Polyvinylchloride; PVDC = polyvinylidenechloride.

as the possible cause of increased adhesion cannot be considered in this case. The surface of cellulose after 1 hr of corona treatment in nitrogen showed very little or no change. At the same time air-corona treatment created considerable surface roughening. In contrast to these observations, short 30 sec air-corona treatments [5] produce essentially no changes in cellulose (cotton fibers).

A corona discharge has been shown to be especially useful for priming paper substrates to increase their adhesion to polyethylene coatings [6]; at speeds of 6–40 ft/sec, excellent adhesion is obtained. The effect is not temporary, since the adhesion is the same or even slightly higher when the coating is applied 24 hr later.

A detailed study has been published recently by Brown and Swanson on the surface energy of corona-treated cellulose [7]. It was concluded that the treatment increased the solid surface free energy very significantly. This change was claimed to be caused by an increase in the polarity of the surface which was observed to be more hydrophilic after the treatment.

The plasma treatment of cellulosic materials can provide enhanced adhesive properties on fibers as well as flat surfaces. Stone and Barrett [8] have reported that the irradiation of cotton yarn in the plasma of a glow discharge caused increased water absorbency. Samples of cotton so treated were no longer soft and pliable, but rough and stiff. Electron microscope studies showed that the normal wax coating was eliminated at certain points and the surface of the fiber was roughened. While such roughened surfaces had no effect on the strength of individual fibers, the breaking strength of the treated cotton yarn was higher. This could easily be explained by the increased adhesiveness of the fiber surfaces. The increase in yarn strength was also found to be dependent on the twist applied during the spinning of the yarn, as shown in Table 4.3. At optimum twist, the treated sample had a yield strength of 1280 g compared with 980 g for the untreated yarn. At lower twist, the difference between the treated and untreated material was as high as 87%, though the absolute value was 1200 g. While it is known that a wet cotton yarn is stronger than a dry one, the treated samples showed no significant increase in strength when wetted.

Thorsen [9] has studied the effects of an air corona on cotton. He found that the removal of the cotton wax and the roughening of the surface were

TABLE 4.3. Breaking Strength and Statistical Data from Single Strand Breaks of Treated and Control AEdl-10A Cotton Yarn

Twist (turns/in.)	Breaking strength (g)[a]					
	Control			Treated		
	$\overline{X}^b$	Sigma[c]	CV (%)	$\overline{X}^b$	Sigma[c]	CV (%)
7.66	620	64.9	10.5	1090	88.6	8.1
8.45	645	64.5	10.0	1207	93.7	7.8
9.07	774	66.8	8.6	1218	108.4	8.9
9.80	814	66.6	8.2	1218	108.1	8.9
11.14	925	77.7	8.4	1259	106.0	8.4
11.95	983	76.5	7.8	1287	98.1	7.6
12.69	953	78.3	8.2	1175	105.6	9,0
13.17	954	81.0	8.5	1169	106.0	9.1
14.89	895	79.2	8.8	1101	99.4	9.0
16.31	880	91.4	10.4	969	91.5	9.4
18.03	840	93.3	11.1	890	78.5	8.8

[a] Mean of 400 determinations.
[b] Values from Uster charts (breaking strength in grams).
[c] Sigma denotes standard deviation.

not necessarily required to obtain better adhesive properties, as was pointed out earlier in the discussion of the treatment of cellulose strips. Many properties dependent on surface adhesiveness showed significant improvement when the treatment was carried out at 93–96°C for 1.4 sec (see Table 4.4). Longer residence time did not give better results and in some cases even caused burning. While a corona discharge (15.57 kV at 2070 Hz) in air produced marked improvements in yarn strength and spinability, the best results were obtained when chlorine was introduced into the airstream. The increase in yarn strength was 10% with air and 18% with a 5/1 air–chlorine ratio. An increase of 51% was observed in yarn abrasion resistance.

In the experiments just described the single fiber strength was unchanged. Thorsen [10], however, studied the frictional properties of cotton fibers treated with an air–chlorine corona and found an increase of 28–32% in the coefficient of friction. Examination of the adhesive forces between fibers also

TABLE 4.4. Comparison of Fibers, Single 13.3's Yarns, and Knitted Fabrics Constructed from Untreated and Corona-Treated Combed Cotton Silver

Temperature (°C)	Residence time (sec)	Air flow (scfm)	Chlorine flow (scfm)	Fiber strength (lb/in.²)	Yarn TS[a] (g)	CV[b] of yarn TS (%)	Elongation at break (%)	Knitted fabric burst strength (lb)	Skein break product (81 yd) Before steaming	Skein break product (81 yd) After steaming
					After steaming					
Control	—	—	—	15.82	710.7	10.6	7.9	141.5	2,100	1,893
49–50	1.4	—	—	17.47	751.8	10.8	7.3	147.8	2,272	1,812
49–50	1.4	0.35	0.07	16.22	789.5	8.6	7.0	144.5	2,357	2,002
49–50	3.2	—	—	18.09	783.8	9.6	7.3	142.6	2,282	1,977
93–96	1.4	—	—	17.37	786.0	11.6	8.0	148.7	2,220	1,962
93–96	1.4	0.35	0.07	23.49	841	8.3	7.7	146.4	2,378	2,050
93–96	3.2	—	—	17.24	822.0[c]	9.6	7.3	147.7	2,237	1,996

[a] Tensile strength.
[b] Coefficient of variation.
[c] Before steaming TS is 854 g.

produced interesting results. Treatment in an air corona permanently increased the adhesion by about 25 %. The addition of chlorine, however, had a very unexpected effect. A 2.8 sec treatment in an air–chlorine corona gave a 323 % increase immediately after treatment. After 5 min the value dropped to about 150%, which did not change further. This suggests that spinning immediately after treatment would be highly effective. The cohesive force versus time curve is shown in Figure 4.1.

The treatment of wool and mohair has also been investigated. These studies have yielded many puzzling results, indicating the necessity for more detailed research in this area. Kassenbeck [11] examined the effects of treating wool in a 500 Hz glow discharge at a pressure of 100 torr. This treatment resulted in a 20% increase in fiber tensile strength, an effect which had not been observed by previous investigators. An increase in strength was also

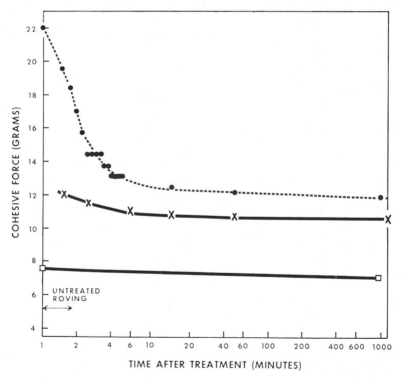

FIGURE 4.1. Cohesive force of cotton roving at various times after corona treatment with electrode voltage 15.57 kV at 2070 Hz, temperature 92–95°C: ●, 2.8 sec residence time, air flow 0.35 scfm, chlorine flow 0.07 scfm; ✕, 1.4 sec residence time, air 0.35 scfm, chlorine 0.07 scfm; ☐, 2.8 sec residence time, air corona.

noted when yarn and fabric were treated. These results are, however, consistent with those reported previously. While the strengthening of yarn and fabric can be explained by an increase in cohesive forces between wool fibers, the increase in fiber strength cannot be explained at this time. In contrast to Kassenbeck's results, wool fibers treated in a high-frequency (13.6 MHz) discharge at a pressure of 1–5 torr showed no increase in strength [12]. It was reported, however, by Roldan [13] that the x-ray diagrams of such treated fibers showed an increase from 45 to 60% in the orientation of alpha-keratin units and an increase from 7 to 15% in the orientation of the amorphous halo. These changes suggested a relaxation of the fiber during the treatment.

Because the surface of mohair fibers is normally smooth, it is difficult to card and felt mohair. These difficulties can be overcome by creating a rougher surface. Thorsen [10] has examined the use of a corona for this purpose. Treatment was carried out at various temperatures between room temperature and 95°C. This indicated a maximum increase in the coefficient of friction around 50°C. Residence times as low as 0.19 sec gave a large increase. More extended exposures provided additional small increases. Figure 4.2 shows that the changes were larger in the presence of chlorine than in air alone. With chlorine, however, the effect started to decrease with longer exposures (0.7 sec or higher), while with air a steady increase was observed up to 2 sec before it reached a maximum. The effect also decreased with increasing electrode voltage, but the decrease was much less at lower frequency. Even 5 sec using 830 Hz did not cause any significant overtreatment. In contrast to the frictional properties, the cohesive forces showed a steady increase with the temperature up to 140°C as shown in Figure 4.3. A practical application of increasing mohair friction by corona treatment was demonstrated in plant trials [14]. It was found that carding efficiency was improved to the degree that it was possible to process mohair with a system designed for wool.

Although plasma treatment has made it possible to obtain significant changes in the frictional and adhesive properties of wool and mohair, the results obtained by different investigators show a great deal of variation. For example, while low-frequency discharges at atmospheric and decreased pressure and high-frequency discharges at decreased pressure result in increased cohesive characteristics and yarn strength, the time dependence of these changes is different. Thorsen's experiments [10] with an air–chlorine corona resulted in a 230% increase in the cohesive forces between wool fibers which quickly decreased to 38% in 95 min (Fig. 4.4). On the other hand, mohair fibers showed an increase of 90% immediately after treatment which did not change with time. High-frequency discharge-treated wool and mohair samples exhibited changes of a similar order, but no time dependence was

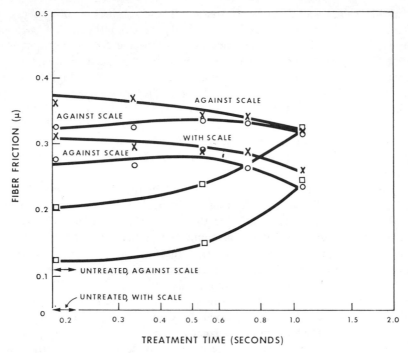

FIGURE 4.2.   Fiber coefficient of friction for mohair after corona treatment at 52°C with and without dilute chlorine gas injection, and at room temperature with dilute chlorine gas injection. Electrode voltage 15.57 kV at 2070 Hz, air flow 0.35 scfm, chlorine flow 0.07 scfm; ×, air–chlorine corona at 52°C; ○, air–chlorine corona at room temperature; □, air corona at 52°C.

observed [12]. At the same time, the low-frequency discharge at decreased pressure increased the strength of the yarn, but the increase became greater with time and it reached a maximum after 2 weeks [11]. The observations reported for the time dependence of the frictional properties of wool and mohair are also unanticipated. For wool treated in a corona [10], the increase in the coefficient of friction was permanent in the direction against the scale, but decayed somewhat in the direction with the scale reaching a stable value only after 6 weeks. The coefficient of friction for mohair treated in a similar manner did not exhibit a similar decay. When a high-frequency discharge was used, both wool and mohair exhibited coefficients of friction which were independent of time.

Thorsen [15] has also made some additional interesting observations. He notes that the frictional coefficient can be decreased to the pretreatment

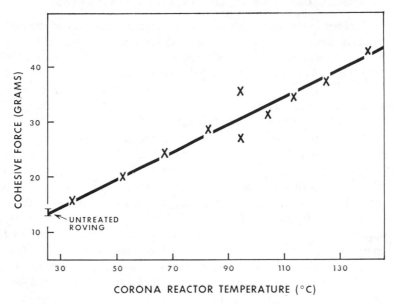

FIGURE 4.3. Cohesive force of wool roving 30 sec after air–corona treatment at various temperatures.

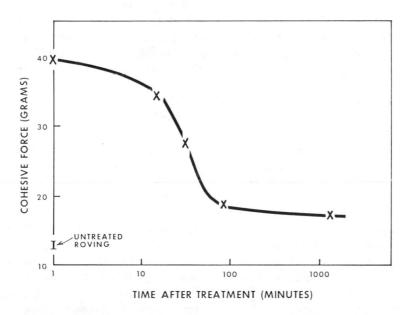

FIGURE 4.4 Cohesive force of wool roving at various times after an 130°C air-corona treatment.

value if the fiber is treated by a cationic surfactant. Removal of the cationic compound restores the original value. The cationic compound apparently attaches to the sulfonic acid groups which were detected at the surface of the wool [16]. This reversal of frictional properties is also seen in felting [17,18] and wicking [15,16] properties. These observations are valid only with the corona treatment. Values obtained after high-frequency glow-discharge treatments did not change with cationic surfactants [12].

A number of other observations have been made concerning the effects of plasma treatment on the surface of natural materials. The surface of corona-treated wool and mohair fibers as in the case of cotton [5] did not show noticeable changes in scanning electron microscopy studies [15]. The hand was somewhat harsher, especially after air–chlorine corona treatment [17,18]. The treatment of wool in a high-frequency glow-discharge at low pressures produced a series of interesting observations. After washing electron micrographs taken of the treated and untreated material showed very few differences [12]. The low-frequency glow-discharge treatment resulted in a bombarded surface, but the high-frequency glow-discharge-treated samples exhibited very few changes [12]. The cysteic acid content increased in both cases [19]. The presence of hexavalent sulfur atoms in high concentration on the surface was confirmed by x-ray photoelectron spectroscopy (ESCA), but a simple washing eliminated the larger part of it while the acquired properties were still retained [20]. In the high-frequency glow discharge, the results were the same whether the gaseous medium was air, oxygen, nitrogen, hydrogen, or even helium [12]. Therefore, it is difficult to theorize about the changes taking place on the surface. The high wettability and increased dyeability does seem to indicate chemical changes in the epicuticle [21]. A small increase in torsion (approximately 6–7%) was reported by Barella [22]. This could indicate cross-linking which generally is expected to occur during glow-discharge treatment. The abrasion resistance is especially improved. One test method resulted in a 50% increase [12], while another kind of investigation reports an astonishing 925% increase. The same report gives a value of 684% for wool treated in a low-frequency discharge [22].

Most of the investigations on wool and mohair were carried out using loose fibers in the form of roving. In general, it can be said that a corona discharge is active only on loose fibers since penetration into yarn or woven fabric is limited under these circumstances [17]. The fiber treated in this manner can be spun into yarn and woven or knitted into fabric. Electrical discharges at decreased pressures are more effective and yarns or even fabrics can be treated with various results. A low-frequency discharge operating at a pressure of 100 torr has been shown to increase the bursting strength of woolen twill fabric from 9.9 to 11.1 kg [11]. The value increased with time just as observed with yarn, but the change was much less in the case of the fabric. The final value was 11.5 kg. The time dependence of the burst strength curves also showed differences. For loose fibers the curve flattened out after 24 hr, while with yarn the curve increased slowly over a period of 2 weeks. Woolen fabric also exhibited increased abrasion resistance after treatment by this method. While the untreated sample lost 700 mg after 5000 revolutions in a standard test, treatments for 5–15 min decreased the loss to 100–120 mg.

Shrinkproofing of Wool and Mohair

When wool or mohair fabric or yarn is washed in the presence of mechanical agitation, a process known as felting shrinkage occurs. The occurrence of this process imposes certain handicaps on the use of wool and mohair and has motivated a search for means to limit the extent of shrinkage. In order to appreciate the source of the problem, it will be necessary to review two of the theories which have been proposed to explain felting shrinkage [23].

Felting shrinkage does not involve the shrinkage of individual wool fibers. Instead, due to mechanical agitation, the individual fibers move in relation to each other and become progressively and irreversibly entangled. The restriction of such movements or the facilitation of return movements can cut down or eliminate shrinkage. According to one theory, shrinkage could be stopped by making the fiber surfaces more hydrophilic thereby allowing the fibers to slip past each other more easily. A second theory proposes that shrinkage is due to the scales present on the surface of the wool fibers. Consistent with this view, a reduction in shrinkage could be achieved if the coefficient of friction measured against the scales were reduced. Since it is known that electric discharges can be used to alter both the wettability and frictional characteristics of wool, it is anticipated that such treatment might cause a reduction in felting shrinkage.

The observation that wool treated in an electric discharge does have improved resistance against felting shrinkage was first reported by Kassenbeck [19]. Wool fibers treated at 100 torr in a low-frequency (500 Hz) discharge were spun into yarn, then knitted into a sweater. After repeated handwashing, only limited shrinkage was observed in contrast to the nontreated control. It was claimed, however, that such treatment could only reduce, but not completely eliminate, felting shrinkage.

Thorsen [18] obtained similar results using a low-frequency discharge at atmospheric pressure. His study indicated that a corona discharge operated at 11.7–13.1 kV and 415 Hz produced no appreciable shrinkproofing in ambient air until the gas temperature rose above 80°C. The dependence of shrinkproofing on temperature is shown in Figure 4.5. The maximum temperature for these experiments was 150°C. Above this temperature it was

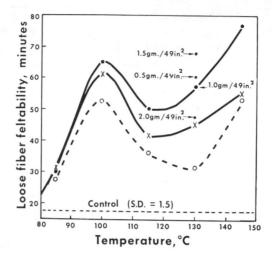

FIGURE 4.5. Effect of temperature and dielectric thickness on loose-fiber feltability. Ambient air, frequency 415 Hz, reaction time 30 sec, 0.25 in. gap, fiber density 1.0 g/49 in^2: ●, 0.063 in. glass, 0.020 in. Mica Mat electrodes, 13.1 kV; ×, 0.063 in. glass, 0.020 in. Mica Mat electrodes, 11.7 kV; ○, two 0.063 in. glass electrodes, 11.7 kV.

expected that extensive damage would occur to the wool due to overheating. The shrinkproofing could also be improved by using an air–chlorine mixture in the corona discharge and the best results were obtained at a 14/1 ratio of the two components. Chlorine alone in the discharge had no effect at all. Nitrogen–chlorine and oxygen–chlorine mixtures gave only half as much shrinkproofing which suggests the contribution of nitrogen oxides to the shrinkproofing process. The addition of water to the air–chlorine mixture further improved the shrinkproofing, but its effect was so complicated that no reaction mechanism could be suggested. Figure 4.6 illustrates that an increase in moisture content creates a small maximum around 25 ppm and a large one at 500 ppm. The shrinkproofing drops below the dry level at 1400 ppm, then reaches another maximum at about 20,000 ppm. In the absence of chlorine, the quantitative effect is smaller, but qualitatively the curve exhibits the same characteristic. The effect of water was also demonstrated by impregnating the wool with various salts before treatment in the corona. Examination of Table 4.5 shows that the use of salts which do not contain water of hydration produced negligible shrinkproofing. Hydrated salts, on the other hand, yielded an increase in shrinkproofing which depends on the

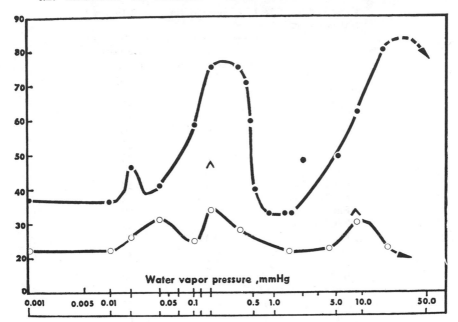

FIGURE 4.6. Effect of water vapor pressure of the air on the air–chlorine reaction and air reaction. Two 0.125 in. glass electrodes, 0.25 in. gap, 18.4 kV, 415 Hz, 93–95°C, reaction time 5 sec, air flow 0.050 scfm, chlorine 0.00364 scfm. Teflon gas preheating coil, center injection; ●, equilibrated 3 hr, air and chlorine; ○, equilibrated 3 hr, air only; ∧, no equilibration, air and chlorine.

salt concentration. A possible explanation of this effect is that the water of hydration is released during plasma treatment in the immediate vicinity of the wool. The effect of the particular salt used is uncertain, but may involve the ease with which the hydrate water is released.

While most of the corona studies were carried out with wool, limited experiments with mohair fibers have indicated similar behavior [24]. Mohair fibers treated in an air corona (18.4 kV at 800 Hz) were spun into yarn and then knitted into fabric. After repeated washings the samples showed very little shrinkage, as indicated in Table 4.6.

The treatment of several worsted and woolen fabrics in a corona discharge [17] failed to produce significant shrinkproofing. Apparently the penetration of the active species responsible for the shrinkproofing effect is more limited in the case of the fabric. This seems to be supported by the different results obtained with worsted and woolen material. The former, which is made of more compact yarn, showed no change after corona treatment, while woolen fabric in which the yarn has a looser structure was partially (50%) shrinkproofed. This level of treatment is not acceptable for

TABLE 4.5. Effect of Impregnating Wool Fibers with Various Salts[a]

Salt	Salt concentration (%)	Loose-fiber felting time (min)
Control	—	38
NH_4Cl	0.2	18
NH_4Cl	2.0	26
NaCl	0.2	26
$Mg(NO_3)_2 \cdot 6H_2O$	0.2	45
$Al(NO_3)_3 \cdot 9H_2O$	0.2	38
$FeCl_3 \cdot 6H_2O$	0.05	58
$FeCl_3 \cdot 6H_2O$	0.1	62
$FeCl_3 \cdot 6H_2O$	0.2	50
$FeCl_3 \cdot 6H_2O$	2.0	27
$Fe(NO_3)_3 \cdot 9H_2O$[a]	0.04	75
$Fe(NO_3)_3 \cdot 9H_2O$	0.1	175
$Fe(NO_3)_3 \cdot 9H_2O$	0.2	115
$Ni(NO_3)_2$	0.2	48
$Cu(NO_3)_2 \cdot 3H_2O$	0.2	50
$Zn(NO_3)_2 \cdot XH_2O$	0.1	47
$Zn(NO_3)_2 \cdot XH_2O$	0.2	66
$ZnCl_2$ dry	0.2	28
$SnCl_4 \cdot 5H_2O$	0.2	41
$SnCl_4 \cdot 2H_2O$	0.2	37
$Pb(NO_3)_2$	0.2	31
$Pb(NO_3)_2$	0.5	37
$Pb(NO_3)_2$	1.5	46
Na_2CO_3	0.2	34

[a] Fibers were treated, after drying, in a cell with one 0.125 in. glass and one 0.015 in. Mica Mat electrode, with 0.25 in. gap, using 17.7 kV, 415 Hz, 91°C for 5 sec. Dry air flow 0.050 scfm, chlorine 0.00364 scfm. Gas preheated in a Teflon tube, center injection. An unimpregnated control sample felted in 38 min.

TABLE 4.6. Felting Shrinkage of Mohair Knitted Fabric

Fabric	Total accumulative length shrinkage (%)		
	Second wash	Third wash	Fourth wash
Untreated	22.0	28.0	29.4
Corona (5 sec)	−4.5[a]	−1.6[a]	4.6
Corona (10 sec)	−8.0[a]	−2.5[a]	−0.6[a]

[a] Negative sign indicates extension, rather than contraction.

practical purposes but indicates the way towards more satisfactory treatments of fabric.

The work of Kassenbeck and of Thorsen and his co-workers has indicated that electric discharges can be used to obtain improvements in resistance against felting shrinkage. However, the conditions used for much of this work are not acceptable for practical application. Many of the treatments require up to 15 min residence time in the reactor chamber. It should be noted, however, that based on a report [25] on the relation between chemical efficiency of a corona discharge and the frequency of altering current used to produce it, Goring concluded [3] that shorter residence times could be achieved by operating at higher frequencies. Thorsen [18] actually studied the effect of frequency on shrinkproofing between 250 and 1500 Hz and found an almost linearly increasing relation as shown in Figure 4.7.

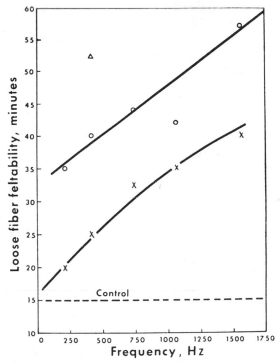

FIGURE 4.7. Effect of frequency, reaction time, and voltage on loose-fiber feltability. Ambient air, 0.063 in. glass, 0.020 in. Mica Mat electrodes, 0.125 in. gap, cell temperature 100°C. An untreated control sample felted in 15 min.: ✕, 10 sec reaction time, 10.9 kV; ○, 20 sec reaction time, 10.9 kV; △, 10 sec reaction time, 21.8 kV.

These findings were the basis on which a successful research plan was formulated [12] at the Western Regional Research Center of the USDA to achieve a practical and acceptable shrinkproofing process for wool. It seemed plausible to assume that the plasma produced in a glow discharge sustained at high frequencies (10 MHz or higher) would contribute to shrink-proofing at short residence times. The electron density of rf and microwave discharges were known to be many times higher than that of the corona. Furthermore, the larger concentration of active particles produced in such discharges was expected to assist in achieving the desired results in a much shorter time.

A low-temperature plasma created at radio frequencies (13.56 MHz) gave highly promising results in the first few experiments [12]. Yarns, both loose and compact, acquired high shrink resistance after 0.3–2.0 sec of treatment. A thorough study was made to determine optimum conditions. This suggested a way for an almost 100% commercially acceptable shrinkproofing process. The relation between the experimental factors and the amount of shrink-proofing, however, was found to be very complicated largely because of the interdependence of the variables [26].

A practical level of shrinkproofing by rf plasma treatment required 10–50 W [12]. Increasing wattage gave better protection against felting shrinkage, but the relation was not linear. Increasing power provided progressively less improvement in shrinkage resistance. A similar relation was found between the residence time and the shrink protection. The upper limit of the residence time was determined by pyrolysis of the wool. The lower limit

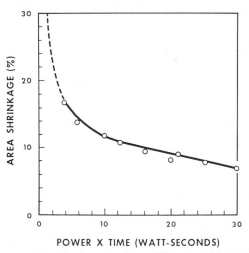

FIGURE 4.8.   Area shrinkage versus power × residence time.

was the maximum speed of the driving mechanism used to pull the yarn through the plasma. With lower residence time, higher power could be used without pyrolyzing the wool. Actually, the product of the power and residence time was found to be a more reliable variable if care was taken to maintain the pressure at the same level. Figure 4.8 indicates a relatively simple, although still nonlinear, relation.

The pressure at which the plasma is formed was also found to be important. With increasing pressure the plasma became confined and hotter. Furthermore, the discharge became unstable at higher pressures, leading to an upper operating pressure of 10 torr for all gases with the exception of the noble gases. Because of the higher temperatures at higher pressures, it was necessary to feed the yarn through the discharge quite rapidly in order to avoid pyrolysis of the yarn. However, as noted in Table 4.7, the shrinkage

TABLE 4.7.   Helium Pressure versus Area Shrinkage

Pressure (mm Hg)	Power (W)	Residence time (sec)	Area shrinkage (%)
3	30	1.5	2.0
15	30	1.5	3.2
50	50	1.0	8.8
180	50	1.0	16.8
180	50	0.7	33.6

protection decreased with increasing pressure.

Since the nature of the gaseous medium could have an effect on the shrinkproofing process, the influence of various gases was also studied. Unexpectedly, the nature of the gas was found to have no bearing on shrinkproofing, as indicated in Table 4.8. The differences between the various

TABLE 4.8.   Effects of Various Gases on Area Shrinkage

Gas	Pressure (mm Hg)	Power (W)	Residence time (sec)	Area shrinkage (%)
Air	2	30	1.2	4.0
O_2	4	30	1.2	3.0
N_2	3	30	1.2	4.3
CO_2	3	60	1.2	2.1
CO_2	3	60	0.7	8.1
H_2	3	30	1.2	4.2
He	3	30	1.5	2.0
NH_3	4	60	1.2	3.6

shrinkage values are too small to infer that the nature of the gaseous molecule had any direct influence on shrinkproofing.

Because the shrinkproofing of wool in an rf discharge shows no apparent dependence on the chemical composition of the gas used to sustain the discharge, it is hypothesized that the active species are the electrons rather than the ions or free radicals. The large difference between the electron densities of a corona and a glow discharge also seems to support this hypothesis, since the latter is more effective than the former. The lack of any significant difference between the surface of wool fibers treated by the plasma in various media also seems to exclude direct action by activated molecular species.

A complete explanation of the means by which electrons could produce shrinkproofing which would be consistent with either of the current theories [23,27] of felting shrinkage is not presently available. It is true that the treated wool surface does become more hydrophilic, but it has also been observed that the deposition of a fluorocarbon polymer on wool in a glow discharge creates a highly water-repellant surface as well as providing a large degree of shrinkproofing [28]. Some changes in the frictional characteristics are also observed after glow-discharge treatment, but the differential friction remains practically the same even after complete shrinkproofing by the plasma [12]. A potential explanation might be found, however, by extension of Makinson's observation on wool shrinkproofed in a corona discharge [29]. This work showed that the cuticle on the treated wool fiber surface softened when wetted. If a similar change could be brought about in a high-frequency glow discharge, this would provide a means for reducing the effects of the scales when the fibers were wet.

Miscellaneous Properties

In addition to those properties discussed above, a number of other fiber properties have been shown to be affected by plasma treatment. Thus Thorsen [5] has noted that an air–chlorine corona can be used to increase the wettability of cotton fibers. A similar change cannot be achieved with an air corona, though. His work also showed considerable evidence that the primary wall of cotton fibers had been affected by corona treatment, the degree increasing with the time of treatment. For example, mercerization (treatment with 18 % caustic) was more complete and there was a greater conversion to the cellulose II lattice. The degree of mercerization provides a good indicator of the status of the permeability of the primary fiber wall. A weakening of the wall will permit an agent such as caustic to swell the secondary wall completely. Conversely, a strengthening of the wall will restrain secondary wall swelling and fiber swelling will be reduced. The apparent weakening or increase in porosity of the primary wall caused by corona treatment also increases the effectiveness of alkaline yarn-setting reagents such as benzyl trimethyl ammonium hydroxide, and improved cotton stretch yarns can be produced.

The Alkali Centrifuge Value, ACV [30], which is a sensitive index of the primary wall status and reflects the swellability of cotton in alkali, also is increased by corona treatment [5]. However, the Iodine Sorption Value, ISV [31], is not affected. If the wax coating on the surface of the cotton fibers is disturbed, this will also affect mercerization and ACV results. Surprisingly, however, scanning and transmission electron microscopy studies show that the wax coating present on the fiber exterior is essentially unaltered by corona treatment.

Further evidence of corona modification of the primary wall of cotton is seen in increased sensitivity to radiation [5]. Fibers exposed to beta rays in a scanning electron microscope (20 kV accelerating voltage) rapidly develop deep cracks generally running parallel to the spiral ridge structure of cotton. These cracks are several thousands of angstroms deep. Crack development also can occur in untreated fiber, but at such a slow rate that it is usually not noticed. Since the fibers were gold-plated for the study, it is probable that the beta rays did not penetrate below the surface of the fibers, at least initially. The heat generated by the beam could have initiated the cracking which then continued due to heat and radiation.

Plasma modification of the epicuticle of natural fibers can also be used to alter the polar and dyeing characteristics of wool and mohair. Belin [32] and Stigter [27] observed that corona and rf glow discharges altered the contact potential of wool. Landwehr [33] studied the electrostatic properties of corona-treated wool and mohair and found that the electropositive wool surface could be made even more negative in a triboelectric series than a polyethylene surface. This increased electronegativity improved the soil repellency of wool and mohair [33], since soils are generally negatively charged. With proper treatment, the electrostatic properties of wool and mohair can be made to match any other insulating material such as leather. Corona treatment produced polar sites on the surface of wool or mohair [34] which were subsequently identified as sulfonic acid groups [16] by ESCA. Apparently, due to these groups the fibers become hydrophilic and exhibit very interesting and commercially important water-wicking properties. Using such fibers, highly desirable mohair paint rollers can be made which hold much more latex (water base) paint. Finally, Kassenbeck [35] reported that wool treated in a low-frequency discharge exhibited increased dyeability. By localizing the contacting of the plasma with the fabric through the use of templates, decorative patterns could be created on wool fabric. When a dyed wool was subjected to rf glow discharge, not only was the usual shrink-proofing effect observed, but the dye became more strongly attached to the fiber [12].

While most of the studies described so far have dealt with fibrous materials, two studies have been made of the effects of plasmas on seeds. Glow-discharge treatment of cottonseed resulted in high water absorption [36]. Corn exposed to radiation germinated faster and more uniformly than untreated controls but high-intensity radiation killed the seeds. Treatment of soybeans gave similar results as well as improved dehulling [37]. Soybeans also absorbed water much more rapidly after treatment. Treated seeds doubled their size in water after a few minutes and after 30 min completely disintegrated. These studies indicate that glow-discharge treatment may be helpful for killing weed seeds admixed with crop seeds, as well as stimulating more efficient germination, and reducing cooking time. It is possible to hypothesize that glow-discharge treatment might even induce certain mutations in the plants.

4.3. GRAFTING ON NATURAL MATERIALS

Although there is a large amount of published work on grafting chemicals to solids by high-energy radiation (see for example reference [38] which deals with grafting to textile fibers), very little attention has been given to the use of electric discharges for this purpose. The majority of what has been published regarding the use of electric discharge for grafting has dealt with the grafting of polymers onto the surface of plastics, metals, and other inorganic materials [2]. These experiments, however, could serve as basis for research to achieve grafting on natural materials through the application of plasma chemistry.

Two methods can be suggested for grafting onto natural materials. In the first, the substrate is exposed to a discharge sustained in the gas or vapor of the monomer to be grafted. A drawback of this approach is that the polymer formed in the discharge may not be grafted to the substrate but only deposited, and so would have a weak bond to the substrate.

A second and potentially better method is to expose only the substrate to the glow discharge. By this means, free radicals would be formed on the exposed surface which could then be used for grafting in two ways. If the treated surface is contacted immediately after leaving the glow discharge with the monomer in either liquid or vapor phase, these free radicals would serve as chain initiators. Naturally precautions would need to be taken to prevent the decay of the free radicals before they could initiate polymerization. Since the polymerization could start only at the surface of the natural material, this method should provide a better graft. Due to the chain transfer mechanism, some homopolymerization might occur simultaneously, but its extent should be much less than when the monomer is introduced into the plasma.

If the exposure of the natural material takes place in an air or oxygen atmosphere, the radicals initially formed might be converted to peroxides, which are frequent catalysts for free-radical polymerization. Such treated surfaces could be exposed later under various conditions to monomers. This would again assure the start of the polymerization from the surface. While this is the simplest method for grafting to synthetic polymer surfaces, it is not widely applicable to natural materials. Most of them, especially protein based materials, are easily oxidized under these circumstances without forming peroxides or, if they do form peroxides, they are highly unstable and rapidly decompose through self-oxidation without initiating grafting.

The grafting of various acrylates to cellulose has been reported by Akutin in a corona discharge [39]. Since corona discharges can be achieved at atmospheric pressure, this might be the best way to graft polymers of volatile monomers. The only drawback is that a part of the polymer is also deposited on the walls of the reactor. In other experiments polyacrylonitrile was grafted to cotton fibers in an rf glow discharge (45 MHz) in argon [40]. The amount of homopolymerization was found to be negligible compared to the results of other methods. Byrne and Brown [41] reported grafting cotton successfully with di- and tetrafluoroethylene obtaining a hydrophobic surface.

The use of a glow discharge to achieve grafting to natural materials is most promising in the case of monomers of very low volatility. These monomers can easily impregnate fabrics, yarns, etc. Then, when exposed to a glow discharge at low pressure, the monomer will be in intimate contact with the activated sites, assuring grafting. This approach has recently been used [28] to show that the impregnation of a commercially available fluorocarbon monomer into wool and the subjection of the wool to an rf glow discharge in argon results in a graft. Some of the polymers formed in this case were homopolymers, since a part of them could be extracted by a fluorocarbon solvent. The rest, however, could not be removed in any way. The grafted wool exhibited an oil-repellent surface as well as felting shrinkage resistance. The same authors studied the grafting of other monomers of low volatility, such as vinyl–silicon and vinyl–phosphorus derivatives, with good results [42].

4.4.   EQUIPMENT   REQUIRED   FOR   PLASMA   TREATMENT   OF   NATURAL   MATERIALS

The equipment used for treating natural materials in various forms in glow discharges is frequently specific to a particular application. Consequently it is difficult to speak in general terms. Nevertheless, it can be said that a

primary concern in all cases is that the material being treated not be over-
heated so that it degrades. Residence times are therefore held to low values
and facilities for cooling are often provided.

Since corona treatment is carried out at atmospheric pressure, the equipment required for this process is very simple. Some modification of a simple
corona reactor is required when other than air is used as a medium. Thorsen
[24] has designed a pilot-scale corona reactor for the treatment of wool and
mohair top with an air–chlorine corona. The reactor is capable of treating
43 lb/hr at a residence time of 2 sec and a power consumption of 4500 W.

Discharge treatments (other than corona) of natural materials must be
carried out at low pressures. In a continuous process this necessitates the
introduction and the removal of the material receiving treatment under
vacuum. This problem has been surmounted in a relatively simple manner
[12].

The idea is based upon a reactor design originally proposed by Stone and
Barrett [8]. They placed a bobbin of untreated yarn in a sealed plastic tube
from which the yarn is led through a tube into the discharge chamber. After
treatment the yarn leaves the chamber and is brought out to atmospheric
pressure through a small diameter polyethylene tube which is stretched
lengthwise to such an extent that a good pressure seal is obtained with the
yarn. This design is based on the assumption that the amount of air diffusing
into the system between the wall of the polyethylene tube and the yarn is less
at the required pressure than the capacity of the vacuum pump. The actual
dimension of the tube can be determined from a semiempirical equation
available for the calculation of the amount of air leakage [43]:

$$V = \frac{k \cdot A \cdot P \cdot (A - D)^3}{L \cdot \mu}$$

where P is the pressure difference between the outside and the reactor
chamber, A is the inside diameter of the tubing, and L is its length; D is the
diameter of the yarn and μ is the viscosity of the gas. The value of the constant
k depends on the units used.

For a given yarn, the dimensions of the capillary tube can be chosen in
such a way that the required vacuum can be maintained with adequate
pumping. By using capillary tubes for both the introduction and the removal
of the yarn, a true continuous design is obtained [12]. It has been shown
that in the case of a yarn with a diameter of 0.4 mm and a capillary tube with
a diameter of 1 mm, approximately 4 ft of tubing was required on each side
of the reactor to maintain a pressure of 4–5 torr in the reactor by a conventional laboratory vacuum pump. This would make the reactor too

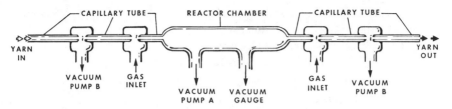

FIGURE 4.9. Reactor design for continuous treatment of yarn.

awkward; therefore, a modified design was tested as indicated in Figure 4.9.

In this case only a foot long capillary tube was required on both sides in order to reach a pressure of 3 torr. The reactor requires two vacuum pumps. The application of a third vacuum pump gives a slight improvement. It is a simple set-up which is ready for use immediately after turning the vacuum pumps on. If the yarn breaks, it can be rethreaded within seconds and it is ready for treatment again. All the auxiliary equipment is outside of the reactor. When the gas inlets are closed, the plasma is created in the residual air. Under this condition a pressure of 35 torr is measured at these points. When a gas flow is introduced at the gas inlets at a somewhat higher pressure, the reactor chamber is quickly purged of air and is filled with the applied gas. In this way not only air, but any other gas, can be used.

The same principle can be used for the treatment of ribbons or fabric. In this case a narrow slit is used instead of a capillary tube with appropriate vacuum chambers [44].

The application of electric discharge to the improvement of the properties of natural materials appears advantageous. Laboratory experiments as described in this chapter tend to support this conclusion. The future, however, is largely dependent on the availability of simpler equipment than has been available up until now. This should be given more attention in future research.

ACKNOWLEDGMENT

The author wants to express his deep appreciation to Harold P. Lundgren and Walter J. Thorsen for their critical review of this chapter and the many valuable suggestions.

REFERENCES

[1] M. Hudis, Chapter 3 of this book.

[2] M. M. Millard, Chapter 5 of this book.

[3] D. A. I. Goring, *Pulp Paper Mag. (Canada)* **68**, T-372 (1967).

[4] C. Y. KIM, G. SURANYI, and D. A. I. GORING, *J. Polymer Sci. C* **30,** 533 (1970).

[5] W. J. THORSEN, private communication.

[6] R. E. GREENE, *Tappi* **48**(9), 80A (1965).

[7] P. F. BROWN and J. W. SWANSON, *Tappi* **54**(12), 2012 (1971).

[8] R. B. STONE, J. R. BARRETT, JR., *Text. Bull.* **88,** 65 (1962).

[9] W. J. THORSEN, *Text. Res. J.* **41,** 455 (1971).

[10] W. J. THORSEN, *Text. Res. J.* **41,** 331 (1971).

[11] P. KASSENBECK, U.S. Patent 2,977,475 (1961).

[12] A. E. PAVLATH and R. F. SLATER, *Appl. Polymer Symp.* Part II (18), 1371 (1971).

[13] L. ROLDAN, private communication, 1970, J. P. Stevens & Co., Inc., Garfield, N.Y.

[14] W. J. THORSEN and R. C. LANDWEHR, *Text. Res. J.* **41,** 264 (1971).

[15] W. J. THORSEN, *Appl. Polymer Symp.* Part II (18), 1171 (1971).

[16] W. J. THORSEN, Proc. 11th Cotton Util. Conf. ARS 72-92, New Orleans, May 1971.

[17] W. J. THORSEN and R. Y. KODANI, *Text. Res. J.* **36,** 651 (1966).

[18] W. J. THORSEN, *Text. Res. J.* **38,** 644 (1968).

[19] P. KASSENBECK, *Bull. Inst. Text. France* **18,** 7 (1963).

[20] M. M. MILLARD, *Anal. Chem.* **44,** 828 (1972).

[21] L. GALLICO, private communications, 1970, Centro Ricerche e Sperimentazione per l'Industria Laniera, Biella, Italy.

[22] A. BARELLA, *Ann. Sci. Text. Belges* **20,** 44 (1972).

[23] *Wool Sci. Rev.* **42,** 2 (1972).

[24] W. J. THORSEN and R. C. LANDWEHR, *Text. Res. J.* **40,** 688 (1970).

[25] J. A. COFFMAN and W. R. BROWNE, *Sci. Am.* **212**(6), 90 (1965).

[26] K. S. LEE and A. E. PAVLATH, 166th National Meeting, Am. Chem. Soc., August 1973.

[27] D. STIGTER, *J. Am. Oil Chem. Soc.* **48,** 340 (1971).

[28] M. M. MILLARD, K. S. LEE, and A. E. PAVLATH, *Text. Res. J.* **42,** 307 (1972).

[29] R. MAKINSON, *Appl. Polymer Symp.* Part II (18), 1083 (1971).

[30] E. HONOLD, *Text. Res. J.* **39,** 1023 (1969).

[31] K. S. FASERFORSCH, *Textiltech.* **3,** 251 (1952).

[32] R. E. BELIN, *J. Text. Inst.* **62,** 113 (1971).

[33] R. C. LANDWEHR, *Text. Res. J.* **39**(8), 792 (1969).

[34] H. P. LUNDGREN and W. J. THORSEN, A.S.T.M. Atlanta, Ga., meeting, September 1968.

[35] P. KASSENBECK, French Patent 1,256,046 (1961).

[36] O. A. BROWN, R. B. STONE, JR., and H. ANDREWS, *Agr. Eng.* **38,** 666 (1957).

[37] C. N. WANG, M. R. JOHNSTON, R. B. STONE, and J. L. GOODENAUGH, USDA, ARS **1970**, ARS 42-178; *C. A.* **73**, 108461 (1960).

[38] R. D. GILBERT, V. STANNETT, *Iso. Rad. Tech.* **4**(4), 403 (1967).

[39] M. S. AKUTIN, N. YA. PARLASKEVICH, N. KOGAN, V. V. RUBINSTEIN, and R. N. GRIBKOVA, *Plast. Massy* (6), 2 (1960).

[40] C. SIMIONESCU, N. ASANDEI, and F. DENES, *Cellul. Chem. Technol.* **3**(2), 165 (1969).

[41] G. A. BYRNE and K. C. BROWN, *J. Soc. Dyers Col.* **1972**, 113.

[42] M. M. MILLARD and A. E. PAVLATH, *Text. Res. J.* **42**, 460 (1972).

[43] L. S. MARKS, *Mechanical Engineers' Handbook*, 5th ed., McGraw-Hill, New York, 1951, p. 939.

[44] A. BRADLEY and J. D. FALES, *Chemtech.* **1**, 232 (1972).

Chapter 5

Synthesis of Organic Polymer Films in Plasmas

M. Millard

5.1. INTRODUCTION

Undoubtedly, the initial observation of the formation of nonvolatile polymeric solid films during the decomposition of organic and inorganic substances in various types of plasmas was viewed as an undesirable result. Almost anyone who has carried out an experiment involving the decomposition or interconversion of organic and inorganic vapors in a plasma has observed the formation of nonvolatile films in the apparatus and encountered the time-consuming task of cleaning or removing these films. During the early 1960s several workers [1] realized that the properties of polymeric films produced in a plasma such as chemical inertness, good adhesion to the substrate, and ease and versatility of preparation could lead to important practical applications. In the last 10 years, many papers have appeared describing the properties of organic polymeric films synthesized by plasma processes, and many applications for these films have been suggested and investigated [1]. Various portions of this work have been summarized in reviews [1–4].

This review will concentrate on the experimental techniques used to synthesize these films in plasma devices, properties of these films, and applications that have been investigated and proposed for these films. Polymeric films have been synthesized in low-temperature ac- and dc-excited electrode and electrodeless devices at low pressures as well as in corona or ozonizer devices at atmospheric pressure. Most of the emphasis in this chapter will be on the use of low-pressure nonequilibrium plasma systems. Only organic polymer films will be covered.

Most of the literature covered in this review is summarized in Table 5.5. This table is included to allow a quick survey of the type of monomers polymerized, the plasma conditions used, the properties of the films that resulted and their intended application.

5.2.   SCOPE OF POLYMER FILM SYNTHESIS IN PLASMAS

Reactivity of Monomers toward Polymerization under Plasma Conditions

One outstanding advantage of a plasma process for the formation of organic polymer films is the wide variety of organic compounds that may be polymerized by this technique. In particular, one is able to consider monomers that are normally inert toward polymerization such as saturated alkanes and aromatics in addition to materials which are known to be conventional monomers. Several reports contain surveys of the types of compounds that are capable of forming polymer films [5–8]. These include saturated and unsaturated aliphatics, aromatics, and a variety of organometallics. Smolinsky and Heiss [7] has given a qualitative rating of the tendency of a variety of monomers with diverse functional groups to form acceptable polymeric films. These materials are ranked in Table 5.1. In this study, organosilanes were found to form the most desirable films while aliphatic halogen compounds were reported not to form films at all. Bradley and Hammes [6] found that vinyl ferrocene and 1,3,5-trichlorobenzene had the greatest tendency to form films while carbon tetrachloride, hexachlorobenzene, and ammonia did not form films. Table 5.2 presents their data on polymerization efficiencies. Based on their own results they concluded that a polymer could be obtained when the monomer contained either a carbon–hydrogen bond, a carbon–carbon double bond, or an aromatic ring unless it was perhalogenated. A similar study undertaken by Yasuda and Lamaze [9] examined the polymerization of 90 monomers in an electrodeless discharge. These monomers were found to fall into two classes according to their effect on the pressure in the flow system. One class of monomers increased the pressure in the flow system when they were introduced into the plasma afterglow and

TABLE 5.1. Monomers Used to Form Polymeric Films in a Glow-Discharge System [7]

Formed best films	Formed films slowly
Dimethylpolysiloxane	Cyclohexane
Triethylsilane	Hexane
Diethylvinylsilane	Chlorobenzene
Vinyltrimethylsilane	1-Chloropropene
Cyclohexene	1-Chloro-2-methylpropane
Heptene-2	Butanol-1
	Proprionaldehyde
Formed very good films	2-Heptanone
2,5-Dimethyl-2,4-hexadiene	
Styrene	**Not suggested as monomers**
Valeronitrile	Cyclopentanone
	2-Buten-1-ol
	Ethyl butyrate
Formed satisfactory films	Tetrahydrofuran
Toluene	Pyridine
Xylene	Tetravinyl tin
Benzene	Perfluorohexane
	Trichloropentafluoropropane
	Did not form films
	Tetrachlorotetrafluoropropane
	Chlorotrifluoroethylene

the other class of monomers caused a decrease in the pressure of the flow plasma system.

The possibility of synthesizing tailor-made films by varying the composition of the gas subjected to the plasma has been demonstrated by several authors. Hollahan and McKeever [10] described the formation of a polymer film when a mixture of CO, H_2, and N_2 was exposed to an rf electrodeless discharge. The nitrogen content of the polymer film was found to be directly proportional to the flow rate of nitrogen gas through the plasma region. This linear relationship is shown in Figure 5.1. A somewhat similar study by Redmond and Pitas [11] has illustrated the effects of varying the composition of a monomer feed containing ethylene and tetrafluoroethylene. Polymerization was initiated using a feed of pure ethylene so that the first layer of polymer resembled polyethylene. Subsequently, the feed composition was changed until only tetrafluoroethylene was present. This yielded a top coating of polytetrafluoroethylene. The composition of the intervening film was found to vary uniformly from polyethylene to polytetrafluoroethylene with no evidence of a discontinuity in the film.

The effects of diluting the monomer with a carrier gas such as argon or other rare gases has also been explored. The results indicated that the

TABLE 5.2. Polymerization Efficiencies of Monomers Deposited on Parallel Plate Electrodes in a Static System [6]

Monomer	Yield (g/kWhr)	Moles (kWhr)
Vinyl ferrocene	300	1.4
1,3,5-Trichlorobenzene	190	1.05
Chlorobenzene	75	0.67
Styrene	69	0.66
Ferrocene	67	0.36
Picoline	65	0.70
Naphthalene	62	0.48
Pentamethylbenzene	61	0.44
Nitrotoluene	55	0.40
Acrylonitrile	55	1.04
Diphenyl selenide	52	0.20
p-Toluidine	47	0.44
p-Xylene	45	0.42
N,N-Dimethyl-p-toluidine	39	0.29
Toluene	38	0.41
Aniline	38	0.41
Diphenyl mercury	30	0.08
Hexamethylbenzene	28	0.17
Malononitrile	25	0.38
Tetracyanoethylene	19	0.15
Thiophene	13.5	0.16
Benzene selenol	13.5	0.086
Tetrafluoroethylene	12	0.12
Ethylene	11	0.39
N-Nitrosodiphenylamine	10	0.05
Thianthrene	10	0.046
Acetylene	9	0.35
N-Nitrosopiperidine	9	0.08
Dicyanoketene ethylacetal	7	0.04
Cyamelurine	6	0.013
1,2,4-Trichlorobenzene	5.5	0.03
Propane	5.2	0.12
Thiourea	4.7	0.06
Thioacetamide	4.4	0.059
N-Nitrosodiethylamine	4.2	0.04
Hexa-n-butyl(di)tin	4.0	0.007
Triphenyl arsenic	3.4	0.19
Methyl mercaptan	1	
Carbon tetrachloride	0	
Hexachlorobenzene	0	
Ammonia	0	

presence of the inert gas can change the properties of the plasma as well as the structure of the film. As an example we can consider the work of Tkachuk and Kolotyrkin [12,13] who studied the polymerization of siloxanes in the presence and absence of argon. Gas phase Electron Spin Resonance (ESR) measurements revealed that higher concentrations of free radicals resulted when a rare gas was present. Furthermore, examination of the films by ir

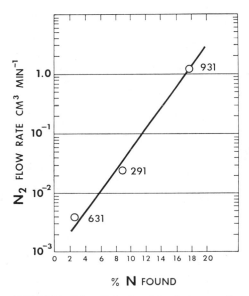

FIGURE 5.1. Relationship between the flow rate of nitrogen gas through the plasma system and the nitrogen content of the polymer formed [10].

spectroscopy indicated that films formed in the presence of argon were more cross-linked than those prepared in the absence of argon.

Finally, it has been reported [14] that the addition of halocarbons can change the rate of polymer film deposition. The yield of polymer obtained by passing styrene vapor through a corona discharge was found to increase when haloforms were added to the styrene prior to polymerization. The largest percentage increase in yield of polymer was observed for low levels of added halocarbon.

The possibility of obtaining films of varying composition by changing the composition of the gas phase is quite attractive and this approach probably merits further investigation.

Properties of Plasma-Polymerized Films

The methods used to characterize organic films deposited in plasma systems and the properties of these films have been summarized in review articles [1,3]. The articles by Wightman and Johnston [15], Neiswender [16], and Thompson and Mayhan [17] as well as others [5,6,18] contain particularly good summaries of the properties of organic polymers formed by this technique and methods for studying these films.

In general, the properties of polymeric films synthesized in plasmas may be summarized as follows. Films are amorphous, pinhole-free, and highly cross-linked. The polymers have superior thermal stability, high melting points, and low solubility. They contain high concentrations of unpaired spins and undergo rapid surface oxidation when exposed to the atmosphere.

Lower molecular weight materials which are more soluble and softer, can be formed under conditions of higher pressure and lower current densities [19,16].

A variety of methods are available for measuring the thickness of polymer films prepared in a discharge and a review of these may be found in reference [20]. In general, it appears that film thickness is limited to the range of 0–2 μ. Several authors have observed that an upper limit exists beyond which the deposited film tends to become brittle and discolored [6,7].

Among the various techniques used to characterize plasma-polymerized films, ir spectroscopy has yielded the most information. Discussions of the ir spectra of such polymers can be found in references [10,15,16,21–24]. The conclusions reached from these spectra are generally similar indicating high degrees of cross-linking, unsaturation, and features similar to those found in polymers prepared by conventional means. The work of Kronick, Jesch, and Bloor [22] represents one of the more detailed ir investigations. In this study, frustrated multiple internal reflection spectroscopy (FMIR) was used to obtain the ir spectra of polymers formed in a static discharge sustained between parallel plate electrodes. A summary of their results is shown in Table 5.3. The general conclusions to be derived from their work are that, independent of whether the monomer is aromatic, olefinic, conjugated or unconjugated, or fully saturated, the solid product is a dense, highly branched, and cross-linked polymer containing considerable unsaturation in the form of both olefinic bonds and free valences. Whereas the production of aromaticity was not observed, this characteristic was preserved in polymers prepared from aromatic compounds. The presence of carbonyl bonds were also observed when the films were exposed to the atmosphere. The latter observation is confirmed by the work of Konig and Brockes [25] who studied the changes occurring in the spectra of films obtained from benzene upon exposure to air. The absorptions due to bonds in the carbonyl

TABLE 5.3. Structure in Glow-Discharge Polymers from Hydrocarbons [22]

Starting vapor	Functional groups in polymers
Pentane	Branches at each pentane molecule, methyl chain ends, ($-CH=CH-$)
Ethylene	($-CH_2-CH_2-CH_2-$), methyl chain ends, ($-CH=CH$), cross-links at saturated carbons
Butadiene	($-CH_2-CH_2-CH_2-$), ($-CH=CH-$), methyl chain ends, and cross-links at saturated and unsaturated carbons
Benzene	($-CH_2-CH_2-CH_2-$), ($-CH=CH-$), ($-C\equiv C-$) or ($-C=C=C-$), methyl chain ends, and pentyl side groups
Styrene	Same as benzene, (C_6H_3)$-CH_2-$
Naphthalene	($-CH_2-CH_2-CH_2-$), ($-CH=CH-$), ($-C\equiv C-$) or ($-C=C=C$), methyl chain ends,

and hydroxyl regions were found to increase in intensity with the time of exposure to the atmosphere. Their spectra are presented in Figure 5.2.

Differential scanning calorimetry (DSC) and thermal gravimetric analysis (TGA) have been used to examine the thermal characteristics of plasma-polymerized films. As a rule, such studies indicate the absence of observable melting points or other phase transitions. An example of the thermal stability of plasma-polymerized films is shown in the work of Wrightman and Johnston [15]. In this study thermograms of films prepared from methane and methylchloride were compared with that for polyethylene. As may be seen in Figure 5.3, the film prepared from methane behaved very similarly to polyethylene in terms of its thermal decomposition. By contrast, 80% of the polymer made from methylchloride remained at 800°C.

Thompson and Mayhan [17] analyzed the soluble and insoluble fractions of plasma-polymerized styrene by both DTA and TGA. The TGA curves for the insoluble polymer showed a weight loss at temperatures less than 75°C. This behavior indicated that the insoluble film contained some volatile components even after solvent extraction prior to analysis. The insoluble polymer exhibited unusual thermal stability and 40% of the initial weight

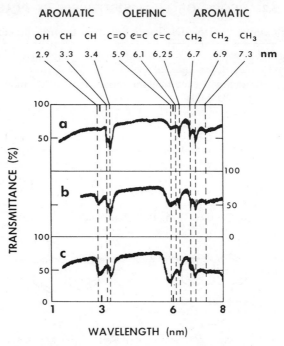

FIGURE 5.2. Variation in ir spectrum of polymer films obtained from benzene with time of storage in air: (*a*) 2 hr; (*b*) 5 days; (*c*) 1 month [25].

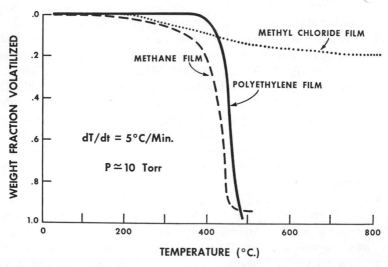

FIGURE 5.3. Comparison of thermograms obtained with polyethylene and with polymer films produced from methane and methyl chloride [15].

184

remained after heating to 700°C. This behavior was compared to a TGA curve for a 10^6 molecular weight standard, which was stable until it decomposed completely at just under 400°C. The fact that a carbon matrix remained after the insoluble portion of the film was heated to 700°C was interpreted as an indication of cross-linking.

Thompson and Mayhan [17] also compared the mass spectra of the decomposition products of a standard polystyrene sample of 2×10^4 molecular weight with the insoluble fraction of plasma-polymerized styrene. The major difference in the two spectra was the observations that nearly all the peaks from the linear polystyrene standard were derivatives of either C_2H_x or C_4H_x while the spectra of the plasma polymerized styrene were derivatives of odd carbon species of C_3H_x, C_5H_x, or C_7H_x. The difference in the mass spectra of these two samples gave further support for the presence of cross-linking in the plasma-polymerized sample.

The presence of unpaired spins caused by trapped free radicals in freshly prepared films has been observed by many authors [15,16,18,26–29]. The source of these radicals is probably due to the incorporation of gaseous free radicals formed in the discharge which are known to exist in abundant quantities in plasmas [22,30–34] and to the formation of radicals from the polymer material through the process of bond rupture caused by the impingement of energetic particles and radiation coming from the plasma. As an example of this work, Morita, Mizutani, and Leda [27] passed styrene, *para*-xylene, and ethylene through a flow plasma system and studied the films formed by ESR. The film from styrene decomposition was sealed in a tube in air and under vacuum and exposed to various heat treatments. The

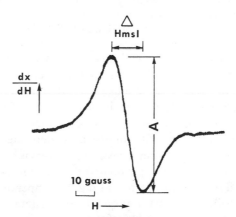

FIGURE 5.4. Typical ESR signal of a thin polymer film formed in a plasma of styrene vapor [27].

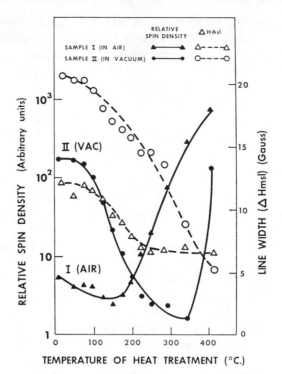

FIGURE 5.5. Relative spin density and line width ($\triangle$ Hmsl) as a function of the temperature heat treatment for the thin polystyrene films formed in a plasma of styrene vapor; ESR measurements are made at room temperature after heat treatment for 10 min [27].

polystyrene film contained 5×10^{17} spins/g at room temperature under vacuum. The samples in air and under vacuum were heated for 10 min at increasing intervals of temperature and the ESR signal observed. The ESR signal in air decreased to 150°C then increased. The ESR signal from the sample in vacuo decreased until ~340°C and then increased. A typical ESR spectrum obtained from a film of discharge deposited styrene is shown in Figure 5.4. A plot of the spin density after various heat treatments is shown in Figure 5.5. The authors postulated that the radical formed at lower temperatures was due to carbon, while at higher temperatures a new radical formed as a result of reaction with oxygen in the system.

The presence of trapped radicals in films formed in plasma systems also accounts for the reactions of these films with oxygen that has been reported by a number of workers [14–16,22,23,35–39].

TABLE 5.4. Some Polybutadiene Film Properties Reported by Bashara and Doty [40]

Sample	I^a (0.5 V)	Area (10^{-6} m²)	I/A (A/m²)	C_t^b (10^9 F)	T_c^c (Å)	C_d/C_t^d	$R_{dc}\,\Omega$ (0.5 V)	R_p^e (10^9 Ω)	Dissipation factor
1	8.0 × 10^{-13}	5.17	1.55 × 10^{-7}	2.65	450	1.05	6.2 × 10^{-11}	2.2	0.027
2	1 × 10^{-12}	5.5	1.8 × 10^{-7}	2.69	480	—	5 × 10^{-11}	2.2	—

[a] I: Current in amperes.
[b] C_t: Total capacitance.
[c] T_c: Thickness from a capacitance measurement using a dielectric constant of 2.65.
[d] C_d: Dielectric capacitance.
[e] R_p: Parallel resistance from a bridge measurement.

The electrical properties of polymer films prepared by plasma techniques have been discussed in review articles by Mearns [3], Gregor [4], and Allam and Stoddard [2]. The electrical properties usually of interest are the resistivity, dielectric constant, and dissipation factor.

Bradley and Hammes [6] have studied the electrical conductivity of 24 polymeric films prepared in a static parallel plate plasma device. At 150°C most of the films gave conductivities between 10^{-6} and 10^{-4} mho/cm; 15 films gave an average of the activation energy for conduction of 1.36 ± 0.06 eV. These data suggest that conduction in these films is determined by some structural feature common to films prepared by this technique.

Bashara and Doyty [40] have investigated electrical conduction in thin films of plasma-polymerized butadiene. Some of the film properties reported in this paper are reproduced in Table 5.4.

5.3.   EXPERIMENTAL METHODS

A wide variety of reactors and experimental configurations have been used to prepare polymer films in electric discharges. These may be classified according to whether the gas is static or flowing and whether internal or external electrodes are used.

Reactors with Internal Electrodes

Figure 5.6 illustrates a typical experimental set-up utilizing internal parallel plate electrodes. This type of apparatus has been used with dc power supplies as well as ac supplies operating over the range of 50 Hz to 2450 MHz. When a dc discharge is used, the polymer films form principally on the electrode acting as the cathode. In the presence of an ac field, films form on both electrodes.

In the absence of a flow of gas through the bell jar, it has been observed [6,19,23,41] that the gas pressure can either rise or fall when the discharge is first turned on. The direction of the pressure change depends upon the monomer used and is a reflection of the relative rates of monomer fragmentation and polymerization. The rate of polymer deposition has been found to increase up to an assymptotic limit as the initial monomer pressure is increased. For a fixed initial monomer pressure, the polymerization rate increases with increasing discharge power or current density. It has also been observed that the deposition rate can be increased by cooling the electrodes. The dependence of the rate on source frequency has not been ascertained. The monomer may be fed continuously through the bell jar in the presence or absence of a carrier gas [21,42]. It has been claimed that operating in the

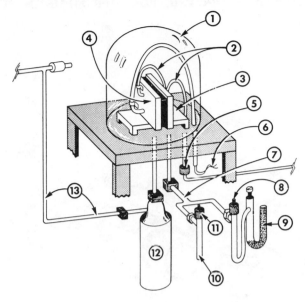

FIGURE 5.6. Typical experimental set up for deposition of films in a bell-jar plasma reactor [42]: (1) Bell jar, (2) cooling lines, (3) sample holder, (4) heat sink, (5) thermocouple, (6) electrical leads, (7) monomer feed lines, (8) air bleed valve, (9) drierite absorber, (10) monomer reservoir, (11) metering valve, (12) dry-ice–acetone trap, (13) vacuum line.

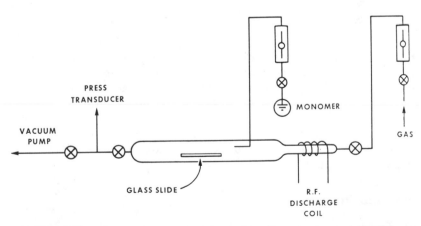

FIGURE 5.7. Schematic representation of a flow system used with an electrodeless discharge [48].

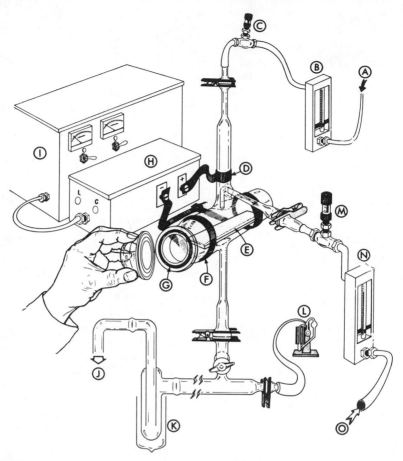

FIGURE 5.8. Flow system for the coating substrates with fluorocarbon polymer films [29]. (A) From gas cylinder, (B) rotometer, (C) flow control valve, (D) high voltage, (E) substrate, (F) ground, (G) o-ring, (H) coupling unit, (I) generator, (J) to pump, (K) liquid nitrogen trap, (L) Mcleod gauge, (M) flow control valve, (N) rotometer, (O) from gas cylinder.

flow-through mode has advantages such as reducing the buildup of contaminants in the film [21].

Electrodeless Reactors

Most of the studies performed with electrodeless discharges have utilized a reactor similar to that shown in Figures 5.7 and 5.8. Several modes of operation have been described. The monomer may be fed directly through the plasma zone [10,15–17,24,43–46]. The monomer and a carrier gas may be fed through the plasma zone [7–9,47]. The monomer may be introduced into the afterglow of an inert gas [29,48].

Using the first configuration, it has been found that the polymerization rate increases towards a maximum value as the discharge power is increased. Above a certain power level, all of the monomer can be converted into polymer. For high-polymerization rates it has been observed that a very fine polymer powder is formed in addition to the deposition of a film. By contrast, excitation of a carrier gas and subsequent introduction of monomer leads to a lower polymerization rate and avoids the formation of a powder. The structure of the polymer films formed in the presence of a carrier gas can differ from those found in the absence of the carrier gas, as was mentioned earlier [12,13].

5.4.   MECHANISMS OF PLASMA POLYMERIZATION

In view of the large number of elementary reactions taking place in an electric discharge, the formulation of a mechanism for polymer formation and the verification of this mechanism are very difficult tasks to accomplish. The mechanisms which have been advanced have usually been based upon qualitative observations and measurements of overall polymer deposition rates rather than on direct identification of the active species present. Only a few cases exist where a postulated mechanism has been used to derive a rate expression which can be fitted to the data. Unfortunately, these expressions are limited in their use to the apparatus for which they were derived since the dependence of the fitted constants on the plasma characteristics was not identified.

By examining the discussions of mechanism which have been proposed, two major approaches to explaining plasma polymerization may be discerned. In the first approach, polymerization is assumed to take place on an electrode or other solid substrate as a result of monomer adsorption and the subsequent bombardment of the monomer by active species and radiation produced in the plasma. The second approach assumes that free radical or ionic species are produced in the gas phase and that these may interact among themselves or with the monomer to produce active species of larger molecular weight, i.e., oligomers. The formation of a film occurs when both the original species and the oligomers diffuse to the substrate surface where they can react further. A second distinction which can be noted is the selection of ions or free radicals as the active species responsible for polymerization [49].

It seems reasonable to believe that both approaches to explaining plasma polymerization have merit and that one may predominate over the other depending upon the conditions and apparatus used for polymerization. The same conclusions may be drawn for the choice of active species. The balance of this section will review some of the work done with hydrocarbon systems which has been aimed at identifying the nature of the active intermediate and will examine the details of several of the proposed mechanisms for plasma polymerization.

Reaction Mechanisms for the Formation of Volatile Products

Extensive studies have been made of the volatile products formed when various hydrocarbons and related materials are passed through a discharge. An excellent review of this work has been presented by Blaustein and Fu [50]. As a result, we will limit the discussion presented here to the work performed with benzene and toluene, two compounds whose polymerization has been extensively investigated.

Streitwieser and Ward [47] have observed the products formed when either a helium or argon carrier gas containing toluene vapor was passed through a 3000 MHz microwave discharge. Approximately 10% of the toluene was converted into products which contained benzene, ethylbenzene, phenylacetylene, and styrene as the major components. Traces of hydrogen, ethylene, and acetylene could also be observed. Based upon a discussion of energy-transfer processes, the authors concluded that the formation of an anion intermediate was needed to explain the products formed.

The reactions of benzene and halobenzenes in a rf discharge have been studied by Stille, Sung, and Vander Kooi [44] and Stille and Rix [45]. In addition to identifying the products produced, emission spectra were recorded of the plasma. The spectrum obtained for a discharge through benzene was found to be quite similar to that reported for phenyl radicals. Further support for the presence of phenyl radicals was obtained from experiments in which deuterium was added to the benzene prior to its passage through the discharge and to benzene which had already passed through the discharge. These experiments showed that approximately 85% of the products had a single atom of deuterium incorporated into them.

Schuler and Stockburger [51], Walker and Barrow [52] (Walker and Bindley [53], and Walker, Bindley, and Watts [54]) have also reported the emission spectra obtained during the passage of benzene derivatives through a discharge. In both cases the principal emitting species was assigned as the benzyl radical. Other organic radicals could be identified as well.

Dinan, Fridman, and Schirman [46] have discussed the decomposition of toluene in an electrodeless rf discharge and concluded that the volatile products could be accounted for in terms of benzyl, phenyl, and methyl radicals. The visible emission spectrum of a toluene discharge was found to contain bonds which could be assigned to benzyl radicals by comparison with the work of Schuler and Stockburger [51]. Experiments with deuterotoluene indicated that deuterium was attached to the aliphatic carbons of the volatile products and that both the aliphatic and the aromatic portions of the polymer product contained deuterium. This difference was interpreted as indicating that the polymeric products were formed by a different mechanism than the volatile products.

Finally, it should be noted that Ranney and O'Connor [55] have investigated the reactions of benzene in a corona discharge at atmospheric pressure. The formation of both the volatile and nonvolatile products could be explained through a mechanism involving phenyl radicals.

As may be seen from this brief survey, a majority of the studies involving benzene or its derivatives points to the formation of the products through a free-radical mechanism rather than an ionic mechanism. This result is not surprising when one considers that discharges are known to be an excellent source of atomic species [33,34] and have been shown to produce simple free radicals from a number of aliphatic hydrocarbons [50,56,57].

Reaction Mechanisms for the Formation of Polymeric Products

The interpretations of plasma polymerization and the proposed mechanisms presented in the literature can be distinguished according to whether electrodes are present within the discharge itself. In the discussion given below we will bring out the main points as seen by several authors and attempt to derive some conclusions from this work.

Williams and Hayes [21] have studied the polymerization of 12 vinyl monomers in a bell jar apparatus containing parallel electrodes which are excited at 10 MHz. A qualitative model was proposed based upon an examination of the dependence of the polymerization rate on the monomer pressure and the current density. The possibility of gas phase polymerization was discussed but its importance ruled out on the basis of the observation that the polymerization rate was independent of pressure above a certain minimum. The almost complete absence of polymer on the apparatus walls and its presence on the electrode surfaces were used to argue that the active species responsible for polymerization were created on the electrode surfaces. It was postulated that the active species were formed when monomer molecules adsorbed on the electrodes were bombarded by either electrons or ions derived from the plasma. The fragments formed in this manner were assumed to initiate polymerization within the adsorbed layer. Further evidence of the importance of monomer adsorption was the observation that the polymerization rate increased with monomer pressure up to a limiting value for a given current density. By decreasing the electrode temperature, the polymerization rate could be increased. This behavior with pressure and temperature was rationalized by suggesting that at low pressures the amount of adsorbed monomer limited the rate of polymerization. As more monomer was adsorbed, the rate increased until all the charged species reaching the electrodes initiated polymerization and the rate became dependent on the current density. This final conclusion was supported by the fact that the polymerization rate was found to be proportional to the current density until a limiting value was reached.

Denaro, Owens, and Crawshaw [19] have reported a study of the polymerization of styrene and related monomers in a bell jar type of apparatus containing internal parallel electrodes excited at 2 MHz. These authors discussed the nature of the plasma and the uncertainties involved in trying to choose between ionic and radical polymerization intermediates.

The authors noted that under conditions of high-current density, the formation of particles of polymer could be observed in the vapor phase. However, the amount of polymer formed in this way was believed to be negligible compared to the amount of material that formed on the electrode surfaces. The initiation of polymer formation was assumed to take place through free radicals produced by collisions between the free electrons present in the plasma and the monomer vapor.

The importance of ions was not excluded but the fact that trapped radicals had been observed in plasma-deposited polymer films was used to support the predominance of a radical reaction.

Subsequent to their formation the radicals were assumed to react according to the following scheme:

$$R\cdot_n + M \xrightarrow{k_1} R\cdot_{n+1} \tag{5.1}$$

$$R\cdot_m + R\cdot_n \xrightarrow{k_2} P_{n+m} \tag{5.2}$$

$$R\cdot_n \xrightarrow{k_3} R\cdot_n \text{ (trapped)} \tag{5.3}$$

where R· represents a free radical, M an adsorbed monomer molecule and P a product molecule. From these reactions the polymerization rate was deduced to have the form:

$$\text{Rate} = r\left[\frac{p}{p + (k_2/kk_1)[R\cdot] + (k_3/kk_1)}\right]$$

where r is the limiting rate of formation of radicals and $M = kp$ where p is vapor pressure of the monomer.

Rearranging this equation the relationship

$$\frac{1}{\text{Rate}} = \frac{1}{p}\left[\frac{(k_2/kk_1)[\text{R·}] + (k_3/kk_1)}{[\text{R·}]}\right] + \frac{1}{r}$$

was obtained and a value of r could be found from a plot of $1/\text{Rate}$ against $1/p$ at each power.

A relationship between rate and plasma power was derived in this way. Rates of polymer deposition were then calculated at various pressures and powers and found to be in fairly good agreement with the experimental values.

In a second paper Denaro, Owens, and Crawshaw [58] have reported a study of the rate of polymerization of α-methylstyrene and allyl benzene. A more general relationship between the rate of production of radicals and the power was assumed. A treatment of the rate data for those systems indicated that step (5.2) or radical recombination was not very important. The agreement between experiment and the calculated rate was good for allylbenzene but poor for α-methylstyrene.

A somewhat different polymerization mechanism has been proposed for those experiments in which an electrodeless discharge was used. An example of such a study is the work of Yasuda and Lamaze [9] who investigated the polymerization of styrene in a reactor similar to that shown in Figure 5.7. The dependence of the deposition rate on the power and pressure were found to be unlike that previously reported for apparatus containing internal electrodes. The deposition rate was independent of the power and increased with the square of the monomer partial pressure. Based upon the observed pressure dependence it was proposed that polymerization occurred in the gas phase.

Thompson and Mayhan [17,24] have also studied the rate of polymerization of styrene in an electrodeless discharge. In this case, styrene was passed directly through the coil used to excite the discharge. The rate of polymerization of styrene in this system was found to depend on the power coupled into the plasma. At power levels of 36 W, particles of polymer were observed to form in the gas phase and fall to the bottom of the reactor.

Careful observation of the films present on the wall of the reactor showed that particles were also present in the film. On the basis of this evidence, polymerization in the gas phase was considered to be important as well as polymerization on the surface. Two experiments were carried out to determine whether ions, radicals, or ion radicals were the primary gas phase intermediates.

In the first of these experiments styrene was polymerized in the presence of 10 and 90% nitrogen dioxide. The rate of styrene polymerization was found

TABLE 5.5. Films Produced by Plasma Techniques

Monomers	Plasma conditions	Film evaluation	Reference	Application, comments
Aromatics: Benzene, toluene, xylene, chlorobenzene, etc.	Tesla discharge, external high-area electrodes	Brown shellac appearing films, elemental analysis included	[35]	Observed some O_2 in films, appear to be high polymers, inert, insoluble
Aromatics: Benzene, xylene, pyridine, methane, heptane	Electrodeless, 1100 kHz	Solid powder, scales, elemental analysis	[60]	Observed emission spectra of intermediates in discharge
Studied 57 hydrocarbons, aromatics	dc large metal electrodes, coatings on cathode	Solid wax-like brittle dark brown films, C, H analysis	[61]	
Hydrocarbons, halocarbons, organometallics	ac, dc, parallel plates, coatings on substrate in contact with electrodes	Thin uniform pinhole-free films, ir analysis, electrical properties	[5]	Thin film capacitors, early application in electronics
Polyvinylalcohol types	Low-energy discharge		[62]	Patent, paper or paperboard coatings
		Films deposited on moving tapes through discharge zone	[63]	Patent
Benzene	Internal electrodes, bell jar type of apparatus	Electrical properties, ir, temperature stability	[64]	Studied variables related to process
Styrene	180 kc/sec rotating parallel cylindrical electrodes	Capacitance and power factor measured, films for electronics application	[36]	Observed free radicals in films, mentioned surface oxidation effect
Organic	Glow discharge		[65]	Patent, coatings on metals
Butadiene	Hot filament glow discharge	Large area, short free films	[66]	Electronics
Vinyl acetate	Silent electric discharge, N_2 atmosphere		[67]	Studied process variables added

$$(CH_3)_2C-N{=}N-C(CH_3)_2$$
with CN substituent on each terminal carbon.

40 Monomers	Flat plate electrodes, 10–50 kHz, films on electrodes	High-temperature stable inert films, conductivity, capacitance	[6]	Micro electronics application extensive discussion of variables and film properties
Acetylene, benzene	Microwave 2456 MHz	Extensive evaluation, analysis, ir, ESR, amorphous, pinhole-free, surface O_2 reaction and effect of annealing discussed	[18]	Characterized volatile and nonvolatile products
Methyl methacrylate		Polymers dissolved in acetone, forming powders and oils on addition of water	[68]	Added metal carbonyls and CCl_4
Styrene, butadiene, acrylonitrile, acrylate esters [69]	ac discharge, parallel plates, bell jar set-up	Solubility, ir data	[42]	Protective coating for cans, plastic films, paper, paper board, etc., studied effect of various process variables
Styrene, acrylates, olefins	16 kHz parallel plates, bell jar set-up	Solubility, ir, thermal stability data	[41,59]	Protective coatings for steel, evaluated potential and economics of process, studied process variables polymerization rate related to adsorption of monomer on electrodes
Styrene, divinylbenzene, allylglycidyl ether, etc.	Electrodeless rf with external longitudinal magnetic field to confine discharge		[70]	Thin film capacitors deposition rate studied

(*Continued*)

TABLE 5.5. (*Continued*)

Monomers	Plasma conditions	Film evaluation	Reference	Application, comments
Styrene	Deposition on Al foil, substrate in contact with internal electrodes, 1.0 torr	Electrical measurements, loss tangent, permittivity measured	[37]	Thin film electronic devices, noted effect of surface oxidation on properties
Hexafluoropropane	Internal plate electrodes ac, low pressure	Film more resistant to heat than poly(tetrafluoroethylene)	[71]	Patent
CF_2X_2, XCF_2CF_2X, $X = H$ or halogen not F	Internal plate electrodes, ac low pressure		[72]	Patent
Styrene, chlorobenzene etc.		Electrical measurements, thermal stability	[73]	Thin film electronics dielectric or insulating films
Propylene	Ozonizer, 788 torr	Infrared elemental analysis, thermally stable films	[74]	
Styrene	Low-pressure discharge, substrate on electrode	Protective coating	[75]	Process for continuous deposition of polymer film on moving steel strip
Pentane, ethylene, butadiene, benzene, styrene, naphthalene	ac parallel plate electrodes, bell jar apparatus	Transmission and FMIR spectra of films presented and interpreted	[38]	Observed high reactivity of films toward water observed radicals in film by ESR
Propylene oxide	Ozonizer type of discharge	Insoluble heat-stable polymer film	[76]	Coating on paper patent
Siloxanes	Electrodes with Al substrate		[77] [78]	Film characterization, ir spectra of polysiloxane films from [77], observe spectra changes from surface oxidation

Monomer	Conditions	Application	Ref.	Comments
Methyl methacrylate	Metal electrodes, 6 kHz	Coating on metal articles	[79]	Observed good bonding, to methyl methacrylate film
Divinylbenzene, butadiene	Electrodes and substrate control grid with pattern used to define area where film deposited	Decorative film with pattern	[80]	Patent
	Device from [80]	Coat various substrates	[81]	Film treated with vapors to stabilize film against atmospheric degradation, patent
Ferrocene, vinylferrocene, styrene, acrylates, polyaromatics	Parallel electrodes, one perforated—allowing film to form pattern on substrate behind perforated electrode	Thin film electronics films	[82]	Films deposited on variety of substrates
Divinylbenzene	Al electrodes, 1 torr	Measured capacity and loss factor	[83]	Capacitors with polymer film as dielectric, electron-microscope indicated no pinholes
Acrylates, styrene, vinyl-benzenes, etc.	Low-pressure discharge in inert gas	Coating on moldings	[84]	Apparatus for continuous polymerization described, patent
Organosiloxanes	Al electrodes in glass chamber	Infrared spectra of films discussed	[85]	Process variables studied, i.e., deposition rate
Acrylonitrile	Silent electric discharge	Measured resistivity, optical absorption, photo current, change with heat treatment of film	[86]	Photo-conductive thin films
Styrene	Metal electrodes rf glow discharge	Solubility, ir of films discussed	[19]	Variables related to deposition studied—surface radicals detected

(*Continued*)

TABLE 5.5. (*Continued*)

Monomers	Plasma conditions	Film evaluation	Reference	Application, comments
Hydroperfluoropropane, dibromodifluoromethane, ethylene	Internal electrode plates high-frequency ac		[87]	Coatings on cutting surface of knives and tools, patent
Butylphthalate, acrylamide silicones, uncured resins	Low-volatile momomers evaporated on rotating drum and polymerized by maintaining glow discharge between drum and stationary electrode		[21]	Protective coatings for steel process variables studied
Methylethylene	Siemens ozonizer	Inert, thermally stable films ir of films given and interpreted	[88]	Coatings to improve adhesive compatibility and printability
Chlorobenzene			[89]	Apparatus to coat polyethylene bottles described
Toluene	ac glow discharge, internal electrodes	Infrared of film, bulk electrical resistance, electrical conductivity loss factor	[26]	ESR measurements indicated addition of argon increased radical concentration in vapor phase
			[90]	Patent, apparatus for deposition of organic films on objects on a carrier in a discharge chamber

Monomer	Conditions	Results	Ref.	Comments
Methylsiloxanes	Silent discharge, Al electrodes	Films examined by ir spectroscopy	[12]	Argon addition in vapor increased cross-linking in polymer and cleavage of monomer, ESR indicated higher concentration of free radicals in vapor phase in presence of argon
Benzene	Low pressure, 13.6 MHz	Insoluble thermally stable yellow film, conductivity data given	[91]	Patent
Butadiene, naphthalene	Internal electrodes, ac glow	Measured photoelectron emission and energy distribution of external photoelectrons from films	[92]	
Methane, methyl chloride	Microwave 2456 MHz	Extensive film characterization, elemental analysis, ir, thermal stability, trapped radicals by ESR	[15]	
Toluene	Electrodeless rf 28 MHz	Cross-linked polymer, ir, NMR data given	[93]	
	Polymers prepared from excitation of two or more substances in the gas phase		[94]	Polyphenylene oxide, sulfone, sulfide, polymetallophenylenes, polymetallosiloxanes
Hexafluoropropylene, tetrafluoroethylene, chlorodifluoromethane	Corona discharge		[95]	Patent

(Continued)

TABLE 5.5. (Continued)

Monomers	Plasma conditions	Film evaluation	Reference	Application, comments
CO, N_2, H_2	Electrodeless rf 13.56 MHz	Elemental analysis, ir, FMIR spectra	[10]	Nitrogen content of polymer found to be proportional to flow rate of N_2; amino acids resulted on hydrolysis of polymer
12 Organic compounds studied	Corona discharge with moving strip substrate	Analytical and ir data given for films	[14]	Studied effect of halocarbons in vapor—halocarbons varied polymer deposition
Benzene	Corona discharge	Polymer fractionated, molecular weight distribution, elemental analysis, and NMR data given	[96]	
	Electrodeless rf 3.69 MHz	Extensive characterization, analysis, thermal stability ir data, detection of unpaired spins by ESR	[16]	Studied variables related to process
Chlorobenzene, fluorobenzene, benzonitrile, acetophenone, phenylacetylene, nitrobenzene, toluene	Internal electrodes 20 kHz, 1 torr	FMIR spectra of films presented and interpreted, spectra support monomer substituted aromatic rings	[22]	Observed bands from emission spectra of discharge in toluene, phenylacetylene, and diacetylene
30 Monomers studied	Electrodeless rf 13.56 MHz bell jar type of apparatus	Measured capacitance and dielectric loss of films	[7]	Dielectric in capacitors, studied variables effecting deposition rate
Hydrocarbons, saturated, unsaturated, cyclic	Low pressure 100–300 MHz	Cross-linked polymers, analysis ir, thermal analysis, electrical resistivity data	[97]	

Monomer	Conditions	Film characteristics	Ref.	Remarks
Ethylene	Low-pressure parallel electrodes		[58]	Extension of work reported in [87]
	Substrate and additional electrode		[98]	Protective coating for metal substrate, patent
Toluene, trifluoroethylene, tetrafluoroethylene	Electrodeless discharge	Cross-linked thermally stable films	[11]	Impermeable coatings in fuel storage containers, gas composition can be changed to continuously vary composition of film
Styrene, methylmethacrylate	Pulsed electric discharge internal electrodes	Thin coatings	[99]	
Siloxanes	Electrodes in glass chamber	Studied ir of polymer deposited in presence and absence of argon and also effect of annealing film, concluded that film deposition in the presence of argon cross-linked film	[13]	
Methylsiloxanes			[100]	Studied changes in discharge process variables as a function of structure of siloxane monomers
Vinyl halides, styrene, acrylates	Internal metal electrodes, 3.14 MHz excitation bell jar set-up	Analysis and ir data given for films	[23]	Observed analytical and ir evidence for surface oxidation, studied effects of process variables
Tetrafluoroethylene	rf glow discharge		[101]	Process variables studied
	Glow discharge	Films contained stable negative charges	[102]	Nonthrombogenic surfaces

(Continued)

TABLE 5.5. (*Continued*)

Monomers	Plasma conditions	Film evaluation	Reference	Application, comments
Xylene, divinylbenzene, ethylene siloxane, methylmethacrylate		Polymeric films with dielectric loss	[103]	Insulators for micro electronic devices
Siloxane	Internal electrodes—glow discharge	Effect of carrier gases such as Ar, O_2, N_2, H_2 on film properties studied as well as changes in electrical properties of films with time	[104]	
About 20 monomers	Various glow discharge deposition process	Survey of glow discharge films and applications	[69]	
Toluene, ethylbenzene	Glow discharge	Liquid products characterized	[105]	
Hexafluoropropene	Ozonizer discharge conditions	Liquid linear branched and unsaturated fluoroalkanes	[39]	Observed free radicals in products by ESR hydrolyzed products for further characterization
Styrene, xylene, ethylene	Internal electrodes 5 kHz, 0.5 torr	Observed ESR signal from films and changes in signal with time and temperature in air and vacuum	[27]	
Toluene	Internal electrodes 20 kHz, bell jar set-up	Observed changes in the water permeation rate through nitrocellulose films exposed to a toluene discharge	[106]	
Styrene	Electrodeless rf	Films on membranes	[48]	Study of effect of carrier gas, pressure, power on the rate of film deposition

204

Monomer	Reactor/conditions	Evaluation	Ref.	Purpose
Triethylsilane	Electrodeless rf 100 W power	Oxygen transfer rates through membrane	[9]	Polymer membranes
90 Monomers, olefins, and various aromatics, etc.	Electrodeless rf at 13.56 MHz	Evaluated reverse osmosis characteristics of films studied deposition rate and classified monomers according to their deposition rate parameters	[8]	Preparation of reverse osmosis membranes
Perfluorobutene-2	Electrodeless rf at 13.56 MHz	Evaluated film thickness, surface wetting properties, and studied free radicals present in films by ESR	[29]	Fluorocarbon protective coatings on glass substrates
Styrene, ethylene, 1,3-butadiene, 1,1-difluoroethylene, methylvinyldichlorosilane, and 5 vinyl monomers	Inductively coupled 4 MHz, 50 W, used reactor capable varying inlet and outlet configuration	Infrared, measured distribution of film in reactor	[24]	Observed formation of particles of polymer as well as film
Styrene	Same as described in [24], extensive study of plasma polymerization	Extensive polymer evaluation DTA and TGA analysis, ir, mass spectroscopy, x ray, SEM of polymer surface	[17]	Observed soluble and insoluble polymer studied kinetics of deposition—presented vapor phase ionic mechanism for polymerization
Butadiene	Internal electrodes 60 Hz, 400 V	Extensive evaluation of electrical properties of films such as capacitance resistance current voltage behavior	[40]	Investigation of electrical conduction in thin polybutadiene films formed in a plasma

(Continued)

TABLE 5.5. (*Continued*)

Monomers	Plasma conditions	Film evaluation	Reference	Application, comments
Styrene, tetrafluoroethylene, benzene, *p*-xylene	Low-pressure ac-generated plasma	Electrical properties such as capacitance/area loss factor, heat and insulation resistance were measured	[107]	Dielectric films in electrical devices such as capacitors
Vinylenecarbonate, acrylonitrile, acrylonitrile, vinylacetate	Convex-concave metal electrodes excited at 400 kHz millipore filter substrate	Salt rejection behavior evaluated	[108]	Reverse osmosis membranes mixtures of monomers were used and the substrate was precoated with siloxanes
Hexamethyldisiloxane, vinyltrimethylsilane, tetramethylsilane	Internal electrodes excited at 13.56 MHz, helium, argon, and hydrogen were used as carrier gases	Characterized the surface of the film by scanning electron microscopy	[109]	Laser light guides in optoelectronic devices
15 Monomers nitrile, vinyl and aromatic	Internal brass electrodes excited at 350 kHz	FMIR spectra and gas permeability	[110]	Permeselective membranes, various substrates used

16 Volatile fluorinated organic monomers	Bell jar reactor excited electrodelessly at 13.56 MHz and up to 300 W argon used with the fluorocarbon	Ir, detected surface oxidation	[111]	Barrier coatings on semiconductor devices; performance of semiconductor devices studied after exposure to plasma conditions
A variety of monomers were studied	Bell jar with internal electrodes operated between 3000–10,000 Hz at 300 V and 20–25 mA; static conditions employed	Ir, elemental analysis, VPC used to analyze gases formed during polymerization, studied decomposition of films on exposure to oxygen discharge	[112]	General study of polymerization of organic monomers in a static discharge
Allylamine	Bell jar excited electrodelessly at 13.56 MHz; argon and nitrogen monomer blends were used in addition to pure monomer	Transmission ir. salt and urea rejection evaluated	[113]	Membranes for reverse osmosis
Ethylene	Bell jar apparatus with internal electrodes excited at 13.56 MHz	Chemical analysis, ir, observed atmospheric oxidation, observed amorphous powder formation as well as film formation	[114]	Studied effect of polymerization conditions on the physical form of the polymer
Variety of monomers	Electrodeless discharge	Salt rejection	[115]	Reverse osmosis membrane, good general review articles

to be unaffected by the presence of NO_2. Since nitrogen dioxide acts as a radical scavenger and its presence did not affect the rate of polymerization, a radical mechanism was felt to be unimportant. For the second experiment, two electrodes were located beyond the coil in the after glow region of the plasma. A dc potential of 350 V was maintained on these electrodes during the polymerization. The plasma was deflected to the negative electrode and most of the polymer film deposited on the negative electrode.

From this evidence it was postulated that positive ions were the most important species for the vapor phase polymerization. The source of these ions was taken to be either the direct conjugation of the monomer or the ionization of fragments formed from the monomer molecule.

From this review it becomes apparent that a comprehensive mechanism for explaining plasma polymerization is not available. If one attempts to piece together the material available, it is possible to conclude that the precursors for polymerization, i.e., ions and radicals, are formed in the gas phase. These species most certainly interact both with themselves as well as with the monomer gas to produce intermediates of increasingly greater molecular weight. This set of reactions can lead to the formation of polymer chains present in the gas phase and to the ultimate formation of the fine polymer powders reported by a number of authors. The preferential deposition of polymer films on electrodes when these are present does not conflict with the proposed interpretation. When electrodes are present within the discharge, it is frequently observed that the plasma glow is most intense in the immediate vicinity of the electrodes (see Chapters 1 and 10). Since the rates of excitation processes are high in these regions, it is reasonable to expect a high concentration of polymer intermediates to be present a short distance from the electrode surface. By diffusing to the electrode surface, these intermediates can become incorporated into a film and can react with adsorbed monomer molecules. The radiation emitted from the discharge zones near the electrodes undoubtedly also plays a role in enhancing the rate of film formation on the electrode surface.

5.5. APPLICATIONS OF PLASMA-DEPOSITED FILMS

The polymer films formed in an electric discharge are finding a variety of applications which derive from their unique properties. Plasma-deposited organic films have found wide application as dielectrics in the electronics industry [2], as protective coatings for metals and other reactive surfaces [59], and more recently in reverse osmosis membrane fabrication [8]. Detailed descriptions of these applications can be found in review articles [1–4].

For convenient reference, the literature on plasma polymerization of organic monomers is presented in tabular form in Table 5.5. Information on the monomers employed, the plasma conditions, and the properties and applications of the films is included. It is apparent from this table that plasma polymerization of organic monomers is achieving the status of an important process for preparing polymer films. The established features of the process should assure it a continued and expanding use.

REFERENCES

[1] V. M. KOLOTYRKIN, A. B. GILMAN, and A. K. TSAPUK, *Russ. Chem. Revs.* **36**, 579 (1967).

[2] D. S. ALLAM and C. T. H. STODDARD, *Chem. Brit.* **1**, 410 (1965).

[3] A. M. MEARNS, *Thin Solid Films*, Vol. 3, Elsevier, Lausanne, 1969, pp. 201–228.

[4] L. V. GREGOR, *Physics of Thin Films*, Vol. 3, in G. HASS and R. E. THUNED (eds.) *Advances in Research and Development*, Academic Press, New York, 1966.

[5] J. GOODMAN, *J. Polymer Sci.* **44**, 551 (1960).

[6] A. BRADLEY and J. P. HAMMES, *J. Electrochem. Soc.* **110**, 15 (1963).

[7] G. SMOLINSKY and J. H. HEISS, *Div. Org. Plastic Chem. Preprints* **28**(1), 537 (1968).

[8] H. YASUDA, Final Report to Office of Saline Water, U.S. Department of the Interior, Contract No. 14-30-2658, Research Triangle Park, N.C. (1972).

[9] H. YASUDA and E. LAMAZE, *J. Appl. Polymer Sci.* **16**, 595 (1972).

[10] J. R. HOLLAHAN and R. P. MCKEEVER, *Advan. Chem. Ser.* **80**, 272 (1969).

[11] J. P. REDMOND and A. F. PITAS, NASA Contract Rep. NASA-CR-94310, 1968.

[12] B. V. TKACHUK and V. M. KOLOTYRKIN, *Vysolomol. Soedin. Ser. B* **10**, 24 (1968).

[13] B. V. TKACHUK and V. M. KOLOTYRKIN, *Ukr. Khim. Zh* **35**, 768 (1969).

[14] P. M. HAY, *Advan. Chem. Ser.* **80**, 350 (1969).

[15] J. P. WIGHTMAN and N. J. JOHNSTON, *Advan. Chem. Ser.* **80**, 322 (1969).

[16] D. D. NEISWENDER, *Advan. Chem. Ser.* **80**, 338 (1969).

[17] L. F. THOMPSON and K. G. MAYHAN, *J. Appl. Polymer Sci.* **16**, 2317 (1972).

[18] F. J. VASTOLA and J. P. WIGHTMAN, *J. Appl. Chem.* **14**, 69 (1964).

[19] A. R. DENARO, P. A. OWENS, and A. CRAWSHAW, *Eur. Polymer J.* **4**, 93 (1968).

[20] B. SCHWARTZ and N. SCHWARTZ (eds), *Measurement Techniques for Thin Films*, Electrochemical Society, New York, 1967, pp. 102–122.

[21] T. WILLIAMS and M. W. HAYES, *Nature* **216**, 614 (1967).

[22] P. L. KRONICK, K. F. JESCH, and J. E. BLOOR, *J. Polymer Sci. A1* **7**, 767 (1969).

[23] A. R. WESTWOOD, *Polymer Prepr. Am. Chem. Soc. Div. Polymer Chem.* **10**, 433 (1969).

[24] L. F. THOMPSON and K. G. MAYHAN, *J. Appl. Polymer Sci.* **16**, 2291 (1972).

[25] H. KONIG and A. BROCKES, *Z. Physik.* **152**, 75 (1958).

[26] L. S. TUZOV, A. B. GILMAN, A. N. SHCHUROV, and V. M. KOLOTYRKIN, *Vysokomol. Soedin. Ser. A*, **9**, 2414 (1967).

[27] S. MORITA, T. MIZUTANI, and M. LEDA, *Japan. J. Appl. Phys.* **10**, 1275 (1971).

[28] R. MANGIARACINA and S. MROZOWSKI, "Trapped Radicals in Organic Deposits," *Proc. Carbon Conf.*, 5*th* **2** (1963).

[29] M. MILLARD, J. J. WINDLE, and A. E. PAVLATH, *J. Polymer Sci.* **17**, 2501–2507 (1973).

[30] A. A. WESTENBERG, *Science* **164**, 381 (1969).

[31] S. N. FONER, *Science*, **143**, 441 (1964).

[32] N. JONATHAN, *Chem. Soc. Ann. Reports A* **66**, 152 (1969).

[33] M. A. A. CLYNE, *Chem. Soc. Ann. Reports A* **65**, 167 (1968).

[34] F. KAUFMAN, *Ann. Rev. Phys. Chem.* **20**, 45 (1969).

[35] J. B. AUSTIN and I. A. BLACK, *J. Am. Chem. Soc.* **52**, 4552 (1930).

[36] M. STUART, *Nature (London)* **199**, 59 (1963).

[37] M. STUART, *Proc. Inst. Elec. Eng.* **112**, 1614 (1965).

[38] K. JESCH, J. E. BLOOR, and P. L. KRONICK, *J. Polymer Sci. A1* **4**, 1486 (1966).

[39] E. S. LO and S. W. OSBORN, *J. Org. Chem.* **35**, 935 (1970).

[40] N. M. BASHARA and C. T. DOYTY, *J. Appl. Phys.* **35**, 3498 (1964).

[41] T. WILLIAMS and M. W. HAYES, *Nature* **209**, 769 (1966).

[42] R. M. BRICK and J. R. KNOX, *Modern Packaging* **123** (1965).

[43] R. A. CONNELL and L. V. GREGOR, *J. Electrochem. Soc.* **112**, 1198 (1965).

[44] J. K. STILLE, R. L. SUNG, and J. VANDER KOOI, *J. Org. Chem.* **30**, 3116 (1965).

[45] J. K. STILLE and C. E. RIX, *J. Org. Chem.* **31**, 1591 (1966).

[46] F. J. DINAN, S. FRIDMANN, and P. J. SCHIRMANN, *Advan. Chem. Ser.* **80**, 289 (1969).

[47] A. STREITWIESER, JR. and H. R. WARD, *J. Am. Chem. Soc.* **85**, 539 (1963).

[48] H. YASUDA and C. E. LAMAZE, *J. Appl. Polymer Sci.* **15**, 2277 (1971).

[49] I. HALLER and P. WHITE, *J. Phys. Chem.* **67**, 1784 (1963).

[50] B. D. BLAUSTEIN and Y. C. FU, *in* A. WEISSBERGER (ed.), *Organic Reactions in Electrical Discharges in Physical Methods of Chemistry 1*, Part 11B, Wiley, 1971, pp. 90, 93–207.

[51] H. SCULER and M. STOCKBURGER, *Spectrochim. Acta* **13**, 841 (1959).

[52] S. WALKER and R. F. BARROW, *Trans. Faraday Soc.* **50**, 541 (1954).

[53] S. WALKER and T. F. BINDLEY, *Trans. Faraday Soc.* **58**, 217 (1962).

[54] S. WALKER, T. F. BINDLEY, and A. T. WATTS, *Trans. Faraday Soc.* **58**, 849 (1962).

[55] M. W. Ranney and W. F. O'Connor, *Advan. Chem. Ser.* **80,** 297 (1969).

[56] F. K. McTaggart, *Plasma Chemistry in Electrical Discharges*, Elsevier, New York, 1967.

[57] A. S. Kana'an and J. L. Margrave, *in* H. J. Emeleus and A. G. Sharpe (eds.), *Advances in Inorganic Chemistry and Radiochemistry*, Vol. 6, Academic Press, New York, 1964, p. 154.

[58] A. R. Denaro, P. A. Owens, and A. Crawshaw, *Eur. Polymer J.* **5,** 471 (1969).

[59] T. Williams, *J. Oil Colour Chem. Assoc.* **48,** 936 (1965).

[60] W. D. Harkins and J. M. Jackson, *J. Chem. Phys.* **1,** 37 (1933).

[61] E. G. Linder and A. P. Davis, *J. Phys. Chem.* **35,** 3649 (1931).

[62] G. J. Argnette, U.S. Patent 3,061,458 (1962); *C.A.* **58,** 1636 (1963).

[63] J. H. Coleman, U.S. Patent, 3,068,510 (1962); *C.A.* **58,** 5898 (1963).

[64] H. Pagnia, *Phys. Status, Solidi* **1,** 90 (1961).

[65] A. P. Cornelius, U.S. Patent 3,108,900 (1963); *C.A.* **60,** 1386 (1964).

[66] E. M. DaSilva and R. E. Miller, *Electrochem. Technol.* **2,** 147 (1964).

[67] Y. Kikuchi, *Denki Kagaku* **28,** 268 (1960).

[68] W. Strohmeier and H. Gruebel, *Z. Naturforsch.* **20b,** 11 (1965).

[69] A. Bradley, *Eng. Chem. Prod. Res. Dev.* **9,** 101 (1970).

[70] R. A. Connell and L. V. Gregor, *J. Electrochem. Soc.* **112,** 1198 (1965).

[71] Y. Kometani, A. Katsushima, and T. Fukui, Japanese Patent 10,989 (1965); *C.A.* **64,** 12838 (1966).

[72] Y. Kometani, A. Katsushima, and T. Fukui, Japanese Patent 10,991 (1965); *C.A.* **64,** 12838 (1966).

[73] B. G. Carbajal, *Trans. Met. Soc. AIME* **236,** 365 (1966).

[74] Z. I. Ashurly, V. G. Babyan, M. A. Bagirov, and C. A. Fatalizade, *Dokl. Akad. Nauk Azerb. SSR* **22,** 29 (1966).

[75] T. Williams and J. H. Edwards, *Trans. Inst. Metal Finishing* **44,** 119 (1966).

[76] M. C. Tobin and W. G. Diechert, U.S. Patent 3,287,242 (1966); *C.A.* **66,** 29570j (1967).

[77] B. V. Tkachuk, V. V. Bushin, and N. P. Smetankina, *Ukr. Khim. Zh* **32,** 1256 (1966).

[78] B. V. Tkachuk and V. V. Bushin, *Ukr. Khim. Zh* **33,** 224 (1967).

[79] Continental Can Co. Netherlands Patent 6,601,363 (1966); *C.A.* **67,** 12657r (1967).

[80] Metal Containers Ltd. Netherlands Patent 6,610,513 (1966); *C.A.* **67,** 22973e (1967).

[81] Metal Containers Ltd. Netherlands Patent 6,610,517 (1966); *C.A.* **67,** 22974f (1967).

[82] B. G. Cargajal and B. G. Slay, Jr., U.S. Patent 3,318,790 (1967); *C.A.* **67,** 22975g (1967).

212   SYNTHESIS OF ORGANIC POLYMER FILMS IN PLASMA

[83] T. Kanazono, M. Takamura, and K. Kojima, *Mem. Inst. Sci. Ind. Res. (Osaka Univ.)* **24**, 65 (1967).

[84] Metal Containers Ltd. Netherlands Patent 6,604,207, (1966); *C.A.* **66**, 76490t (1967).

[85] B. V. Takachuk, V. V. Bushin, V. M. Kolotyrkin, and N. P. Smetankina, *Vysokomol. Soedin. Ser. A,* **9**, 2018 (1967).

[86] T. Hirai and O. Nakada, *Japan. J. Appl. Phys.* **7**, 112 (1968).

[87] H. R. Watson, British Patent 1,106,071 (1968); *C.A.* **68**, 106152v (1968).

[88] Z. I. Ashurly, V. G. Babyan, M. A. Bagirov, and E. Y. Volchenkov, *Vysokomol. Soedin. Ser. B,* **10**, 356 (1968).

[89] M. Erchak, Jr., U.S. Patent, 3,387,991 (1968); *C.A.* **69**, 28664t (1968).

[90] F. Grasenick, Austrian Patent 258,664 (1967); *C.A.* **69**, 54868m (1968).

[91] C. Simionescu, N. Asandei, and F. Denes, Romanian Patent 50976 (1968); *C.A.* **70**, 20527a (1969).

[92] P. L. Kronick, *J. Appl. Phys.* **39**, 5806 (1968).

[93] F. J. Dinan, S. A. Friedmann, and P. J. Schirmann, *Advan. Chem. Ser.* **80**, 289 (1969).

[94] E. L. Bush, J. P. W. Ryan, and J. H. Alexander, British Patent 1,146,550 (1969); *C.A.* **70**, 107086e (1969).

[95] S. W. Osborn and E. Broderick, French Patent 1,524,571 (1968); *C.A* **70**, 115704v (1969).

[96] M. W. Ranney and W. F. O'Connor, *Advan. Chem. Ser.* **80**, 297 (1969).

[97] C. Simionescu, N. Asandei, F. Denes, M. Sandulovici, and Gh. Popa, *Eur. Polymer J.* **5**, 427 (1969).

[98] C. D. Fisher, U.S. Patent, 3,457,156 (1969); *C.A.* **71**, 82742w (1969).

[99] E. H. Manuel, U.S. Patent, 3,471,316 (1969); *C.A.* **71**, 114299a (1969).

[100] V. V. Bushin and B. V. Tkachuk, *Tr. Vses, Soveshch. Kriotronike.*, 3rd ed., 1967, p. 73.

[101] H. U. Poll, *Z. Angew. Phys.* **29**, 260 (1970).

[102] P. L. Kronick and M. E. Schafer, U.S. Clearinghouse Fed. Sci. Tech. Inform., PB Rep. #192062, 1969.

[103] R. Aires, French Patent 1,584,542 (1969); *C.A.* **72**, 89249r (1970).

[104] L. S. Tuzov, V. M. Kolotyrkin, and N. N. Tunitskii, *Vysokomol Soedin. Ser. A,* **12**, 849 (1970).

[105] A. N. Singh and I. S. Singh, *Indian J. Pure Appl. Phys.* **8**, 119 (1970).

[106] P. L. Kronick and M. E. Schafer, *J. Appl. Polymer Sci.* **13**, 249 (1969).

[107] P. J. Ozawa, *IEEE Trans. Parts Mater. Packag.*, PMP-5, 112 (1969).

[108] K. R. Buck and V. K. Davar, *Br. Poly. J.* **2**, 238 (1970).

[109] L. F. Thompson and G. Smolinsky, *J. Appl. Polymer Sci.* **16**, 1179 (1972).

[110] A. F. Stancell and A. T. Spencer, *J. Appl. Polymer Sci.* **16**, 1505 (1972).

[111] S. M. LEE, *Insulation/Circuits*, June 1971, p. 33.

[112] A. BRADLEY, *J. Electrochem. Soc.*, **119,** 1153 (1972).

[113] J. R. HOLLAHAN and T. WYDEVEN, *Science*, **179,** 500 (1973).

[114] H. KOBAYASHI, A. T. BELL, and M. SHEN, *J. Appl. Polym. Sci.*, **17,** 885 (1973).

[115] H. YASUDA, *Appl. Polymer Symposia* No. 22, 241, M. A. GOLUB and J. A. PARKER (eds.), Wiley, New York, 1973.

Chapter 6

Semipermeable Membranes Produced by Plasma Polymerization

Theodore Wydeven and John R. Hollahan

6.1. INTRODUCTION

The method of plasma polymerization for preparing composite hyperfiltration membranes offers a combination of advantages which cannot be obtained by conventional methods of casting membranes. By plasma polymerization, thin membranes that reject dissolved solids such as ions and organics can be prepared from a variety of conventional (unsaturated) and unconventional (saturated) organic monomers. Thin hyperfiltration membranes yield a high water flux and high salt rejection. For both homogeneous ionic and nonionic polymer membranes, high water flux or permeability is normally accompanied by low salt rejection, as shown in Figure 6.1 [1,2]. Controlling the deposition time with the plasma method is a convenient way of controlling the membrane thickness.

The ability to select the porous substrate upon which the plasma-polymerized hyperfiltration membrane is deposited permits the choice of substrates with minimum compaction during hyperfiltration. Compaction of most polymeric hyperfiltration membranes causes a decrease in water flux with time. Furthermore, the thin films prepared by plasma polymerization are highly cross-linked [3] and undoubtedly achieve maximum density during preparation; therefore, the membrane itself does not undergo further compaction.

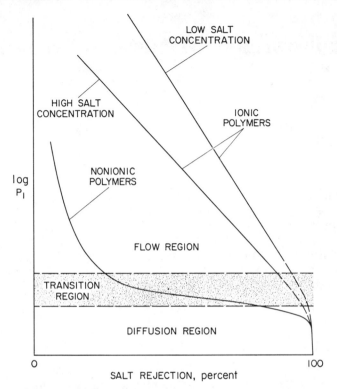

FIGURE 6.1. Schematic representation of reverse osmosis characteristics of ionic and noionic polymer membranes; P_1 = water permeability, Rejection = $(C_{feed} - C_{product})/C_{feed}$ where C = concentration [1,2].

Another advantage of the plasma method is that the membranes are deposited in the clean environment of a partial vacuum; dust particles from the surrounding environment are therefore prevented from collecting in the film during preparation. Dust particles can cause membrane imperfections so that conventional casting methods may yield membranes of reduced salt rejection. Films prepared by plasma polymerization are generally found to be pinhole-free.

Plasma-polymerized membranes are prepared dry and can be stored indefinitely in the dry state; in this respect, they differ from most cast membranes, such as cellulose acetate, which generally must be stored in water. Dry membranes are also easier to handle than most cast membranes in assembling a water purification or processing unit.

Although it has not yet been demonstrated, the plasma method should lend itself to the preparation of membranes in a variety of configurations such as hollow fibers, tubules, and flat sheets. Hollow fibers are an especially attractive possibility because they can be used for obtaining a large membrane surface to volume ratio, thereby achieving a high water processing rate with a small water purification unit. As a result of the potential advantages the plasma-polymerization method offers for preparing hyperfiltration membranes, research has begun in a few laboratories to develop further the technique [1,2,4,5]. The potential for preparing thin selective membranes of high gas permeability by plasma polymerization also makes the method attractive for the separation of gas mixtures; some preliminary studies have shown the application of plasma-polymerized membranes to this kind of problem [6].

6.2. APPARATUS FOR MEMBRANE FORMATION

A variety of deposition systems have been utilized for preparing plasma-polymerized semipermeable membranes. Buck and Davar [4] used a reactor with internal curved electrodes for preparing hyperfiltration membranes. The electrodes were located 0.6 cm apart. A Millipore filter substrate made of a cellulose ester was firmly attached to one of the electrodes to assure uniform coating. The electrodes were connected to a 40 kHz rf power source.

A membrane was formed by first evacuating the deposition chamber, sealing off the chamber from the vacuum pump, and then admitting the monomers to the desired partial pressures. A discharge was established at 400 V and allowed to operate for 90 sec at a current density of 1.8 mA/cm² before repeating the evacuation-deposition cycle. Typical monomer partial pressures for preparing a membrane were 80 N/m^2 for vinylene carbonate and 7 N/m^2 for acrylonitrile.

Yasuda and co-workers [1,2] used an electrodeless glow-discharge system (Fig. 6.2) for preparing hyperfiltration membranes. The substrates upon which

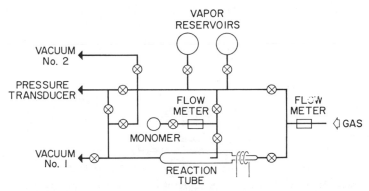

FIGURE 6.2. An electrodeless glow-discharge system [1].

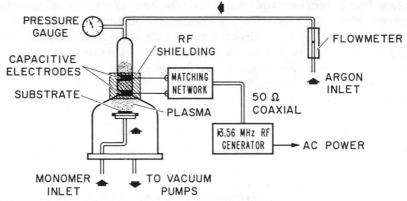

FIGURE 6.3.   Plasma-polymerization system.

the plasma-polymerized membranes were deposited were located against the inside wall of the reaction tube. Yasuda studied a variety of preparative conditions, including some with a flowing plasma gas and some with a stagnant gas. However, one common feature of his experiments was the utilization of a rf (13.56 MHz) electrodeless glow discharge with inductive coupling. The power used for preparing membranes ranged from 30 to 150 W and the deposition or polymerization time from 60 to 3000 sec. Typically, monomer pressures ranged from 2.7 to 170 N/m^2.

Hollahan and Wydeven [5] used the deposition system shown in Figure 6.3 for preparing plasma-polymerized hyperfiltration membranes. Membranes were prepared with the plasma gas flowing over a porous substrate. The plasma-polymerized membrane was deposited on the shiny side of a cellulose ester type of Millipore filter of either 0.1 or 0.025 μ average pore size. The monomer pressure was 13 N/m^2; when using an additive gas, e.g., nitrogen or argon, this gas was also at 13 N/m^2. Deposition times ranged from 500 to 2300 sec and these times yielded membranes ranging in thickness from 0.8 to 1.6 μ. The net rf power to the reactor was generally 40 W.

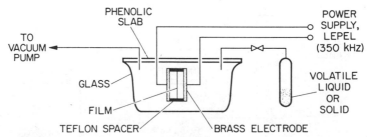

FIGURE 6.4.   Schematic of apparatus for film coating by plasma deposition [6].

Plasma-polymerized membranes for gas permeability and selectivity studies were prepared by Stancell and Spencer [6] using the reactor shown in Figure 6.4. The deposition chamber was continuously pumped during membrane preparation and the desired monomer pressure was achieved by adjusting the valve attached to the monomer reservoir. The polymer coated both sides of the porous substrate with this reactor configuration. The power for the plasma was supplied by a 350 kHz capacitive generator at the 100 W level.

6.3. HYPERFILTRATION CHARACTERISTICS OF PLASMA-POLYMERIZED MEMBRANES

Buck and Davar [4] were the first to report success in preparing hyperfiltration membranes using an rf glow discharge. Their best hyperfiltration performance, shown in Figure 6.5, was obtained with a polymer of vinylene

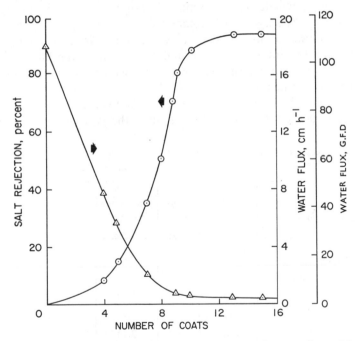

FIGURE 6.5. Variation of salt rejection and water flux with number of coats of polyvinylene carbonate deposited by glow-discharge polymerization on coarse side of Millipore VS backing membrane. Salt rejection and water flux measured after one day in test cell; pressure = 1400 psi = 9.6×10^6 N/m^2; GFD = gallons per square foot per day.

carbonate and 8% acrylonitrile deposited on the coarse (or dull) side of a Millipore cellulose ester filter. Several coats of polymer were applied before a highly salt-rejecting membrane was formed. Each coating yielded a polymer ~0.07 μ thick. Buck and Davar also noted the interesting result that water flux increased with up to 8 days of hyperfiltration with a membrane prepared from vinylene carbonate and 8% acrylonitrile. Most polymeric hyperfiltration membranes show a decrease in water flux with time due to compaction.

Yasuda recently reported [1,2] on a parametric study of the formation of composite hyperfiltration membranes from a large number of organic monomers using an electrodeless glow discharge (Fig. 6.2). As a result of his studies with several different monomers, a few general conclusions were drawn concerning plasma-polymerization mechanisms. These mechanisms are discussed in Chapter 5.

Yasuda also separated the organic compounds he studied by plasma polymerization into two categories. One category (Category A) contained those compounds which are predominantly polymerized in the plasma; the second category (Category B) comprised those compounds which are primarily decomposed in the plasma. Compounds in each category had certain structural features in common and also yielded hyperfiltration membranes of common characteristics. Compounds which generally decomposed in the plasma contained oxygen (e.g., —CO—, —CO—O—, —O—, —OH), chlorine, aliphatic hydrocarbon chains, or cyclic hydrocarbon chains. These groups and structures tended to be absent in the *polymers* prepared from

TABLE 6.1. Reverse Osmosis Results[a] of Plasma-Polymerized Polymers from Various Monomers with Porous Polysulfone Films as Substrates

Monomer	Salt Rejection, %	Water Flux, GFD	Preparative Conditions[b]		
			Monomer Pressure, μ Hg	Power, watts	Deposition Time, min
4-vinylpyridine	95	1.6	360	60	7
5-vinyl-2-methylpyridine	82	14.6	500	30	3.5
4-picoline	98	6.4	540	30	8
4-picoline	95	13.6	560	30	3
2-vinylpyridine	88	3.2	380	30	2.5
pyridine	62	10.4	560	30	5
4-ethyl pyridine	98	9.6	520	30	7

[a] 1.2% NaCl at 1200 psi.
[b] These membranes were prepared by feeding monomer into a closed system at a constant flow rate until the pressure reached a predetermined level. There was no monomer feed-in or pumping during glow discharge.

TABLE 6.2. Reverse Osmosis Results of Plasma-Polymerized Polymer from Hydrophilic Type A Monomers

Monomer	Substrate	NaCl Concentration, %	Applied Pressure, psi	Salt Rejection, %	Water Flux, GFD
4-vinylpyridine	porous glass	1.2	1200	90	0.49[a]
	porous glass	1.2	1200	96	0.81[a]
	porous glass	1.2	1200	87	0.51[a]
	porous glass	1.2	1200	96	0.72[a]
	polysulfone	1.2	1200	97	3.7
	polysulfone	1.2	1200	98	7.0
	polysulfone	1.2	1200	89	4.0
	polysulfone	1.2	1200	95	1.6
	Millipore-VS	1.2	1200	99	38.0
	Millipore-VS	1.2	1200	98	4.0
	Millipore-VS	1.2	1200	95	24.0
	Millipore-VS	1.2	1200	97	7.4
N-vinylpyrrolidone	polysulfone	1.2	1200	91	10.6
4-picolin	polysulfone	1.2	1200	98	6.4
4-ethyl pyridine	polysulfone	1.2	1200	98	9.6
4-picolin	polysulfone	3.5	1500	96	7.6
4-methylbenzylamine	polysulfone	3.5	1500	96	2.2
n-butylamine	polysulfone	3.5	1500	94	2.7
4-vinylpyridine	polysulfone	3.5	1500	97	4.9
4-vinylpyridine	porous glass	3.5	1500	96	0.80[a]
3-5 lutidine	polysulfone	3.5	1500	99	12.0

[a] Water flux of porous glass tube is approximately 1.0 GFD.

these compounds. Compounds which tended to polymerize without decomposition in the plasma also produced superior hyperfiltration membranes. Compounds in this category (A) were also relatively hydrophilic, an apparent requirement for a semipermeable membrane which allows transport of water. Among the compounds investigated by Yasuda and also by Hollahan and Wydeven [5], nitrogen-containing compounds, particularly amines, fell into Category A. Tables 6.1 and 6.2 show the salt rejection and water flux of membranes prepared [1,2] from some nitrogen-containing monomers.

The polymerization conditions were not always the same when preparing the membranes used in obtaining the data in Table 6.2. The different preparative conditions may account for some of the variability in flux and rejection values. However, Yasuda, and also Hollahan and Wydeven, found it difficult to obtain reproducible membranes with some monomers; work is continuing to determine the causes for lack of reproducibility.

It was also noted [1,2] that certain characteristics of the porous substrate were important in preparing composite hyperfiltration membranes by plasma polymerization. Of particular concern was the degradation or decomposition of the substrate in the plasma or glow discharge, the adsorption characteristics of the substrate and the substrate pore size. Of the porous substrates studied, including the Millipore filter (made of cellulose esters), and filters made of polysulfone or porous glass, the Millipore filter with a 250 Å nominal pore size yielded some of the best results (see Table 6.2). The Millipore filter did not degrade appreciably in the plasma, and its small pore size allowed for the deposition of thin semipermeable membranes. With the Millipore substrate, the prior adsorption of contaminants, particularly water vapor, and the adsorption of large amounts of monomer during membrane preparation were not as troublesome as with the porous glass substrate.

Yasuda viewed the process of forming a semipermeable membrane on a porous substrate as the building up of a layer of polymer on the pore walls until closure of the pore by the polymer was achieved. This model requires that the polymer thickness be at least equal to or greater than the radius of the largest pore in the substrate to obtain an imperfection-free hyperfiltration membrane. Yasuda, as did Buck and Davar, also found an increase in water

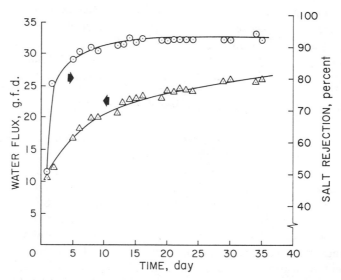

FIGURE 6.6.   The change of salt rejection and water flux with time (1.2%, NaCl, 1300 psi). Membrane of plasma-polymerized 4-vinylpyridine on Millipore VS filter.

TABLE 6.3. Flux and Rejection Characteristics of a Polymer Membrane, Synthesized from Plasma-Polymerization of Allylamine, as a Function of Time during Reverse Osmosis

Time, hr	Membrane Constant[a], A, cm sec^{-1} atm^{-1} × 10^6	Flux, GFD	Percent Rejection[b] NaCl	Urea
1.5	3.21	2.30	78.0	33.0
12.7	4.64	3.28	87.0	28.9
38.0	6.32	4.46	88.7	25.3
61.7	6.52	4.60	89.3	23.7
86.6	6.96	4.91	89.3	22.7

[a] Calculated from $J_1 = A(\Delta P - \Delta \pi)$, where J_1 is the solution flux, cm^3/cm^2/sec^1; A is the membrane constant, cm/sec^1/atm^1; and $(\Delta P - \Delta \pi)$ is the pressure difference minus the osmotic pressure difference across the membrane, atmospheres.
[b] Feed solution composition = 10 g/l NaCl + 10 g/l urea; pH = 7.79; temperature = 19.5°C; pressure = 600 psig; and average linear flow rate across membrane = 17 cm/sec.

flux and salt rejection with time for plasma-polymerized membranes (see Fig. 6.6).

Hollahan and Wydeven [5] recently reported that plasma-polymerized allylamine monomer forms a useful hyperfiltration membrane. The same authors also studied the rejection of dissolved urea as well as NaCl. Typical flux and rejection data for a plasma-polymerized allylamine membrane is shown in Table 6.3. The membrane was formed on 0.1 μ Millipore filter during a 600 sec polymerization period. Again, the data shows an increase in *both* water flux and salt rejection with time. However, urea rejection decreased instead of increasing with time. The difference in salt and urea rejection behavior may be indicative of different rejection mechanisms for these two solutes.

Table 6.4 illustrates the effect of deposition time on water flux and urea

TABLE 6.4. Flux and Rejection Properties as a Function of Deposition Time[a]

Deposition Time, sec	Membrane Constant, A, cm sec^{-1} atm^{-1} × 10^6	Flux, GFD	Percent Rejection NaCl	Urea
600	6.32	4.46	88.5	25.3
2300	0.864	0.581	97.9	41.8

[a] All depositions made on 0.025 μ MF Millipore filter media; same test conditions as in footnote *b*, Table 6.3.

and salt rejection. It is apparent from these data that longer deposition times lead to reduced water flux, presumably due to increased membrane thickness and improved salt and urea rejection. Improved rejection may be ascribed to nearly complete closure of the larger pores in the Millipore filter with longer deposition times.

6.4. GAS PERMEABILITY AND SEPARATION WITH PLASMA-POLYMERIZED MEMBRANES

Stancell and Spencer [6] recently reported the results from a study of the gas permeability and selectivity of some plasma-polymerized membranes. Their objective was to achieve both a high gas selectivity and high permeability by depositing very thin but selective membranes onto relatively permeable

TABLE 6.5. Hydrogen and Methane Permeabilities of Silicone–carbonate Copolymer Plasma Treated with Various Monomers[a]

Monomer	Permeability $\times 10^7$ at 25°C, cc(S.T.P.)-cm/cm^2-sec-atm		Permeability ratio H_2/CH_4
	H_2	CH_4	
None	9.2	10.5	0.87
Nitrile type			
Cyanogen bromide	2.74	0.075	36.6
Nicotinonitrile	3.96	0.11	36.0
Benzonitrile	7.26	0.22	33.0
3-Butenenitrile	8.87	0.39	21.7
Methacrylonitrile	9.04	0.55	16.4
Chloroacetonitrile	13.0	4.97	2.6
Acrylonitrile	15.3	4.37	3.5
Vinyl type			
Acrylic acid	8.15	0.33	24.5
Allyl bromide	8.37	0.66	12.6
Styrene	8.46	0.68	12.4
Divinylbenzene	9.85	1.33	7.0
Aromatic			
Naphthalene	8.1	0.44	18.4
Benzene	9.57	1.33	7.2
Mesitylene	9.43	1.72	5.5
1,3-Di(trifluoromethyl)-benzene	9.92	5.98	1.7

[a] Treatment time 1 min, pressure 1 torr.

TABLE 6.6. Hydrogen and Methane Permeabilities of Poly(phenylene oxide) Plasma Treated with Cyanogen bromide and Benzonitrile

| | Plasma treatment | | Permeability $\times 10^7$, cc(S.T.P.)-cm/cm^2-sec-atm | | | Permeability ratio |
Monomer	Time, min	Pressure, torr	H_2	CH_4	T, °C	H_2/CH_4
None	—	—	5.88	0.25	25	23.5
None	—	—	8.47	0.46	45	18.4
None	—	—	8.53	0.55	64	15.5
Cyanogen bromide	0.5	0.5	3.41	0.015	25	297
Cyanogen bromide	0.5	0.5	5.06	0.035	45	161
Cyanogen bromide	0.5	0.5	7.32	0.11	64	68
Benzonitrile	1.0	1.0	4.47	0.066	25	68

substrate films. The plasma-deposited film thicknesses were judged to be less than 1 μ. The polymer substrates selected were polyphenylene oxide (PPO), silicone–carbonate copolymer and cellulose acetate butyrate.

One of the gas pairs chosen for study was hydrogen and methane because of the significant difference in their molecular sizes. The permeability of each individual gas was determined and used to calculate the permeability ratio. Permeability data for different plasma-polymerized films deposited on a silicone–carbonate copolymer are shown in Table 6.5. Comparison between the untreated substrate and various films deposited on the substrate reveals a significant increase in the permeability ratio, H_2/CH_4, or selectivity with the coated substrate. The authors suggested that the increase in the H_2/CH_4 permeability ratio with the treated substrate was due to the difference in molecular size between hydrogen and methane. The average collision diameter is 3.8 Å for methane, compared to 2.9 Å for hydrogen [7]. The nitrile type of monomer gave the best selectivity as a group, while the aromatic monomers were the least selective.

The PPO substrate yielded the permeability results shown in Table 6.6. It is apparent that the cyanogen bromide polymer on the PPO substrate yielded a significantly higher H_2/CH_4 ratio at 25°C than did the same polymer on silicone–carbonate substrate. The permeabilities of hydrogen and methane increased with temperature as expected (see Table 6.6); however, the permeability ratio decreased due to the greater effect the coating has on the activation energy for methane permeability than on the hydrogen permeability.

Stancell and Spencer also studied the permeability of butene-1 and isobutene, an isomer pair having comparable size and solubility in polymers. Therefore, these molecules were expected to have permeability ratios near

TABLE 6.7. Butene-1 and Isobutene Permeabilities as a Function of Substrate and Plasma Treatment[a]

Substrate	Monomer	Plasma treatment		Permeability ×10^7 at 50°C, cc(S.T.P.)-cm/cm²/sec-atm				Permeability ratio Bu/i-Bu
		Time, min	Pressure, torr	Butene-1	Δp, psi	Isobutene	Δp, psi	
Poly(phenylene oxide)	none	–	–	35.3	66	17.1	68	2.1
	styrene–BrCN	6.0	0.7	0.66	66	0.06	70	1.0
Silicone–carbonate copolymer	none	–	–	1330	67	1345	70	1.0
	benzonitrile	0.5	0.7	20.5	65	18.3	67	1.1
	styrene–BrCN	2.0	0.7	29.1	67	22.0	69	1.3
Cellulose acetate	none	–	–	58.9	65	41.7	70	1.4
Butyrate	styrene	10.0	0.7	5.1	68	2.4	68	2.1

[a] The Δp values used in the permeability measurements are tabulated since the solubility of the butene-1/isobutene pair in the membrane can be very high and can cause permeability to vary with Δp.

unity and plasma-deposited coatings should have had little effect on modifying membrane permeability. These expected findings were confirmed by experiment, as shown in Table 6.7.

6.5. SUMMARY AND CONCLUSIONS

The plasma-polymerization approach to produce thin films which are permselective to liquids and gases offers several advantages over conventional methods. To date only a limited number of experiments have been performed. Many potential conventional and unconventional monomers have not yet been investigated. There is not yet an adequate picture developed for the mechanism responsible for the rejection of salt and/or organic dissolved solids. Perhaps a model similar to the solution diffusion model for salt rejection applies to these polymeric membranes formed from a plasma. For some organics, a geometric exclusion model may apply in which solutes are rejected simply on the basis of molecular size relative to the average pore size in the membrane. Whether the permeate is hydrophilic or hydrophobic may also be an important consideration [8]. Hydrophobic organic solutes are generally believed to exhibit poor rejection by membranes. Very little experimental work has been done on the structure/performance correlation between the monomer, resulting polymer, and hyperfiltration (rejection) properties of these membranes.

Another uncertainty in plasma-polymerized film properties relates to the large amount of oxygen uptake exhibited by the polymer shortly after the films are produced. The significance of this observation is unknown as far as hyperfiltration is concerned. For example, the authors found that when the allylamine monomer (percent composition—C, 63.11; H, 12.35; N, 24.53) is polymerized under good vacuum conditions, a film of the percent composition C, 52.88; H, 6.92; N, 17.39; O, 22.81 is produced. Another example is the observation that after polymerization of benzene in a glow discharge [9], the elementary composition of the resultant polymer corresponded to approximately $C_9H_7O_2$, or 23% oxygen. The large amounts of oxygen presumably arose from free-radical deactivation by reaction with atmospheric oxygen to form hydroperoxides or other oxygen-containing moieties. Free radicals are known to be present in polymers after plasma polymerization or irradiation [10] and their lifetimes may be many hours [11]. The uptake of oxygen presumably renders the membrane, at the surface at least, more hydrophilic, which could potentially enhance its reverse osmosis properties.

The fact that both flux and salt rejection appear to increase with time during reverse osmosis probably relates to the swelling of the initially dry membrane by water as it permeates and saturates the polymer film. As the polymer network expands, more water is also passed due to the increased concentration of water in the swollen membrane. Since water is passed at a much faster rate relative to the salt flux, the overall net effect is greater salt rejection. In the one case of organic rejection studied [5], the rejection of urea decreased with time; this observation is consistent with the formation of larger pores due to polymer network expansion. Thus the organic and ionic salt rejection mechanisms are very likely different processes.

The plasma-polymerization technique for the formation of hyperfiltration membranes is very promising. This flexible technique should allow the production of films virtually tailor-made to satisfy a given set of requirements. The fact that very thin films (on the order of hundreds of angstroms thick) can be formed by plasma polymerization leads to the possibility of high rejection *and* high solvent flux being achieved after deposition on a suitable substrate. The biomedical implications of plasma-polymerized membranes for separation processes are as yet unexplored and may represent a promising potential.

REFERENCES

[1] H. YASUDA, and C. E. LAMAZE, *J. Appl. Polym. Sci.*, *17*, 201 (1973).

[2] H. YASUDA, Conference on Polymeric Materials for Unusual Service Conditions, Ames Research Center, NASA, Moffett Field, Ca. Nov. 29–Dec. 1, 1972.

[3] C. J. WENDEL and M. H. WILEY, *J. Polymer Sci. A-1* **10**, 1069 (1972).

[4] K. R. BUCK and V. K. DAVAR, *Brit. Polymer J.* **2**, 238 (1970).

[5] J. R. HOLLAHAN and T. WYDEVEN, *Science* **179**, 500 (1973).

[6] A. F. STANCELL and A. T. SPENCER, *J. Appl. Polymer Sci.* **16**, 1505 (1972).

[7] J. O. HIRSCHFELDER, C. F. CURTISS, and R. B. BIRD, *Molecular Theory of Gases and Liquids*, Wiley, New York, 1964, pp. 1110, 1111.

[8] J. E. ANDERSON, S. J. HOFFMAN, and C. R. PETERS, *J. Phys. Chem.* **76**, 4006 (1972).

[9] J. AUSTIN and J. BLACK, *J. Am. Chem. Soc.* **52**, 4552 (1930).

[10] V. M. KOLOTYRKIN, A. B. GIL'MAN, and A. K. TSAPUK, *Russ. Chem. Revs.* **36**, 579 (1967).

[11] T. SEGUCHI and N. TAMURA, *J. Phys. Chem.* **77**, 40 (1973).

Chapter 7

Applications of Low-Temperature Plasmas to Chemical and Physical Analysis

John R. Hollahan

7.1. INTRODUCTION

General Operations with the Plasma Technique

The laboratory application of low-temperature oxidizing plasmas to sample preparation for chemical or physical analysis has proved highly successful and provides the analyst with unique capabilities. The use of a flowing oxygen plasma at about 1.0 torr which contains an abundance of atoms and excited states to degrade organic materials or modify substances in some manner prior to analysis has obvious advantages when compared to alternate techniques. "Wet-ashing" with oxidizing acids, nitric or perchloric, has been successfully employed to degrade substances. However, this procedure can alter or destroy the physical structure of the sample, such as physiological or crystalline phases. Further, in oxidizing acids there are many chemical species that can interfere or complicate trace analysis. The dry gas plasma ashing process keeps contamination effects to a minimum. The muffle or combustion furnace usually operates at 600–900°C, much too high to retain readily volatilized elements or easily decomposable phases. The low-temperature plasma-oxidation process typically prepares samples at room temperature to 250–300°C. The reaction of an oxygen plasma directly in a cold trap at −196°C has been reported. With an oxidizing low-temperature plasma, more commonly one established by electrodeless microwave or rf excitation, the analyst is able to conduct some basic preparation processes such as:

1. Low-temperature (25–300°C) oxidation or dry-ashing of organic, biological, or polymeric materials

2. Removal of organic matrices, leaving the inorganic residue or structure for subsequent chemical or physical analysis, by x-ray diffraction, neutron activation, emission or absorption spectroscopy, electron or optical microscopy, etc

3. Reaction of samples or substrate materials with the plasma to produce gaseous products that are subsequently trapped or analyzed by gas chromatographic or mass spectrometric methods

4. Preconcentration of organic materials (such as polymer foams, filter media, plant materials, etc.) to reduce their volume which may be high relative to the concentration of trace constituents before chemical or physical analysis

5. "Cleaning" of surfaces for regeneration of their optical or contaminant-free condition

A hydrogen plasma has been used successfully in certain cases to accomplish some of the processes above under nonoxidizing conditions.

These basic procedures are conducted in a variety of disciplines, ranging, for example, from coal science to cell biology. Some industries have virtually "standardized" the plasma sample preparation technique for certain analysis procedures.

Some of the initial analytical applications of electrodelessly excited gases were described by Gleit and Holland [1], and much of the early literature deals with the preparation of organic or biomedical samples for trace metal analyses of both stable and radioactive isotopes. Since the early 1960s, the extension of excited oxygen plasma to a variety of analytical problems has been achieved. It was the rather empirical observation in the late 1950s that when graphite or cellulosic filter media were placed in a 1.0 torr rf oxygen plasma sustained at around 100 W the materials degraded oxidatively with time at just a little above room temperature due to the action of oxygen atoms and other excited species. Further, the degradation proceeded in a very gentle fashion, i.e., there was no sputtering or mechanical disturbance of delicate inorganic structures even on a microscopic basis

These observations led to the development of commercial instruments, particularly in the late 1960s and early 1970s, for plasma-ashing or other analytical or research uses. In addition, these same instruments were applied to industrial processing on a smaller scale, such as in the polymer industry (Chapter 3) and the microelectronics industry (Chapter 9), where an oxidation process is important in the manufacture of small components usually of high intrinsic value.

General Conditions of the Plasma for Analytical Applications

Oxidizing Plasmas

The composition and mechanism of atom formation is an rf oxygen plasma has been studied recently be Bell and Kwong [2], and good discussions are given in works by McTaggart [3] and Blaustein [4]. The primary reactive species in an rf or microwave discharge have been determined to be $O_2 \Delta_g^1$ molecules and O ^{3}P oxygen atoms, their individual concentrations ranging from 10 to 20%, depending on the experimental conditions [5], and wall and other catalytic effects. Both the atoms and excited molecules react with organic substrates or gas phase species [6], and literally thousands of papers have been published on $O_2 \Delta_g^1$ and O ^{3}P atom reactions. There is also a distribution of ion concentrations in the glow of the plasma as well as free electrons. Ion and neutral species concentrations for an rf oxygen plasma have been discussed by Bell and Kwong.

Mechanisms of reactions of O atoms and excited electronic states with other gas phase species comprise a subject of exhaustive kinetic investigation. In these defined systems the mode of interactions, the products formed, and their energies and rates can be determined fairly accurately. However, in very heterogeneous systems, i.e., oxygen plasma interacting with coal or biological tissue, for example, one cannot expect to obtain the same degree of interpretable data to reveal the mechanism of oxidation or of the conversion process to inorganic residues for these complex samples.

In addition to oxygen oxidizing plasmas of air, carbon dioxide, and water have been investigated for their efficacy in degrading organic materials or a sample preparation for analysis. Air plasma has often been quite successful in certain applications; however, rates of oxidation can be significantly slower than with pure oxygen. Water and carbon dioxide plasmas have been investigated with no general advantage being apparent in the use of these forms of mildly oxidizing plasmas.

Nonoxidizing Plasmas

When carbonaceous, hydrocarbon, or, in some cases, protein samples are placed in a hydrogen plasma, the organic material usually degrades. The products are generally water, hydrocarbons (methane, ethane), nitrogen oxides, etc. [3]. The reactivity of hydrogen atoms with oxygen-, carbon-, or nitrogen-containing entities in the substrate is sufficiently high to permit degradation, although at generally lower rates than that exhibited by oxygen plasma. The principal advantage of this nonoxidative mode of low-temperature (ambient, 300°C) degradation is the obvious avoidance of oxidation of constituents in the sample to higher oxidation states or volatile products that might destroy the desired analytical component or structure. Few sample preparation techniques exist to degrade or remove organic materials involving a nonoxidizing approach, and the hydrogen plasma appears an inviting although limited and mostly unexplored medium.

7.2. EQUIPMENT AND TECHNIQUES

Samples and Sample Preparation

In a typical experiment, the sample to be treated with the gas plasma is placed on a microscope slide, glass plate, Petri dish, or a small glass boat and then inserted into the reactor chamber of the plasma system (Fig. 7.1). The vacuum is established at a rate that will not disturb the powdered or fragile samples. After the background pressure has been established with the mechanical forepump at 10^{-2}–10^{-3} torr (a diffusion pump is not necessary in most cases), the oxygen flow rate is regulated via the needle valve at the flow meter to give an operating pressure of 0.5–1.5 torr. For most laboratory system geometries, this usually corresponds to 50–300 cm^3/min (STP) of oxygen flow. The plasma is initiated by proper impedance matching with inductive and capacitive controls by increasing the power from the high-frequency rf or microwave generator to the desired level to provide the necessary sample surface or bulk temperature. After a period of time (which may range from several seconds for filter media to several hours for biological or polymer samples), the sample is removed for analysis or testing. Commercial plasma-ashing instruments combine most of the components in Figure 7.1 in one unit.

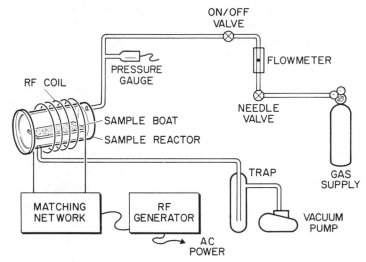

FIGURE 7.1. System components for plasma-ashing samples prior to analysis.

Sample types for plasma oxidation usually fall into one of the following categories: Powder, flat sheet or coupon, microtomed section, film, foam or porous material, microscopic specimens, and in some cases liquids of low vapor pressure. In most cases, powdered samples, say of mineral, plant, or biological origin, are usually prepared by drying and crushing, mechanical pulverizing, or ball milling. Powdered samples are often more desirable since the increased surface area usually promotes the oxidation or ashing rate.

Sample size before ashing may range from 0.5 to 10.0 g for a chamber 4–6 cm in diameter and a flat-bottomed boat container, 3 × 6 cm, where the powdered sample is evenly spread. If the powdered sample is too thick, the ash that develops in time may prevent further oxidation. Hence, it may be necessary to remove the sample boat and stir or redistribute the ash and sample mixture several times for complete ashing. Attempts to stir a sample under plasma conditions or to utilize a rotating chamber that tumbles the sample have been somewhat successful. Unfortunately, some types of sample materials tend to aggregate in the presence of the plasma and distribution of the sample becomes quite difficult.

Completeness of ashing may be determined in several ways. A constant final weight may be determined after repeated stirrings and ashing periods. Another test of ashing completeness that is often successful and quite convenient is the color of the gas under discharge conditions. When oxidizable material is present, CO and CO_2 are formed most commonly. The emitted radiation from the plasma is blue; pure oxygen is pink under discharge conditions. Thus, when the sample has been depleted of oxidizable material, the plasma will change from blue to pink. This observation, though not always foolproof, is often satisfactory.

Flat samples, may be derived from filter media, microtomed plant or animal sections, natural samples such as leaves from plants, films of polymers, contaminant films, etc. In these instances, the surface area is much smaller in comparison to powdered samples and thus ashing times may be quite different even for the same material. It is important, therefore, that the physical state of the sample be accurately described when plasma-ashing times or behavior are cited. Flat samples are often ashed in this form because the residual ash structure or inclusions are of interest, such as distribution of fine particles, catalysts, additives, etc., and these are to be studied or analyzed in situ. In biomedical sections of either tissue or bone, the organic material is removed, leaving undisturbed the physiological distribution of minerals, trace metals, or inorganic structure in the residual ash. High-temperature incineration often compromises the ash integrity through sintering or fusion processes. This often results in a loss of valuable structural information to the physiologist, pathologist, or microbiologist.

Liquid or semisolid samples, such as oils and greases, can be successfully ashed if these materials have a low vapor pressure relative to the usual 0.5–1.0 torr vacuum of the plasma. With some liquid organic samples, preevaporation of the light ends (which are of no analytical interest) renders the sample amenable to the vacuum of the plasma. For samples that may be predominantly aqueous, evaporation completely to dryness is usually required before ashing. With most pure liquids, no surface ash develops until near completion, and the ashing proceeds in an unimpeded manner because a fresh surface is constantly being regenerated. The ashing times for oils and greases at low temperatures in the plasma can be many hours compared to high-temperature combustion. One has to assess the tradeoff between allowable ashing times and the desired analytical results. Ashing of microscopic samples for ultrastructural analysis is described in Chapter 8. In these types of samples, the gentleness of the ashing process and the preservation of the ash structure are manifestly apparent.

Reactors and Sample Containment

Samples may be supported on or in Pyrex, silica, or ceramic dishes, boats, or flat plates. Figure 7.2 depicts some useful reactor and sample container configurations that have been used. In Figure 7.2a, the sample boat is placed in an L-shaped chamber. This reactor was among the earliest used in commercial versions of plasma-ashing equipment. The plasma is excited by an

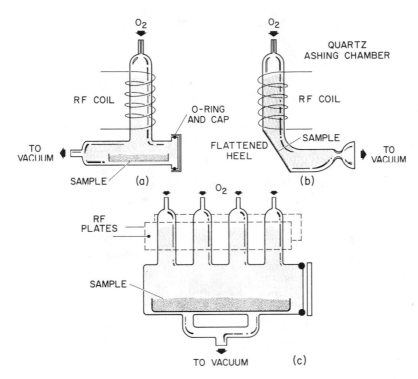

FIGURE 7.2. Examples of plasma reactors which have been used in oxidation processes.

inductive coil and streams over the sample. The plasma glow or the progress of sample ashing is observed through the glass window.

This reactor (Fig. 7.2b) was modified so that a trace analysis of samples was obtained from a liquid suspension and contamination was minimized in the handling procedure [7]. Samples were pipeted directly into the silica chamber (thus avoiding a boat) and allowed to dry on the flattened portion of the tube. When ashing was completed, the sample residue was dissolved in HCl and analyzed.

Figure 7.2c depicts a plasma reactor for large samples, 100–200 g [8]. The reactor is approximately 16 in. long, with a diameter of 5 in. A manifold of tubes entering the reactor is simultaneously excited by capacitive exciter plates. The plasma fills the reactor quite uniformly. In such a reactor, one large sample can be processed, such as polymer foam, large area filter media, etc. for preconcentration before analysis. Additionally, a quantity of samples individually contained in boats could be simultaneously processed after any

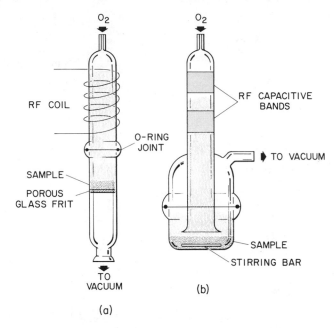

(a)   (b)

FIGURE 7.3. Other versions of sample containment: (a) Plasma is drawn through sample, (b) plasma drawn over sample contained in a cup.

cross-contamination effects have been verified to be inconsequential. This reactor has also been applied to surface treatment of polymers surfaces to improve adhesion [9] or biocompatibility [10].

In Figure 7.3a, a powder sample is supported on a glass frit. The apparent advantage is that the plasma penetrates further into the bulk of the sample. Figure 7.3b depicts a reactor in which the sample is contained in the bottom of a cup, rather than in a boat. A stirring bar is also provided.

A type of reactor reported [11] to give high uniformity of reaction with samples throughout its volume is shown in Figure 7.4. In this design, gas inlet and exit tubes extend internally the full length of the reactor. Gas enters the chamber through orifices and is pumped out an exhaust tube near the front bottom. The inner and outer chamber design is unique. The sample boat or material is placed inside the inner chamber (Fig. 7.4) then is inserted in the outer chamber. The flowing gas is pumped through slots located at the bottom of the inner chamber. In this manner, a type of laminar flow is established over the full length of the reactor, providing high uniformity of plasma reactivity throughout the chamber. The chamber is excited by an inductive rf coil and the plasma is generated directly where the samples are

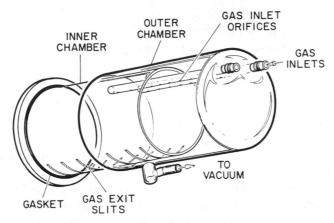

FIGURE 7.4. Plasma reactor with an inner and outer chamber useful for sample handling. Uniform ashing is achieved by the inlet tubes distributing oxygen the full length of the chamber. Inductive coil is not shown around the outer chamber (Tegal Corp.).

located rather than some distance away from the samples. At equivalent power levels and other constant plasma conditions, generally, the walls of the latter reactor design will be at a higher temperature than the walls of the chamber containing samples as in Figure 7.2c. This appears to have some effect on oxidation rate, in the case of graphite samples at least, in that oxidation rate decreases as the power is increased beyond an optimum value that heats the walls. Sample temperature attainment and oxidation rate are discussed further below.

Reactor Oxidation Performance

In assessing the oxidation capabilities of a particular reactor configuration for a given sample type, one approach is to take a model compound or material, for example, a high-purity graphite rod or powder to represent a carbonaceous material, and determine the oxidation rate (taken as the rate of weight loss per unit time) as a function of power, flow rate, pressure, gas type, etc. When the weight loss maximizes or levels off as these parameters are varied, this serves as an indicator for reactor performance. This, of course, is a simplification of how the reactor will operate in real situations with more heterogeneous samples such as, for example, those of biological origin. Since it may be assumed that the oxidation rate will be proportional to the atomic oxygen concentration, as well as sample temperature, a more accurate and physical determination of atomic oxygen concentration as a function of plasma parameter is by means of gas titration techniques. (For good discussions of plasma reactors for a variety of end applications, see the works of Blaustein [4] and McTaggart [3].)

Sample Temperature

The surface temperature of samples (which can be quite different from the bulk temperature) has been successfully measured during plasma oxidation by use of the noncontact ir radiation pyrometer. Several commercial versions of this instrument are available, all based on the same general operating principle, by which emitted ir radiation characteristic of the surface temperature is detected usually by highly sensitive thermistors. For application to plasma-ashing, the sample surface is usually sighted by the optics of the pyrometer through the reactor window onto the solid sample and the surface temperature is read directly and dynamically in °C on the pyrometer meter.

Other methods of measuring temperatures of solid samples have included the use of commercially available pigment dots attached to the sample boat or the sample itself, if convenient. These indicating dots change color at specific temperature levels. This temperature measurement technique has been extensively used in the semiconductor industry. The method suffers from the irreversible nature and once only use of the color-changing dot.

There have been some experiments in which thermocouples or thermometers have been imbedded in the bulk of the sample material. However, this requires accommodating the reactor geometry to this type of sample temperature measurement and ensuring proper shielding of the probes to prevent extraneous heating or pickup from the high-frequency field.

Solid sample materials generally increase in surface and bulk temperature for most conventional reactors as the power and flow rate increase, if steps are not taken to heat or cool samples by external means.

Gleit [12] found that the surface temperature of the walls of discharge flow tubes can affect the plasma-oxidation rate for carbonaceous compounds. As the wall temperature increases the oxidation rate decreases, as reflected in Figure 7.5 for graphite oxidation. These data were obtained in an rf discharge flow experiment consisting of a reaction tube 100 cm long, 3.5 cm id., borosilicate glass. In these experiments, the graphite sample temperature was held constant by external means while the wall temperature was varied. Hence, excessive rf power can act adversely in certain instances in terms of oxidation rate. In many types of plasma experiments, the rf power required to accomplish the chemical process is at a level just above that necessary to sustain the plasma.

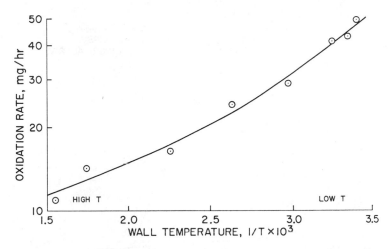

FIGURE 7.5. Effect of wall temperature on the oxidation rate of graphite.

Sample Surface Area and Plasma-Oxidation Rate

Figure 7.6 shows the oxidation rates for graphite rod, carbon felt, and graphite shavings from the rod [1]. The oxidation rate is shown to depend markedly on the physical state of a given type of material. Initially, the oxidation rate is highest for the sample having the highest surface area in contact with the plasma. At longer times mineral residue has accumulated, presenting a barrier to the plasma, and the oxidation rate falls dramatically for the highest surface area. This behavior was also found in biological tissue samples. Therefore, the completion of oxidation, or ashing, after a certain period of time depends on the physical state of the material for a given type of sample. In process or quality control operations where many samples may be plasma-oxidized before analysis, it is important to ascertain (by preliminary investigation) the oxidation behavior of the material so that the analyst is assured complete oxidation is achieved.

Plasma-Oxidation Rate of Organic Compounds

Hozumi and Matsumoto [13] have quantitatively studied plasma parameters governing the oxidation rate of organic substances in low-temperature oxygen and air plasmas operating at 14 MHz. The oxidation rate of several organic compounds was determined under plasma conditions given in Table 7.1. An effort was made to correlate the structure and oxidation rate. Most of the compounds were aliphatic and aromatic carboxylic acids. Acid anhydrides of low molecular weight exhibited extremely high rate of weight

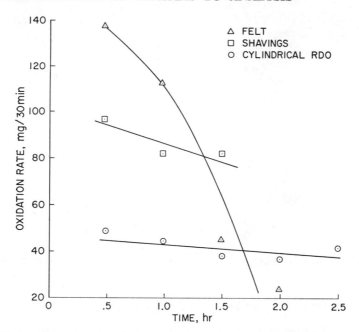

FIGURE 7.6.   Effect of surface area on oxidation rate, maximum temperature 100°C.

losses because their volatilities were higher. The weight loss observations appear to indicate that some differences in the oxidation rates due to the number of carboxylic groups and the lengths of the alkyl chains exist, and a higher oxidation rate of unsaturated compounds over that of saturated ones was found in general. The weight gains observed must be interpreted as oxide products being formed in the solid state which are stable entities toward further oxidation. A mass spectrometric study of the reactant gases found no fraction greater than $m/e = 50$, indicating practically complete combustion was taking place in the plasma/solid reaction.

The data in Table 7.1 are specific and characteristic of the design of the plasma experiment; however, the design is very basic and one that is commonly encountered: a discharge flow tube (35 mm OD) with an inductive coil exciter and the sample placed at a specified point downstream from the center of the coil region.

Figure 7.7 shows the essentially linear dependence of the oxidation rate of sucrose on power up to 30 W. The plasma-oxidation rate generally increases with rf power and then levels off. Undoubtedly, had oxidation rate data in Figure 7.7 been extended to higher powers, the curve would have leveled off. Further addition of power serves only to heat the sample additionally and,

TABLE 7.1. Oxidation Rate of Some Organic Compounds in Oxygen Plasma[a]

Compound	Loss of weight (mg/hr)
Sucrose	42
Oxalic acid	53
Oxalic acid, anhydrous	130
Calcium oxalate	< 1
Succinic acid	18
Succinic anhydride	76
Tartaric acid	35
Fumaric acid	49
Maleic acid	33
Maleic anhydride	400
Sorbic acid	142
Myristic acid	+8
Stearic acid	+21
Benzoic acid	30
Phthalic acid	43
Phthalic anhydride	102
Salicylic acid	41
Bakelite	33
Polyethylene	19
Polyvinylchloride	5
Teflon	1

[a] Sample weight: *ca* 150 mg; +: increase of weight; high-frequency power: 20 W; pressure: 1 mm Hg; flow rate: 15 ml/min; sample location: 10 cm from center of coil.

after a point, this leads to incipient thermal oxidation. A meaningful experiment is one in which the sample temperature is held constant by external means while the effect of increasing the power coupled into the gas load on oxidation rate under specified plasma conditions is determined.

Figure 7.8 shows a typical example of the dependence of sucrose-oxidation rate on the flow rate of oxygen entering the reaction tube. From a review of published data, it is clear that, in general, for a given reactor geometry, exciter configuration, and vacuum-pumping conditions, the oxidation rate increases with increased oxygen flow until more oxygen is present than

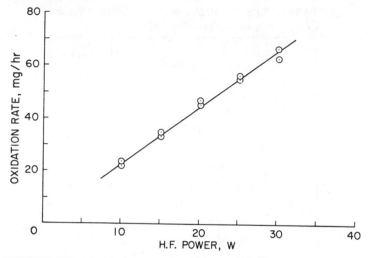

FIGURE 7.7. Oxidation rate of sucrose as a function of rf power; pressure, 1 torr; flow rate, 15 cm³/min; sample location, 10 cm downstream from end of inductive coil.

energy density (power) to cause excitation. At this point, a further increase in the oxygen flow rate does not significantly promote the oxidation rate. Most data presented in the literature are taken under conditions where the pressure is not controlled independently of the oxygen flow rate, i.e., throttling by a valve just before the pump. Hence, as the oxygen flow rate increases, the system pressure also rises.

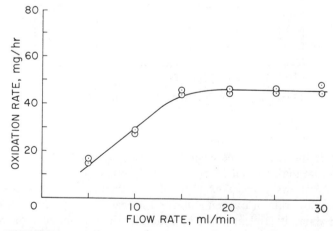

FIGURE 7.8. Oxidation rate of sucrose as a function of flow rate: pressure, 1 torr; rf power 20 W; sample location, 10 cm.

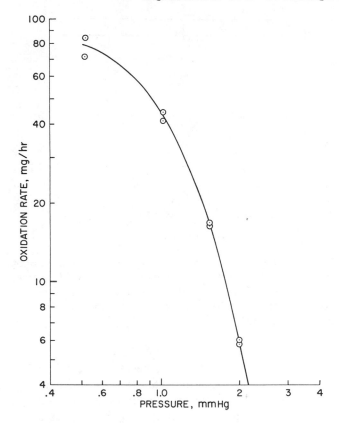

FIGURE 7.9. Oxidation rate of sucrose as a function of gas pressure: rf power, 20 W; flow rate, 15 cm³/min; sample location, 10 cm.

At higher discharge pressures, atomic oxygen recombination occurs to a greater extent, reducing the effective concentration for reaction. The oxidation rate falls dramatically (Fig. 7.9) when the pressure exceeds about 0.5 torr in the case of sucrose oxidation under the experimental conditions used. The point at which the oxidation rate will decrease, in general, with increasing pressure—all other plasma conditions held constant—will vary with the reactor geometry and efficiency since the latter will determine to some extent the production and utilization of atoms (see Bell and Kwong [2]).

Figure 7.10 shown the dependence of sucrose-oxidation rate on sample location relative to the discharge coil for two different power levels at a

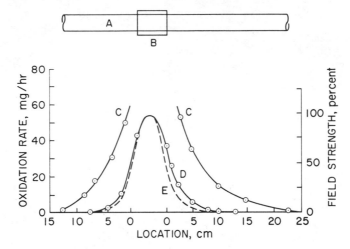

FIGURE 7.10. Distribution curves of oxidation rate of sucrose along plasma tube: pressure, 1.5 torr, flow rate, 15 cm³/min; (A) plasma tube; (B) rf coil; (C) rf power, 20 W; (D) rf power, 10 W; (E) percent distribution of electric field strength [13].

specified pressure and flow rate [18]. As expected in the coil region where the electric field is greatest, the oxidation rate is correspondingly high and falls quite rapidly as one moves a few centimeters away from the center of the coil. However, if the sample temperature is to be kept as low as possible while still maintaining a sensible oxidation rate, as in the case for element retention before analysis, then the sample may be placed well outside the exciter region, as was the case with some of the reactor geometries described previously (Figs. 7.2 and 7.3). The temperature variation with position of the sample (Fig. 7.10) is undoubtedly following curvex profiles similar to C and D.

It must be stressed that the foregoing discussion on oxidation behavior as a function of plasma conditions is necessarily qualitative since accurate statement of power, pressure, and flow-rate values is related to and characteristic of the specific design of the plasma system used in the experiment and, of course, the nature of the sample material used as a model to represent biological or carbonaceous systems. Qualitatively, however, the data should reflect plasma-oxidation behavior toward oxidizable substrates that are predominantly organic and generally encountered in analytical applications of plasmas. The reader is referred to the details in the literature cited for the specific plasma conditions used, although in the majority of cases these conditions are not quantitatively presented.

7.3. APPLICATIONS OF PLASMAS TO CHEMICAL ANALYSIS

With the necessity of environmental control and analytical monitoring of pollutant species, either solid or gaseous, the low-temperature plasma-oxidation technique has been more extensively applied in recent years to isolate trace constituents from a variety of background matrices. Such problems as analysis for metals, especially heavy metals, in foodstuffs, drugs, and plant materials, airborne pollutant particulates, human and animal tissues have all been reported in the literature. In the following sections, some typical examples of plasma analytical applications will be reviewed to represent what can be achieved with this useful approach.

Recovery of Trace Constituents from Largely Organic Matrices

In oxidation studies [1] on a wide variety of representative organic substances including muscle tissue, fat, ion exchange resins, solid wastes, cellulose, polyvinyl chloride, charcoal, and biological tissues, it was found that a number of trace elements that had been isotopically tagged could be retained in the residue after plasma oxidation at less than 100°C in contrast to losses by volatilization observed by muffle furnace combustion of the same materials at higher temperatures (Table 7.2). In every case, the plasma-ashed sample residue was completely soluble in mineral acids; the activity after ashing was counted in a well counter that could accommodate the entire sample boat. Note that some volatilization occurred with Ag^+, Au^{3+}, Hg^{2+}, and I^- isotopes.

Sanui [14] reports complete retention of cations of Na, K, Mg, and Ca after plasma oxidation of biological membranes. These metals or their cations can be difficult to analyze on a low-level submicromolar basis because possibilities always exist for external contamination from reagents, atmosphere conditions, containers, etc. The plasma-ashing approach minimized contamination when special sampling techniques and quartz reactor boats were used, and the only reagent—oxygen—could be easily obtained in a high state of purity. It was concluded from this study that, when used in conjunction with atomic absorption spectrophotometry for the analysis, plasma oxidation of biological materials provides a unique and superior method for measuring submicromolar amounts of biologically important cations with reasonable speed and accuracy. The same degree of success was reported for a variety of biological specimens for trace cation analysis including whole blood, cultured algae cells, muscle, chloroplasts, homogenates of developing bone, and liver microsomes.

TABLE 7.2.   Recovery of Radioactive Tracers after Plasma Oxidation

Nuclide	Sample	R.f. discharge, 1.5 hr		Muffle furnace	
		Boat	Trap and chambers	24 hr, 400° C	3 hr, 900° C
Sb^{124}	$SbCl_3$ + blood	99	0	67	9
As^{76}	$HAsO_2$ + blood	100	0	23	0
Cs^{137}	$CsCl$ + blood	100	0	. . .	. .
Co^{60}	$CoCl_2$ + blood	102	0	98	30
Cu^{67}	$CuCl_2$ + blood	101	0	100	58
Cr^{51}	$CrCl_3$ + blood	100	0	99	56
Au^{198}	$AuCl_3$ + blood	70	30	19	0
I^{131}	NaI + filter paper	31	69	. . .	. .
I^{131}	$NaIO_3$ + filter paper	100	0	. . .	. .
Fe^{59}	$FeCl_3$ + blood	101	0	86	27
Pb^{210}	$Pb(NO_3)_2$ + blood	100	0	103	13
Mn^{54}	$MnCl_2$ + blood	99	0	99	79
Hg^{208}	$Hg(NO_3)_2$ + blood	92	8	< 1	0
Mo^{92}	$(NH_4)_2MoO_4$ + blood	100	0	100	83
Sc^{75}	Alfalfa	99	0	. . .	. .
Ag^{110}	$AgCl$ + blood	72	28	65	21
Na^{12}	$NaCl$ + blood	100	0	. . .	. .
Zn^{65}	$ZnCl_2$ + blood	99	0	100	30

Other examples of elements successfully retained by low-temperature plasma oxidation are summarized in Table 7.3. It is evident that the background matrix can vary widely for given types of trace elements and that plasma oxidation as a preparative technique applies to many subsequent analytical operations.

Retention as a Function of Chemical State and Environment

The manner in which the trace element or chemical entity may be bound, either chemically or physically, such as in the form of a halide, oxide, hydride, chelate, etc., often critically dictates whether that constituent will be retained under the plasma-oxidative conditions, rf power level, temperature, etc. Thus certain elements which, in some cases, are difficult to retain in the residual ash even after plasma-ashing, such as I, Hg, and As, may be successfully retained in some instances, depending on the chemical environment in which the element is bound. For example, it was found that the presence of other metals greatly affected the retention of a given metal under plasma-ashing conditions [15]. Table 7.4 shows the case for selenium retention

TABLE 7.3. Examples of Trace Constituent Isolation and Analysis after Plasma Oxidation

Element or Phase	Matrix or Sample Material	Analysis	Temperature or Power Delivered to Reactor	Retention, %	Reference
Na, K, Mg, Ca	Liver (rat) micro-somes	Atomic absorption	100°–125° C	98–100	7
Ag, Cd, In, Ta, Rh, rare earths	Organic filter media, quartz	Neutron activation, radiochemistry	100 W	90–98	17
Si minerals	Plants and drugs	Microscopy	20 W	Complete mineral retention	18
Sb, As, Cs, Co, Cu, Cr, Au, I, Fe, Pb, Mn, Hg, Mo, Se, Ag, Na, Zn	Filter media, whole human blood	Radiochemical	100° C	99–100[a]	1
Cu	Glyceride oils	Colorimetric	80°–200° C	100	19
Pb, V, Cd, Cr, Cu, Mn, Ni, Zn, in pollutant particles	Glass filters	Atomic absorption	300 W	Complete	20
Pb, in airborne particulates	Glass filters	Atomic absorption	25°–200° C	Complete	21
Alkai, alkaline earth, transition metals, halides	Biological tis-sue (fish)	X-ray fluorescence	80° C	Assumed complete	22
Ra, Th, U	Marine algae	Neutron activation	NS[b]	Assumed complete	23
Ag, Al, Ba, Be, Bi, In, Mn, Mo, Ni, Pb, Co, Cr, Cu, Fe, Ga, Sn, Sr, Ti, V, Zn	Phytoplankton	Arc spectroscopy	NS[b]	Complete	24
Th, U, Sr^{90}, I	Urine, synthetic diets	Liquid chroma-tography	150	90% +	25
Cd, Pb	Whole blood	Atomic absorption	28 W	Complete	26
H^3	Biological tissues	Liquid scintillation	50 W	Complete	27

[a] Except for Au, Ag, Mg, and I where 30–70% retention was obtained.
[b] NS not specified.

TABLE 7.4. Effect of Other Metals on the Volatilization of Selenium

Elements Added to Matrix	Per Cent Recovery of Selenium
Se	43
Se + Cu	72
Se + Cu, As, Hg	100

after plasma-ashing as a function of the presence of other metals added to a matrix of pure cellulose.

This same study reported the retention of trace amounts of soluble compounds of Cu and As in a matrix of cellulose powder with subsequent analysis by atomic absorption. The level of rf power delivered to a reactor of the type in Figure 7.2a was examined as a function of recovery of the elements. Table 7.5 shows some selected results

TABLE 7.5. Retention of Copper and Arsenic as a Function of Radio-Frequency Power Input to Plasma-Oxidation Chamber

RF Power, W	Oxidation Time, hr.	Per Cent Recovery	
		As	Cu
7	23	104	100
14	16	79	98
28	8	55	100
55	5	12	101

Mercury soluble salts in the cellulose matrix were appreciably volatilized under all conditions for the particular sample composition used in this study [15]. In Table 7.5, the tradeoff between oxidation times and element retention is evident in the case of As.

Container Contamination of Samples

The type of container used for oxidation at high temperatures can be a source of contamination for samples to be analyzed for trace elements (Table 7.6), providing further justification for a low-temperature plasma-oxidation approach. Even wet ashing is seen to be a source of contamination in this particular study of the comparison of oxidation methods for biological samples [16].

Inorganic Materials Analytical Preparations

The fields of application of plasmas to inorganic materials analyses include geochemical, metallurgical, marine, and soil sciences [28]. With the exception of coal, most of these materials have a high ratio of inorganic/organic phases, and the analyst often desires to remove the low-level organics before analysis, to separate phases at low temperature without destroying their chemical or physical nature, and to minimize volatilization and sintering.

TABLE 7.6.   Trace Element Analysis of Bovine Tissue Ashed by Different Methods (ppm)

Tissue	Element	LTA[a] 150° C	Platinum Crucible 500° C (48 hr)	Porcelain Crucible 600° C	Wet ashing
Brain	Ca	< 2	< 2	5	10
	Cu	12	12	30	22
	Fe	42	37	57	94
	Mg	360	380	530	745
	Zn	59	60	60	73
Heart	Ca	< 2	< 2	14	96
	Cu	12	16	25	19
	Fe	76	88	95	155
	Mg	580	520	630	695
	Zn	99	101	100	120
Hemoglobin[b]	Ca	21	23	36	235
	Cu	7	2[b]	15	10
	Fe	3550	(38)[b]	2800	7300
	Mg	40	< 0.2[b]	620	88
	Zn	11	8	16	22
Kidney	Ca	< 2	< 2	4	36
	Cu	16	19	36	21
	Fe	80	88	81	179
	Mg	460	360	565	565
	Zn	109	105	116	61
Liver[b]	Ca	< 2	< 2	10	30
	Cu	165	160	235	175
	Fe	100	4[b]	27	180
	Mg	310	300	580	370
	Zn	110	...[b]	115	115

[a] Low-temperature ashing.
[b] Note incomplete ashing in muffle furnace.

The latter criterion is a key requirement when fine particles, their size and distribution, are subject to analysis.

Geochemical Applications

With geochemical specimens, for example, oil shale or marine sediment, it is often desired to determine the total organic-carbon–containing phases without altering the residual inorganic chemical phases. A plasma of oxygen can be used to oxidize the organic carbon with quantitative collection of the carbon dioxide on an adsorbate, such as ascarite. Materials containing carbonate phases incinerated at high temperatures decompose, producing CO_2 from the inorganic phases as well as CO_2 from the organic carbon. With the plasma-ashing technique, the interference as to the source of CO_2 is usually completely avoided. For high accuracy, it is necessary to ensure that the product gas is entirely CO_2 and not a CO/CO_2 mixture under these plasma-oxidizing conditions. If a mixture is produced, a secondary catalytic oxidation unit may be used to oxidize CO to CO_2.

That mineral phases may be retrieved chemically unaltered from a large background matrix has been amply demonstrated in the case of coal samples [29,30]. Quantitative analysis of five commonly occurring coal minerals—quartz, calcite, gypsum, pyrite, and kaolinite—was successfully developed using ir spectrometry down to 200 cm^{-1}. The samples had been plasma-ashed at 150°C, as opposed to the 700°C usually encountered in muffle furnaces. For analysis of multicomponent mixtures, it is essential that no interference between phases occurs, such as exchange reactions, oxidation–reduction, etc. In the above example, 1 g coal samples were completely depleted in organic matter, as evidenced by a disappearance of typical ir C–H absorption bands, in 70–90 hr; the only significant chemical alteration was the complete conversion of gypsum to the hemihydrate. The authors conclude that although the technique was developed for coal analysis, the combination of plasma oxidation and ir analysis should apply to a wide variety of mineralogical problems.

Gluskoter [31,32] has plasma-oxidized high volatile bituminous coal samples and found that the residues were composed primarily of calcite, pyrite, illite, chlorite, and kaolinite. The normal residue after high-temperature ashing in a muffle furnace is calcium oxide, ferric oxide, and dehydrated clay minerals. It was observed that practically all the mineral constituents of the coal remained unaltered after plasma-ashing at <190°C. Some minerals are dehydrated, but there were no irreversible changes in any of the clay minerals detected. Structural identification was followed by x-ray diffraction before and after ashing.

Marine sediments from ocean floors, estuaries, and fresh water sediments may be depleted of organic matter with an oxygen plasma to isolate nondestructively forms of marine shell life, diatoms, skeletal material, mineral aggregates, etc., from the organic phases [33]. Such analyses are revealing for geologic studies, possible valuable mineral indicators, or identification of sources of pollution, or the lack of it, from analysis of organic residues and the inorganic composition.

It has been proposed that an rf oxygen (or hydrogen) plasma reacting with various soils (perhaps extraterrestrial) could potentially offer product gas compositions indicative of either a positive identification of living microorganisms or the degree of complexity of living organisms. In fact, the product gas distribution and abundance would perhaps uniquely "fingerprint" an organism's presence or a precursor of living matter. Soils devoid of life or nonself-replicating organic matter would be expected to have very different product gas compositions after plasma reaction than soils containing forms of life.

In another study [34] on life detection methods in soils or meteorites, it was suggested that microorganisms, independent of their fossil remains, might leave a record of their metabolic activity by alteration of the rocks and minerals of their environment. Fungal microorganisms grown on basalt were found to solubilize significant quantities of the Si, Fe, Al, and Mg of several examples of rock by metabolic processes, as evidenced by chemical analysis and the ir absorption spectra of some of the residual rock and fungal material. Changes in the rock composition, which ranged, for example, from 0.3 to 31 % of Si solubilized, were determined by plasma oxidation for 18–24 hr of the organic matter, leaving the inorganic residue chemically unaltered for analysis.

Metallurgical Applications

Aside from microscopy (Chapter 8), one of the primary examples of the applications of plasmas to preparing metallic specimens for analysis is the removal of carbon from extracted oxide residues of steels. Recent metallurgical research on the influence of nonmetallic phases on the mechanical properties of steel has shown that the size of oxide inclusions can affect these properties more significantly than does the composition or total amount of the oxide in steel [35]. Therefore, it is critical that the inclusions be isolated from the extraction filter media using a technique by which (*a*) no particles are lost during the isolation, (*b*) no agglomeration of the particles occurs, (*c*) no chemical attack on the particles occurs, and (*d*) the filter medium used is quantitatively degraded in the oxygen plasma. The plasma dry-ashing approach successfully met these criteria and, in fact, for nearly all grades of plain carbon steels, the size distribution was determined more accurately than previously possible, either with chemical extraction, muffle furnace ignition, or with quantitative metallographic methods [36]. Ashing times as low as 1 hr were sufficient to isolate the small-diameter (about 2–50 μ) particle inclusions.

Early work in isolation of inclusions from steel for determination of their metal carbide and nitride content was complicated by the coisolation of amorphous and graphitic carbon, the former resulting the chemical decomposition of Fe_3C (cementite) in the steels. The oxidation of amorphous and graphitic carbon caused thermal response during differential thermal analysis that interferred with those due to other constituents. It was found [37] that a low-temperature oxygen plasma efficiently removed the coextracted carbon with little, if any, detriment to the inclusions of interest. Extracted MnS and FeS showed some reaction with the atomic oxygen, but are over 90% recovered following plasma oxidation.

Biomedical Applications

Tooth and bone biological materials are largely inorganic with some amount of tissue in their structure. Microtoming samples of bone, possibly after biopsy, produces thin sections that can be treated with an oxygen plasma to remove the organic matter before biomedical or biological examination. Again, the fine structural detail of the bone material (spicules, cavities, etc.) is generally preserved, even on a microscopic level. With this type of sample preparation, diseases of the bone or teeth can potentially be followed and related to the physiological structure after plasma-ashing. As with other inorganic systems, high-temperature incineration and chemical treatment can compromise a structural detail that may be diagnostic of a bond disorder. The distribution of elements or mineral phases in bone sections can also be followed with an electron microprobe (Chapter 8) and related to malfunctions or diet deficiencies, particularly in children.

Powdered bone material has been effectively plasma-ashed at $-75°C$, in a cold trap reactor, to $140°C$; the results of x-ray diffraction analyses of bone and noncrystalline calcium phosphate showed no detectable phase change in crystallinity [38]. Ashing for 24 hr appeared to completely remove the organic matter before analysis that included x-ray diffraction and microscopy. The utility of the approach for bone, enamel, and dentin was evident.

REFERENCES

[1] C. E. Gleit and W. D. Holland, *Anal. Chem.* **34,** 1454 (1962).

[2] A. T. Bell and K. Kwong, *AIChE. J.*, **18,** 990 (1972).

[3] F. K. McTaggart, *Plasma Chemistry in Electrical Discharges*, Elsevier, Amsterdam, 1967.

[4] B. D. Blaustein (ed.), "Chemical Reactions in Electrical Discharges," *Adv. Chem. Series* (80), Am. Chem. Soc., Washington, D.C., 1969.

[5] F. Kaufman and J. R. Kelso, *J. Chem. Phys.* **32,** 301 (1960).

[6] M. L. Kaplan, *Chem. Tech.* 621 (Oct. 1971).

[7] H. Sanui, *Anal. Biochem.* **42,** 21 (1971).

[8] J. R. Hollahan, U.S. Patent 3,428,548 (Feb. 14, 1969).

[9] J. R. Hollahan, *J. Sci. Instr.* (*J. Phys. E.*) **2,** 203 (1969).

[10] J. R. HOLLAHAN, B. B. STAFFORD, R. D. FALB, and S. T. PAYNE, *J. Appl. Polymer Sci.* **13,** 807 (1966).

[11] TEGAL CORPORATION, Technical Literature, Richmond, Ca.

[12] C. E. GLEIT, "Chemical Reactions in Electrical Discharges," *Adv. Chem. Series* (80), Am. Chem. Soc., Washington, D.C., 1969, p. 232.

[13] K. HOZUMI and M. MATSUMOTO, *Japan Analyst* **21,** 206 (1972).

[14] H. SANUI, *Appl. Spectros.* **20,** 135 (1966).

[15] C. E. MULFORD, *Atomic Absorption Newsletter* **5,** 135 (1966).

[16] J. R. HOLLAHAN, *J. Chem. Ed.* **43,** A401, A497 (1966).

[17] C. E. GLEIT, P. A. BENSON, W. D. HOLLAND, and I. J. RUSSELL, *Anal. Chem.* **36,** 2067 (1964).

[18] K. UMEMOTO, and K. HOZUMI, *Chem. Pharm. Bull.* **19,** 217 (1971).

[19] G. R. LIST, R. L. HOFFMANN, W. F. KWOLEK, and C. D. EVANS, *J. Am. Oil Chem. Soc.* **45,** 872 (1968).

[20] T. KNEIP, M. EISENBUD, C. STREHLOW, and P. C. FREUDENTHAL, *J. Air Pollut. Cont. Assoc.* **20,** 144 (1970).

[21] C. D. BURNHAM, C. E. MOOR, T. KOWALSKI, and J. KRANIEWSKI, *Appl. Spec.* **24,** 411 (1970).

[22] U. COWGILL, G. E. HUTCHINSON, and H. C. W. SKINNER, *Proc. Nat. Acad. Sci.* **60,** 456 (1968).

[23] D. N. EDGINGTON, S. A. GORDON, M. M. THOMMES, and L. R. ALMODOVAR, *Limn Ocean* **15,** 945 (1970).

[24] J. P. RILEY and I. ROTH, *J. Marine Biol. Ass. U.K.* **51,** 63 (1971).

[25] C. TESTA, *Anal. Chim. Acta.* **50,** 447 (1970).

[26] T. R. HAUSER, T. A. HINNERS, and J. L. KENT, *Anal. Chem.* **44,** 1819 (1972).

[27] H. E. DOBBS and G. M. LAND, International Conference on the Use of Radioisotopes in Pharmacology, Geneva, Sept. 1967.

[28] I. E. DEN BESTEN and J. J. MANCUSO, *Chem. Geol.* **6,** 245 (1970).

[29] P. A. ESTEP, J. J. KOVACH, and C. KARR, *Anal. Chem.* **40,** 358 (1968).

[30] R. A. FRIEDEL (ed.), *Spectrometry of Fuels*, Plenum Press, New York, 1960, Chap. 18.

[31] H. J. GLUSKOTER, *Fuel* **44,** 285 (1965).

[32] H. J. GLUSKOTER, *J. Sedimentary Petrol.* **37,** 205 (1967).

[33] G. VILKS, Bedford Institute Contribution No. 30, Atlantic Oceanographic Laboratory, Bedford Institute, Dartmouth, Nova Scotia.

[34] M. P. SILVERMAN and E. MUNOZ, *Science* **169,** 985 (1970).

[35] R. G. SMERKO and D. A. FLINCHBAUGH, *J. Metals* **20,** 43 (1968).

[36] D. A. FLINCHBAUGH, *Anal. Chem.* **41,** 2017 (1969).

[37] W. A. STRAUB and R. G. THEYS, United States Steel Corp., personal communication, 1972.

[38] P. D. FRAZIER, F. J. BROWN, L. S. ROSE, and B. O. FOWLER, *J. Dental Res.* **46,** 1098 (1967).

Chapter 8

Use of Chemically Reactive Gaseous Plasmas in Preparation of Specimens for Microscopy

Richard S. Thomas

8.1. INTRODUCTION

Low-temperature plasmas of electrically excited oxygen, and occasionally of other gases, have increasingly been used in recent years to ash bulk samples of mineral-containing organic material for physical and chemical analyses of their mineral content. In replacing high-temperature furnace ashing or wet-chemical digestion methods, the plasma treatment has proven valuable because of its demonstrated gentleness and cleanliness. It has been able to avoid volatile mineral losses, phase changes, fusion and other, undesirable, high-temperature dependent alterations which can occur in the furnace, and it also avoids trace contamination of the samples which may result from introduction of aqueous reagents. These advantages have been discussed and documented in Chapter 7.

Substitution of low-temperature plasma procedures have conferred similar advantages in ashing- and digestion-type methods of preparing minute, mineral-containing specimens for microscopic analyses of various sorts. The virtual absence of volatilization, fusion, etc., during ashing has frequently allowed preservation of not only the original chemical composition of the mineral material, but also its microscopic morphology and fine structure. Because the plasma is a gaseous rather than a liquid reaction medium, the plasma treatment has also avoided the structure-disrupting, nonspecific extractions, diffusion, viscous shear, and surface tension effects which can occur during wet digestions. These can be particularly devastating to microscopic specimens because of the small size of the structures involved.

A frequent disadvantage of plasma-ashing in macroscopic applications has been the slowness of specimen decomposition by comparison to previous methods. Specimens have often required many hours to several days for complete ashing. This difficulty stems from the characteristic of plasma treatment that it is surface-limited (see Chapter 7). In most microscopic applications there has typically been no such problem because of the huge surface area of microscopic specimens relative to their mass and volume, and ashing for microscopy has usually required only a few minutes to 2 or 3 hr at most.

The surface-limited nature of the plasma-solid reaction has actually been the major advantage in another class of microscopic applications of plasma—namely, surface-etching. Common problems in studying the microscopic structure of many bulk materials have been the difficulties of obtaining thin preparations required by the microscopy technique without altering the native structure of the specimen, and also obtaining structure-related image contrast. Etching has not infrequently been used in place of mechanical sectioning to thin specimens for microscopy and has been commonly used to differentially erode surfaces and develop structure-related surface relief to provide the needed image contrast. In many cases, chemically reactive plasma treatment has provided a gentler, cleaner, more reliable alternative to earlier methods which could alter the subsurface bulk properties of the material or otherwise produce artifacts.

The surface-limited nature of the plasma treatment has also led to numerous macroscopic practical applications in surface modification of organic and inorganic materials (see Chapter 3). Surfaces have been made harder, stronger, cleaner, more adhesive, more wettable, etc., by plasma treatment. Many of these surface modifications have their parallel in special preparatory techniques for microscopy.

Yet another practical, macroscopic application of chemically reactive plasmas has been the synthesis and surface deposition of various sorts of thin films produced in the plasma by activated-gas reactions (see Chapters 5 and 9). This application also has found special use in microscopy.

The present review attempts to survey all of this wide range of microscopic preparatory applications of chemically reactive plasmas, and the presentation has been conveniently divided into sections dealing with ashing, etching, and surface modifications and thin films. Inevitably, and by choice, the coverage has been uneven, however, because of the author's lack of first-hand acquaintance with many of the types of application. Greatest emphasis is given to ashing, and particularly to microincineration, an area in which the author has worked.

Microscopy in the present review includes a range of techniques: Various sorts of optical microscopy (OM) such as phase contrast, polarization optical, reflection, etc.; scanning electron microscopy (SEM); transmission electron microscopy (TEM); selected area electron diffraction (SAED); electron probe x-ray microanalysis (EXM). These should be more or less familiar to the reader. Occasionally in some of the metallurgical work reviewed, the less well-known technique of electron emission microscopy (EEM) has been used. In the EEM instrument, the specimen is heated or exposed to ionic bombardment or other radiation to cause the emission of electrons, which are focused to an image in a manner similar to that of TEM. Contrast in the fairly low-resolution images results from the varying emission of electrons from different regions of the specimen caused by differing physical or chemical states of the surface. The abbreviations, OM, SEM, TEM, SAED, EXMs and EEM, will be used throughout this chapter to refer to the various type, of instrumental techniques.

8.2. APPLICATIONS TO ASHING

Low-temperature plasma-ashing of specimens for subsequent microscopic examination of a mineral residue has generally been used in one or the other of two ways. In the one, the object has been to purify and often concentrate particulate mineral material dispersed in a relatively large volume of organic or carbonaceous matrix, e.g., filter material, so as to get the "needles in the haystack" all together, convenient for microscopic examination and suitable for other types of analysis which may demand relatively pure and/or concentrated samples. Alternatively, the ashing may have aided in dispersing the mineral particles to improve the suitability of the specimen for microscopic analysis. In this first type of application, preservation of particle morphology and original physical and chemical composition has been demanded but the original location of individual particles or ash structures in the initial organic matrix was of no consequence. For want of a better name, we will call this type simply *particle analysis*.

In the second type of application, retention and detection of the mineral at its original site in the organic specimen has been the most important objective, although often it has also been important to preserve the original composition of the mineral for analysis as well. This latter type, which most commonly has dealt with biological specimens, has been called *microincineration*. In the examples of plasma-ashing gathered here, there have been a few overlaps of the two types, but largely they have been distinct and are conveniently considered separately.

Nearly all of the ashing applications of plasma have been done in commercially available, flowing-gas, rf electrodeless discharge devices, or precommercial prototypes of the devices, and the activated gas has almost invariably been oxygen. Unless otherwise indicated, this may be assumed for the examples described below.

Ashing for Particle Analysis

The examples in this section may constitute only a small fraction of what truly exists in the literature, for examples have been difficult to recognize. Microscopic analysis of particles—identification and classification on the basis of size, shape, and other physical and chemical properties of individual microscopic particles—has frequently been only an unheralded part of larger analytical endeavors using bulk specimens and macroscopic techniques. Similarly, use of plasma-ashing may have been considered only a refinement of technique, not necessary to mention in title or abstract. It may be no happenstance, however, that a major fraction of the recognized examples has been in the field of airborne particulate sampling. The heterogeneity of the particulates has forced the use of microscopy for meaningful analyses. Their low incidence in recovery filters has necessitated some sort of ashing for their concentration, and the necessity of unaltered morphology and composition for their identification has strongly recommended plasma-ashing.

Recovery of Airborne Particulates

Turkevich (1959; see also Streznewski and Turkevich, 1959; Turkevich and Streznewski, 1958) was seemingly the first to report use of oxygen plasma-ashing for recovery and microscopic analysis of airborne particulates, and this may also have been the first report of low-temperature oxygen plasma-ashing for any purpose. The plasma consisted of atomic oxygen produced in a high-voltage discharge, as described by Harteck and Kopsch (1931). Fragments of cellulose or synthetic fiber filters containing inorganic aerosol submicron particles and also carbon soot particles were mounted on silicon-monoxide–filmed TEM grids and ashed in place, depositing the inorganic particles on the film. A particular advantage of the plasma treatment was that it easily burned away the confusing carbon particles together with the filter. In connection with this work, Turkevich and Streznewski conducted a pioneer TEM study of the attack of atomic oxygen on carbon (see sections on Microincineration, and Etching of Carbon). Turkevich (1959) in his use of plasma-impervious silicon monoxide films for TEM specimen supports, anticipated subsequent developments in adapting the microincineration technique to TEM uses, and he in fact suggested (but did not explore) the possibility of using atomic oxygen for microincineration of biological tissue sections.

The Tracerlab Company† pioneered the commercial development of low-temperature plasma-ashing devices (Gleit 1963; Gleit and Holland 1962), and they have used their equipment for in-house contract research programs involving microscopic characterization of atmospheric particles, especially radioactive, nuclear weapons debris. Much of this work has been described only in contract progress reports and in an unpublished review (Holland 1970), but the latter may be obtained by writing the company.‡ Brief accounts have been given by Bersin, Hollahan, and Holland (1966); Gleit (1964, 1966); Hollahan (1966); Nathans, Thews, and Russell (1970) and Stafford (1969). They have plasma-ashed a wide variety of cellulose and synthetic polymer filter materials with no difficulty, and also collodion- or oil-coated impactors and graphite cartridges and even whole alfalfa leaves to recover the collected particles. On occasion, hydrogen or carbon dioxide plasmas have been used successfully in place of oxygen, but combustion rates have generally been lower. Typically, the filters have been ashed onto glass slides for OM examination of the residue, which could then be incorporated into a collodion membrane cast onto the surface and transferred to grids for TEM examination, and also for combined SEM and EXM analysis (Holland and Abelmann, 1967). Figure 8.1 shows an ashed IPC (cellulose coated with an organic adhesive) filter containing nuclear fall-out particles. In the absence of fusion and

† Reference to a company or product name does not imply approval or recommendation of the product by the U.S. Department of Agriculture to the exclusion of others that may be suitable.
‡ LFE Environmental, Analysis Laboratories Division, 2030 Wright Ave., Richmond, Ca. 94804.

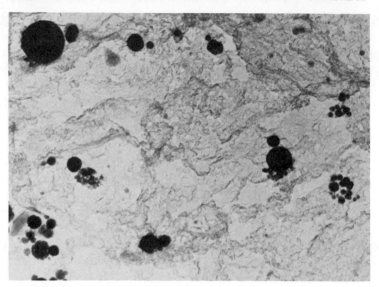

FIGURE 8.1.   TEM micrograph of collodion membrane with incorporated ash of IPC filter containing nuclear weapon debris particles (× 15,000) (print kindly supplied by W. D. Holland).

coalescence which can occur at high temperatures, the residual delicate filamentous ash of the filter material is easily distinguished from the spherical fall-out particles. Identification of radioactive particles has been aided by several techniques of microautoradiography applied to the ashed residues, and particle size classification by sedimentation in viscous liquids has also been useful. Careful statistical comparisons of particles with or without exposure to the plasma have shown that the ashing treatment has no effect on the particle-size distributions. Besides fall-out particles, the Tracerlab group has also studied other health-hazardous airborne particulates such as quartz and asbestos, and submicron aluminum oxide particles from rocket exhaust.

Quite recently, Mueller, et al. (1972) have also reported the beneficial use of oxygen plasma in ashing air filters for microscopy. They evaluated several methods for recovering airborne asbestos, and found that a procedure using plasma-ashing gave the most faithful, quantitative representation of the populations of particle sizes and morphologies. After ashing the filter (cellulose ester material), the particles are sonically dispersed in a collodion solution which is cast into a membrane of known surface area. Samples of this membrane with the included particles are then mounted appropriately for OM and TEM examination at magnifications ranging from 430 to 20,000×.

Selikoff and co-workers (see citations below) have for some years been concerned with the recovery and microanalysis of airborne particulates, especially asbestos, from lung, and they were among the earliest to use the low-temperature oxygen plasma method for tissue-ashing. Their first use of the method was to improve the quality of ash patterns from sectioned lung tissue so that inspired microscopic mineral particles could be visualized in situ (Berkley, et al. 1965; see Microincineration). The gentler ashing procedure proved beneficial not only to the tissue structure but also to the particles themselves; so it was later adopted routinely for the recovery of asbestos from bulk specimens. Berkley, et al. (1965) reviewed the inadequacy of previous methods. As discussed by Berkley, Langer, and Baden (1967); Langer, Rubin, and Selikoff (1970); and Langer, Selikoff, and Sastre (1971), some of the various species of asbestos particles, particularly chrysotile, are chemically altered by acids, alkalies, and heat and also undergo fragmentation during conventional procedures of wet-chemical digestion or high-temperature ashing. The same is also true of the so-called asbestos bodies which are formed in lung tissue by a biological encapsulation of the original particles. These alterations are largely eliminated by the plasma-ashing procedure. An indication of the improvement is seen in results reported by Langer, et al. (1971), comparing incidence of asbestos bodies in autopsied "normal" lungs which were either plasma-ashed or subjected to an alkaline wet-chemical digestion. On the basis of equivalent sample sizes, the average incidence of recognizable asbestos bodies was almost thirtyfold greater in the low-temperature ashed samples (author's calculation from Langer's data). Besides simple morphological recognition and enumeration of the asbestos bodies and fibers by OM, more recent work (Berkley, et al., 1967; Langer, 1970; Langer, et al., 1970, 1971, 1972) has also employed TEM, SAED, and/or EXM. The latter examinations have in some cases been performed on particles within the entire ash or in dried aqueous concentrates of the particles, but more commonly, the very low concentration of the particles has required individual selection of a particle by OM and micromanipulator transfer and mounting for subsequent analysis.

Langer, et al. (1971*a*) have also used low-temperature plasma-ashing to recover incipient airborne particulates—microscopic diatom fragments, aluminum silicate fibers, and other inorganic particles—from cigars. Of interest were the potentially health-hazardous changes in these mineral particles which could occur when they were subjected to high temperature as the cigars were smoked. The particles recovered by low-temperature ashing were unaltered, as verified by OM, TEM, SAED, and EXM, and provided the necessary control specimens. In another nontissue microscopic application, Berkley, et al. (1965) briefly described oxygen plasma-ashing of Millipore air-sampling filters.

Keenan and co-workers (Keenan, 1968; Keenan and Kupel, 1968; Keenan and Lynch, 1970) have reviewed techniques and instrumentation for recovery and analysis of airborne mineral fibers and particles and briefly discussed oxygen plasma-ashing and also OM, TEM, and EXM. They have been using the ashing technique on an experimental basis to decompose lung and other biological tissues "without changing the crystal lattice" of incorporated mineral particulates, and they were presumably applying the OM, TEM, and EXM capabilities in their laboratory to these. No specific details were given, however. In the same laboratory, Nenadic and Crable (1970) have used oxygen plasma-ashing of the filtrate from enzymatically digested lung tissue to differentiate between coal dust and inorganic particles before gravimetric analyses. Again, the report does not mention specifically that microscopy was applied to the residues, but it does seem likely.

Recovery or Dispersion of Various Other Particulates

Flinchbaugh (1969, 1971; Smerko and Flinchbaugh, 1968) has used oxygen plasma-ashing on particulate residues chemically extracted from steel for a purpose similar to that of Nenadic and Crable (1970)—to eliminate carbon particles from a population of inorganic inclusions without altering the morphology and composition of the latter. The mineral particles in this case were iron and manganese oxides which would have been further oxidized if high temperature was used to burn out the carbon. The purified particle suspensions liberated from plasma-ashed filters were characterized not only by OM (size, shape, color, birefringence, etc.) but also in a Coulter counter and by other nonmicroscopic methods which would have been confused by the presence of carbon particles. The overall procedure for Coulter counter determination of extracted oxide-particle size distributions was tested with model systems of oxide particles, subjected or not to the plasma treatment. Both the Coulter counter data and direct microscopic examinations verified that plasma-ashing was innocuous and caused no losses, sintering, or other alterations in the particles.

The ability of low-temperature plasma-ashing to easily remove carbon without altering mineral constituents has been demonstrated and applied in a number of chemical or crystallographic studies on inorganic components of coal (Estep, et al., 1968, 1969; Gluskoter, 1965, 1967; Karr, Estep, and Kovach, 1968; see also Chapter 7). This earlier work did not apparently involve microscopic methods even though the mineral materials in coal occur mostly as microscopic inclusions (e.g., see Kemezys and Taylor, 1964). Quite recently, however, Gluskoter and Lindahl (1973) have used SEM combined with EXM to study individual mineral particles liberated by low-temperature ashing and sorted by density gradient and by micromanipulation under OM. The particles, of the mineral sphalerite, ZnS, were shown qualitatively by the SEM-EXM to contain small amounts of cadmium. Quantitative microchemical analysis (atomic absorption after dissolution) on groups of sorted particles showed that the Cd/Zn ratios in the particles were quite similar to those in the entire coal samples from which the particles were isolated. Evidently, the presence of sphalerite accounted entirely for the occurrence of this potentially volatile, health-hazardous element, cadmium, in the coals.

Vilks (1970) has found plasma-ashing very helpful in separating microscopic calcareous or siliceous skeletal material from marine plankton which are largely organic in nature. Morphological identification and enumeration of these minute sea shells is required for various investigations in marine ecology. Previous methods entailed tedious hand-sorting under a microscope, flotation fractionation in a dense liquid, or high-temperature ashing. In all cases the recovery was biased against small, easily overlooked or entrapped species, or species with fragile, easily broken shells. This difficulty was largely overcome by the gentle and complete low-temperature ashing. Typical examples of the fragile, isolated shells are seen in Figure 8.2.

Grasenick and co-workers (see sections on Etching, and Surface Modification and Thin Films) have developed special equipment and made numerous microscopic applications of low-temperature plasma treatments. In an application appropriate for inclusion here, Ziegelbecker (1965) has used oxygen plasma incineration to remove viscous oil used as a mulling medium to disperse inorganic particles in preparation for identification, counting,

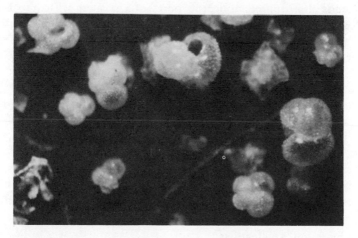

FIGURE 8.2. Typical OM field of sieved ($60\,\mu$ pores), water-insoluble ash from marine plankton sample ($\times$ 100) (reproduced with permission from Vilks, 1970).

and sizing þy TEM. More commonly, the mulling medium has been washed away in a suitable, volatile solvent (Bradley, 1961; Crable, et al., 1967), but this entails the possibility of also washing away some of the particles, thus distorting the particle size distribution. The particles dispersed in the oil were spread on a glass surface, and after removing the oil, a replica or pseudoreplica was made for transfer to TEM grids and viewing. The replication step was necessary because some of the particles were too large for direct TEM viewing. We (Thomas and Greenawalt, 1968) have used a simpler version of the technique to disperse submicron mineral particles for SAED. The particles were mulled in a viscous collodion solution and cast in a film (see Bradley, 1961, and similar technique of Mueller, et al., 1970). After mounting the particle-containing film on TEM grids filmed with silicon monoxide membrane, the collodion was burned away in oxygen plasma.

In a recent, general reference work on handling and identification of microscopic particles, McCrone and Delly (1972) also mention the utility of low-temperature ashing in removing unwanted, nonvolatile mounting or dispersing media (e.g., Aroclor) from OM preparations of mineral particles. The technique has also been useful in eliminating unwanted, biological portions of a sample, e.g., pollens, spores, mold, fungus, hair, dandruff, paper, seed hairs, etc., and other organic materials such as synthetic fibers. In particle-recovery applications, membrane filters are ashed very quickly by the plasma treatment to deposit contained mineral particles on a clean slide for OM examination, but the plasma gasification of the filter is often so rapid that particles are blown away. McCrone (private communication, 1973) recommends first impregnating the pores of the filter with mineral oil to slow down the ashing rate. The authors caution against indiscriminate use of the plasma-ashing technique on unknown mineral particle samples, however, since some metals and lower oxides—e.g., Mg, Zn, Fe_3O_4, Cu_2O— may be oxidized, and other minerals such as gypsum and borax may be dehydrated.

In a hard-to-classify application, which strictly speaking is neither recovery nor dispersion of particles per se, Ong, Lund, and Conrad (1972) have used oxygen plasma-ashing in a technique to *create* particles for microscopic analysis. A small (1 ml) aliquot of blood is low-temperature ashed to minimize elemental losses, the ash is then dissolved, and trace metals such as nickel, copper, and zinc are selectively electroplated onto a small electrode tip to form a microscopic deposit which is subsequently analyzed by EXM.

Microincineration Applications

Careful, high-temperature burning of biological specimens to produce ash patterns demonstrating microscopic, in situ mineral localization was first described by Raspail in 1833, and the technique has been much used in biological investigations since that time. Hintzsche (1956) wrote a monograph on the subject with almost 400 references, and later Kruszynski (1963, 1966) published a comprehensive bibliography with even more references and also an extended review article. The use of plasma for low-temperature micro-incineration has been a fairly recent development, however. Turkevich (1959) apparently was the first to suggest use of oxygen plasma (actually atomic oxygen) for biological microincineration, as mentioned in the section on Particle Analysis, and this author (Thomas, 1962, 1964) may have been the first to actually try it. At the same time, Moscou (1962), following earlier work of Spit (1960) and Jakopic (1960) on oxygen plasma-etching of organic materials, applied the method to a technological object, rubber. Moscou regarded his procedure as an etching method, but in terms of technique it was identical to biological microincineration.

The earlier work reviewed by Hintzsche and nearly all of that by Kruszynski was confined to OM studies. This author (Thomas, 1969) has reviewed the more recent development of the technique for TEM, and also some applica-tions including both high-temperature and oxygen plasma incineration. Additional reports of high-temperature microincineration for TEM (e.g., see Martin and Matthews, 1970; Matthews, et al., 1970; Neff, 1972; Sampson, et al., 1970; Shen, 1972) and also low-temperature plasma micro-incineration have appeared since then, but of these only the latter are reviewed here.

Successful microincineration using high temperatures in a furnace has almost invariably necessitated use of relatively thin specimens which have produced two-dimensional ash patterns (sometimes called spodograms). Owing to shrinkage, melting, and coalescence, and other mechanisms causing ash migration, fineness of detail in the pattern has generally been not greater than the thickness of the specimen. This restriction to thinness has been somewhat relaxed in the case of plasma microincineration. With favorable specimens having sufficient mineral substance, and so distributed that a coherent ash skeleton either preexisted or could form during ashing, it has been possible by use of the plasma technique to obtain detailed three-dimensional ash structures.

Three-Dimensional Ash Structures

Reviews and advertisements from Tracerlab describing the capabilities of low-temperature plasma-ashing provided an early published example of three-dimensional ash structure preservation on a macroscale (see Bersin, et al., 1966; Hollahan, 1966; Tracerlab advertisement, 1966). Figure 8.3 shows a sample of polyurethane foam before and after plasma-ashing. The

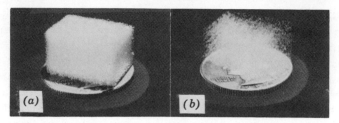

FIGURE 8.3. Sample of polyurethane foam before (*a*) and after (*b*) oxygen plasma ashing (photograph reproduced with permission from LFE Environmental, Analysis Laboratories Division (formerly Tracerlab), Richmond, Ca.).

microcellular membranes of the foam were consumed by the plasma but the slight amount of mineral matter they contained formed into connecting microfilaments and strands, as could be shown by OM, to maintain the gross form of the original specimen.

We have obtained good preservation of ash structure in three dimensions on both a macro- and microscale in recent work on rice hulls (Thomas and Jones, 1970; Thomas, Basu, and Jones, 1972, 1974). Success here has been owed in large measure to the high content of bioincorporated silica—about 20%—which has already formed to some extent a coherent ash skeleton in the unashed material. Figure 8.4*a* shows a macrophotograph of a native hull and a corresponding preparation after plasma-ashing. Higher magnification SEM views are seen in Figures 8.4*b* and 8.4*c*. In order to get rid of the last vestiges of carbon, it was necessary to heat the ash briefly to high temperature after the plasma-ashing, but this had no effect on the ash structure since the latter had already been stabilized by the plasma treatment. By comparison, hulls which were incinerated initially at high temperature became grossly warped and finer structures in the ash were fragmented. The ashing work was done in connection with chemical studies on extraction of silica from the hulls, and the opportunity to correlate SEM views of three-dimensional ash structure with TEM views of ashed thin sections aided greatly in interpretation of chemically modified hulls (see Two-Dimensional Ash Patterns).

On a much finer scale, plasma incineration was able to preserve the three-dimensional ash structure of bacterial spores in one of the earlier demonstrations of this capability (Thomas, 1964). Figure 8.5*a* shows a TEM micrograph of whole spores of *Bacillus cereus*, and an especially well-preserved ash residue from one such spore is seen in Figure 8.5*b*. The preparations were shadowed to display the third dimension. Similar results were also obtained on *Bacillus megaterium* spores. Good results here were again at least partly due to the relatively high mineral content—ash weight represents 12–13% of this material. The objective of this work was to determine the overall

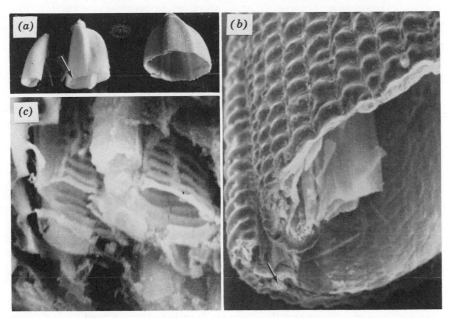

FIGURE 8.4. Plasma-ashed rice hulls: (*a*) Macrophotograph of native hull cut in half (right) and two fragments of similar half hull after ashing (left). The fragile ash remained intact after incineration but fragmented subsequently during handling (× 6). (*b*) SEM micrograph of portion of ash indicated by arrow in (*a*) (× 90). (*c*) SEM micrograph of region indicated by arrow in (*b*) (× 2600). (Reproduced with permission from Thomas and Jones, 1970.)

distribution of mineral substances in spores in connection with certain biochemical and physiological studies (resistance to high-temperature sterilization). The unexpectedly good results on whole spores, showing a coat and internal core ash, were most welcome since unlike thin section results (see Two-Dimensional Ash Patterns), they pertained to chemically intact preparations which had lost none of their mineral content before ashing. Figure 8.5*c* shows for comparison the poorly preserved structural residue obtained from whole spores ashed at high temperature. These latter results by themselves would be rather difficult to interpret.

Also on a fine scale, the delicate three-dimensional ash skeleton of a rotifer jaw prepared by plasma incineration has been shown in TEM by Koehler (1971). The work has been part of an extensive study of the mouthpart structure of these microscopic aquatic animals.

In specimens of low mineral content, and this means most biological materials, good preservation of ash microstructure in three dimensions is unlikely because the widely dispersed mineral is unable to form a coherent

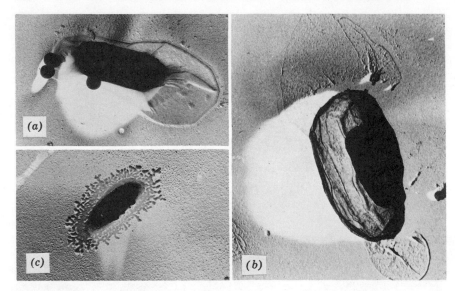

FIGURE 8.5. TEM micrographs of *Bacillus cereus* spore. Spores were sprayed from aqueous suspension onto silicon monoxide membranes, air dried, and uranium-shadowed after treatment (if any). Shadows are shown in bright contrast: (*a*) Typical native (not ashed) spore. The spore body retains its normal, three-dimensional shape but the surrounding, loose, exosporal membrane is flattened by surface tension during drying. Nearby spherical particles are 0.264 μ polystyrene latex added for calibration (× 15,000). (*b*) A spore after oxygen plasma-ashing. A central ash body, displaced to one side during the ashing process, is seen within the delicate, three-dimensional ash of spore coat. A trace of ash from the exosporal membrane can be seen on the support film (× 24,000). (*c*) A typical spore after ashing in a furnace at 500°C. Patterns of small droplet-like particles evidently represent both the exosporal membrane and the collapsed spore coat (× 19,000). (Reproduced with permission from Thomas, 1964.)

skeleton as the supporting organic material is removed. Possibilities exist, however, to artificially and selectively introduce mineral material into the specimen so that its ash can form such a skeleton. Tolgyesi and Cottington (1970) described a fortuitous example of this in their work on human hair, which in the native state has a very low mineral content (see Two-Dimensional Ash Patterns). The hair had been infiltrated with aluminum salts in an experimental treatment to improve the physical properties of bleach-damaged fibers and was ashed by the oxygen plasma technique in preparation for chemical analyses of the bound aluminum. Unexpectedly, the ash residues retained the three-dimensional form of the original fibers and became the object of a brief SEM study.

Drum (1968, 1968a) has invented techniques of artificial silicic petrifaction of plant material for demonstration of structure by OM and TEM. He uses chemical digestion rather than plasma-ashing, however, and his structures have been somewhat fragmented. Lewis (1971), on the other hand, is developing an artificial petrifaction technique specifically for use with the plasma method. The problem requires dissection of neuronal tissue so that delicate nerve nets buried within may be viewed by SEM. Following classical methods of nerve staining developed by Golgi and Cajal, it proves possible to fill the nerves with precipitated silver which effectively forms a metal replica of the nerve cells and their processes and connections. Plasma-ashing removes the organic substance of the tissue but the filamentous, three-dimensional replica survives. The detailed faithfulness of the replicas requires improvement but the method shows good promise. In a preliminary version of the technique, wet-chemical digestion, as used by Drum, was tried, but this disrupted the delicate replicas.

Boyde and Wood (1969) also mention, but with no details, that they are experimenting with artificial petrifaction of bio-organic materials to permit plasma dissection of specimens for SEM viewing. They have explored plasma-etching of entirely organic materials but find that the intrinsic etch resistances of various organic substances differ so slightly that no valid structural dissection results. Structure-specific incorporation of mineral substances would be used to create sufficient differences in intrinsic etch resistance.

Techniques for fine dissection of SEM specimens are greatly needed, and it would seem that the general approach as outlined by Boyde and Wood, and developed in one instance by Lewis may eventually find wide application. Plasma-inert infiltration agents could include not only mineral materials but also such substances as silicones which are quite resistant to oxygen plasma destruction. Injection of silicone rubber latex into the vascular system of tissue or air spaces of lung, with subsequent wet-chemical digestion of the organic material, has been used to demonstrate complex arrays of blood vessels, bronchioles, etc., under SEM (Nowell, Pangborn, and Tyler, 1970, 1972).

Two-Dimensional Ash Patterns from Thin Biological Specimens

Good preservation of three-dimensional ash structure by plasma incineration has been limited to especially favorable specimens. With more typical specimens having slight, sparsely distributed mineral content, even the plasma incineration technique has demanded thin specimens which would form only two-dimensional ash patterns. Besides minimizing migration of ash which can occur even without high-temperature fusion, etc., the thin specimen has also eliminated geometrical confusions inherent in thick objects, and so has been better able to display fine details in the ash.

Although some of the work with thin specimens has been concerned with physical or technological materials, and some of it has been confined to OM studies, most investigations have used TEM to look at biological materials. It is the fine mineral details of the latter material, and the high-resolution capabilities of the latter instrument which have best been able to show the improved ashing capabilities of the new plasma techniques. EXM has also been used in many cases to help overcome the chemical blindness of the OM or TEM.

OM and EXM of lung tissue.   Berkley, et al. (1965) applied oxygen plasma microincineration to sectioned lung tissue, in what may have been the first use of the plasma technique in OM studies of biological material. This gentle incineration greatly aided in the in situ visualization of newly inspired asbestos and other mineral particles in the tissue, which because of their small size and refractive indices nearly matching that of the embedded tissue, were essentially invisible in the intact sections.

The earliest of these microincineration experiments were done in a 60-cycle oxygen glow discharge. This produced some high-temperature charring of the sections, however, and complete ashing was apparently difficult to achieve. Better results were obtained using the then newly available Tracerlab LTA-500 Low-Temperature Asher. Complete ashing of the tissue could be demonstrated by its dissolution in dilute HCl, and foreign mineral bodies, insoluble in the acid and remaining in situ, became strikingly visible on a structureless background. Alternatively, the visibility of foreign mineral particles could be improved within the intact ash pattern by using the reversible technique of optical clearing. Figure 8.6a, from a case of silicosis, shows a silica particle within the ash pattern of a lung tissue section. No mounting medium was used, and both tissue ash and particle are equally visible. Figure 8.6b, from a case of asbestosis, shows asbestos fibers and an asbestos body also within a lung tissue ash pattern. In this specimen, however, a mounting medium with refractive index matching the tissue ash "cleared" the latter. Only the asbestos, of differing refractive index, remains visible.

In more recent work (Berkley, et al., 1967; Langer, et al., 1970, 1972) asbestos fibers and bodies within tissue ash patterns have been not only localized by OM, but also analyzed by EXM. Specimens are limited to cases of severe asbestosis, however. Otherwise, the very low occurrence of the fibers in sections necessitates too much tedious and expensive searching in the instrument, and it becomes preferable to first isolate and concentrate the fibers, as was described under Particle Analysis.

OM of bovine tissues.   Kindig (1969) has also used oxygen plasma micro-incineration in OM studies of biological tissues, specifically bovine aorta,

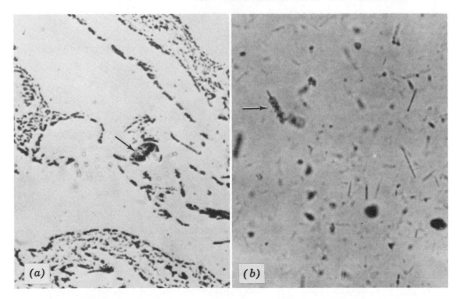

FIGURE 8.6. OM micrographs of plasma-ashed sections of lung tissue containing foreign mineral particles: (*a*) From a case of silicosis; no mounting medium was used on the ash pattern and the tissue ash is readily visible; arrow indicates a silica particle (× 90). (*b*) From a case of asbestosis; the tissue-ash pattern was "cleared" with a mounting medium of matching refractive index, and only the foreign particles remain visible; arrow indicates a coated asbestos body; the other, needle-like objects are uncoated asbestos fibers; phase contrast, × 1000 (approximate). (Reproduced with permission from Berkley, et al., 1965.)

kidney and stomach lining. Other than the instrument used (Coleman model 40 RF Reactor), technical details and description of the work are scant, however. The published ash patterns were viewed by darkfield and/or polarization OM and compared with brightfield views of unashed sections, stained by the Von Kossa method for demonstration of calcium (see Casselman, 1959). The ash patterns appear to show considerably more material than evidenced by the stain, but this would not be surprising since the staining method demonstrates only ion-exchangeable sites of calcium binding, and not necessarily all of the calcium, nor, of course other insoluble or structure-bound mineral which would be visualized in the ash patterns.

OM of leaves and other plant tissue. Hollahan (1966) has described the plasma-ashing of alfalfa leaves which "were completely ashed at about 150°C in a nondisturbing fashion so that the resulting ash still contained the structural integrity of the leaf, the inorganic residues of the leaf, the original veining, etc." This excellent preservation of ash structure had not been sought, but was rather a serendipitous finding in the course of procedures to recover airborne particulates from the leaves.

Quite recently, Umemoto and Hozumi (1971, 1971a–f; 1972, 1972a; see also Hozumi, 1971; Hozumi and Matsumoto, 1972; Hozumi, Hutoh, and Umemoto, 1972; Umemoto, Hutoh, and Hozumi, 1973) have conducted an extensive series of pharmacognostical, taxonomic investigations by low-temperature plasma-ashing of leaves and other plant tissues, making good use of the well-organized ash. It has long been recognized that the size, shape, optical properties, and distribution of inorganic crystals and other mineral bodies in plant tissues can be of sufficient constancy and uniqueness in different species that these characteristics can be used for taxonomic purposes (see Esau, 1965). Ashing the tissue by the plasma technique greatly increases the visibility of these mineral bodies, but unlike high-temperature ashing, it does not alter their distribution or character. Tissue carbon is removed but carbonaceous calcium oxalate crystals, for example, are not altered (Hozumi and Matsumoto, 1972; Umemoto and Hozumi, 1972). By comparing the ash patterns of known minerals in known plants—e.g., silica in bamboo, (Umemoto and Hozumi, 1971b), calcium carbonate in rubber plant (Umemoto and Hozumi, 1971a), calcium oxalate in hops and wintergreen (Umemoto and Hozumi, 1971e, 1972a)—with the ash patterns of crude drug preparations (Hozumi, et al., 1972; Umemoto, et al., 1973; Umemoto and Hozumi, 1971e) it has been possible to identify the previously unknown, specific plant sources of the drugs.

As specimens for ashing, prepared cross-sections of leaves and stems have sometimes been used, but more commonly the specimen has been simply a fresh or dried but rehydrated piece of whole leaf, fastened down to a glass slide with cellophane tape. The damp specimens were found to adhere to the slide during ashing, producing a flatter configuration allowing less potential migration of the ash. The moisture also apparently causes a beneficial reduction in the rate of ashing (Hozumi, et al., 1972; Umemoto and Hozumi, 1971). The authors stress the need to ash slowly, at lower power, to prevent heating effects, and they recommend using air, rather than oxygen, to slow down the ashing rate (Umemoto and Hozumi, 1971c). Their rf plasma-ashing device is of their own construction rather than a commercial model, but they have thoroughly tested its ashing performance (Hozumi and Matsumoto, 1972). After incineration, the ash pattern is fixed and partially cleared under a coverslip with Canada balsam and viewed by either brightfield or polarization optics at magnifications up to about 250 ×. At these relatively low magnifications the preformed mineral bodies are the most obvious features of the ash patterns but the authors mention that ash residues of cell walls, chloroplasts, nuclei, etc. can also be detected. (Umemoto and Hozumi, 1971, 1971c).

Comment on TEM techniques.   For visualization of the really fine details of cellular mineral structure, it is necessary to turn from OM to the much higher resolution capabilities of TEM. The present author has discussed elsewhere (Thomas, 1969) the numerous technical details, problems and possibilities of the microincineration method adapted for TEM, and space limitations will not permit a repetition at length here. The following list will at least indicate some of the points considered in the previous review:

1. Choice of suitable specimen grids and preparation of silicon monoxide support films
2. Preparation and mounting of ultrathin sections or other specimens without artifactual gain or loss of mineral substances
3. Use of thin carbon films of known (spectrophotometrically measured) thickness as standard test objects for determination of burning rates in different instruments and under different plasma conditions
4. Determination of specimen temperature during ashing, and possibility of melting the ash
5. Retention of inorganic elements during ashing
6. Determination of complete ashing
7. TEM examination of the same specimen before and after plasma-ashing
8. Effect of section or specimen thickness and contact with support film on quality of ash pattern
9. Mechanism of ash-granule formation and detailed interpretation of ash-pattern fine structure; effect of mineral concentration
10. Comparisons of high-temperature and low-temperature plasma-microin-cineration results
11. Deliquescence of ash, and possible reaction of ash with support film
12. Chemical identification of mineral constituents within ultramicro regions of ash patterns
13. Quantitative determination of ash mass within ultramicro regions of ash patterns

Many of these points have also been discussed by Boothroyd (1968), Frazier (1971), Hohman (1967), and Hohman and Schraer (1972). Although all of these authors have been concerned primarily with the TEM technique, many of the technical points are also relevant to applications for OM and EXM as well. Among the points listed above, items 2, 9, and 12 are more than mere technical details and in fact represent major problems of the technique which have not yet been fully solved. Because of these problems the TEM micro-incineration technique must be considered still in a state of technical development and exploration.

With regard to point 2, preparing specimens, the problem is that suitable thinness of most specimens is only achieved by thin-sectioning them, and the standard techniques for both OM and TEM were never designed to prevent loss of easily displaced mineral substances or in fact to exclude the artifactual introduction of extraneous mineral materials. The best special technique for preservation of native mineral content is probably frozen sectioning, and this method has been used successfully for some years in preparing specimens for OM and EXM (e.g., see recent review by Läuchli, 1972). Use of the method to prepare the much thinner sections required for TEM has encountered great difficulties, however, which are only now being solved (e.g., see Appleton, 1972; Christensen, 1971). This will probably be the TEM method of the future, however. In the meantime it has been necessary to confine TEM microincineration experiments to problems involving insoluble or structure-bound mineral constituents within specimens, or alternatively, to limit interpretation of results on the basis that unknown quantities of original mineral substances may have been lost.

Concerning point 9, interpretation of ash-pattern fine structure, it is necessary to recognize at the outset that even a structureless specimen containing mineral material as an unstructured continuum will produce a fine-structured ash pattern after plasma incineration. Figure 8.7 shows in TEM the ash pattern from a 1 μ thick section of methacrylate containing 0.3% phosphorus present as dissolved triphenyl phosphate. The ash residue, presumably a polyphosphoric acid, did not deposit as a continuum but rather as a fine-structured, though random, pattern of grains and strands. It seems probable that the pattern resulted largely from subtle inhomogenieties in the methacrylate which were differentially etched in the plasma (see Etching of Organic Polymers) and caused the methacrylate–phosphorus

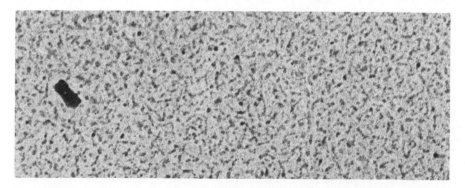

FIGURE 8.7. TEM micrograph of 1 μ thick section of methacrylate containing 0.3% dissolved phosphorus (as triphenyl phosphate), after plasma-ashing. Dense particle on left is a fortuitous contaminant of unknown nature ($\times$ 24,000).

continuum to subdivide into ridges and islands and ultimately grains and strands. Because of the low initial concentration of the mineral substance it seems likely that it but little affected the etching of the methacrylate until the last stages of ashing. In this author's current view, ash patterns in general, of specimens mineral-rich as well as mineral-poor, should probably be seen as the last and ultimate stage of an etching process, and the fine structure of the ultimate ash should be interpreted as a true picture of original mineral distribution only so far as the latter was principally responsible for the etch pattern. Although one may presume that mineral-rich specimens are differentially etched in response to their mineral content, the uncertainties of the etching process in all cases introduces some uncertainty in final interpretation, particularly when 100 Å details are in question.

The third problem, point 12, stems from the fact that the TEM micro-incineration technique is chemically blind; it demonstrates only a class of substances—the incineration-resistant mineral materials—and not individual elements or compounds. In most investigations, and certainly in biological studies, it is precisely the latter demonstration that is most required. The solution to the problem can be difficult in TEM applications to complex specimens because of the excellent spatial resolution, on the order of 100 Å, which the TEM technique has sometimes achieved, and which exceeds that of other microanalytical techniques which are chemically specific. A coward's solution to the dilemma is obtained by confining one's studies to miner-alogically simple specimens (see *TEM of rice hulls*, below) or specimens made simple by previous treatment (extraction, infiltration, see *TEM of cotton*, below). A somewhat wider ranging solution is the combination of a relatively low-resolution analytical technique such as SAED or EXM with high-resolution TEM visualization on specimens which are of only moderate cytochemical complexity and hence amenable to interpretation (see *mitochondria*, and *virus particles*, below). Improved microanalytical techniques are becoming available—e.g., high-resolution EXM (see review by Hall, 1971)—but for the present and near future, the TEM microincineration technique applied to the usual, cytochemically complex specimen is able to visualize finer mineral structures than can be identified chemically (see *spores*, *egg shell gland*, *bones*, etc., below).

TEM of bacterial spores. Our own work with oxygen plasma micro-incineration was initiated in 1962, using a preproduction prototype of the Tracerlab LTA-500 instrument, to improve the quality of thin section TEM ash patterns of bacterial spores we had previously obtained by high-tempera-ture incineration (Thomas, 1962, 1964). A serendipitous result, the preserva-tion of three-dimensional ash structure from entire spores (Fig. 8.5) was described above. Low-temperature ashing of ultrathin sections rather than whole spores could not show the true overall distribution of mineral substances in the spores owing to mineral losses during section preparation. It could, however, provide a closer look at the remarkable structure of the spore coat ash which was probably responsible for its three-dimensional coherence, and it first showed the high-resolution capabilities of the plasma-ashing technique.

A typical thin section plasma ash pattern is seen in Figure 8.8*a* and shows two highly coherent layers of ash representing the outer coat of the spore. In sectioned spores before ashing, three layers were visible but the innermost apparently left no ash. The internal core of the spore is represented by the central circular area with reticular ash structure. (This was a pattern of melted droplets in high-temperature incinerated sections.) This core in unashed sections appeared fairly structureless and homogeneous and so its reticular ash texture might well be an artifact as discussed above (*Comment on TEM techniques*). Certainly artifactual is the annular empty space between the core area and the coat residues. This was occupied by the so-called cortex structure of the spore, but it did not survive the procedures for thin-sectioning. The separation of the coat layers is also an artifact. Almost certainly not artifactual however, is an organized fine structure which was seen in the innermost of the two coat ash layers. The structure is better seen at higher magnification in an ashed, isolated fragment of the layer, as shown in Figure 8.8*b*. It consists of organized arrays of mineral fibrils with regular lateral spacing of about 100 Å. This layer, especially, may have been responsible for the durability of the three-dimensional ash seen in Figure 8.5*b*.

The chemical nature of this organized mineral structure is a tantalizing question which still remains unanswered owing to the limitation of analytical methods, as discussed above (*Comment on TEM techniques*). More recently it has been demonstrated by degradative treatments other than plasma-ashing, however (Holt and Leadbetter, 1969; Rode and Williams, 1966), and a fibrillar material with high ion-binding capacity has recently been isolated from spore coats and may represent the unashed precursor of the fibrillar mineral structure (Kondo and Foster, 1967; see also review by Tipper and Gauthier, 1971).

TEM and EXM of virus particles. Early work on particles of *Tipula* iridescent virus (TIV) (Thomas, 1965, 1967), subsequently updated with additional experiments (Thomas, 1969*a*; Thomas and Corlett, 1969, 1973, 1974) provided another example of very fine ultrastructural detail in plasma ash patterns. In this case, the fortunate nature of the specimen and the combined application of TEM and EXM allowed definite chemical identification of the ash.

Figure 8.9*a* shows typical examples of the unashed virus particles, which are icosahedral in shape and about 1300 Å in diameter. The particles contain 12–13% DNA. Previous characterization (Thomas, 1961; Thomas and

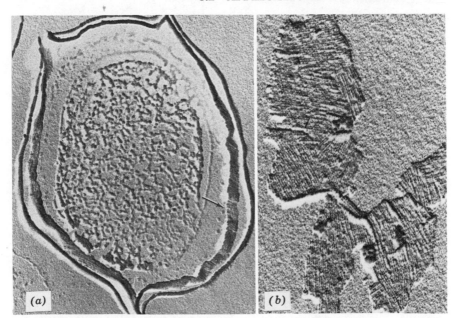

FIGURE 8.8. TEM micrographs of thin-sectioned *Bacillus megaterium* spore (formalin-fixed, methacrylate-embedded) and spore coat fragment after plasma-ashing, and uranium-shadowing: (*a*) Section, approximately 600 Å thick; arrow points to fine-structured inner ash layer of spore coat ($\times$ 48,000). (*b*) Isolated fragment of fine-structured coat layer ($\times$ 63,000). (Reproduced with permission from Thomas, 1964.)

Williams, 1961) had shown that this DNA is contained within a central 900 Å diameter core structure of the virus but its exact location within the core remained unknown. Chemical characterization of the isolated DNA showed that it contained about 30% phosphoric acid. This, and the likely possibility that this structure-bound phosphorus constituted nearly all mineral substance in the virus, suggested that ash patterns might be used to finely localize the DNA within the core.

Preliminary experiments using high-temperature microincineration produced little or no residue from the virus, but with low-temperature plasma-ashing, a very definite and reproducible ash resulted. See Figure 8.9*b*. Further information on the origin of this ash relative to initial structure was obtained from ultrathin-sectioned crystalline pellets of the virus. TIV particles spontaneously form microcrystals when they are highly concentrated under appropriate conditions. Figure 8.9*c* shows such a thin section, about 400 Å thick (1/3 virus diameter) before ashing. The coat and core structures are easily distinguished. When such sections were partially ashed, the coat structure was preferentially etched away, leaving only the cores (not shown,

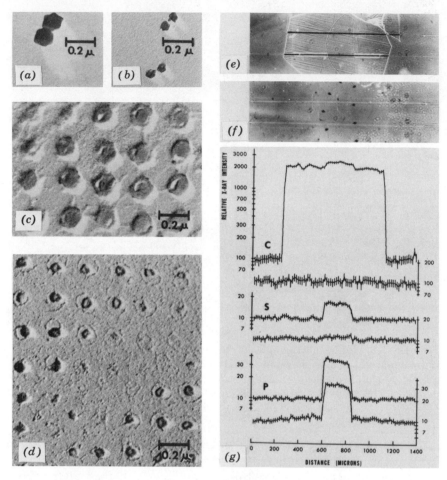

FIGURE 8.9. TEM and EXM of native and plasma-ashed TIV: (*a*) Native TIV particles, freeze-dried on silicon monoxide membrane and uranium-shadowed. (*b*) Similar to (*a*) but plasma-ashed. (*c*) Thin (400 Å) section of TIV crystal (ethanol-fixed, methacrylate-embedded) mounted on silicon monoxide membrane, embedding removed and section shadowed. A slight, artifactual separation of the viral coats from the cores occurred during the embedding procedure. (*d*) Similar to (*c*) but thinner (200 Å) section after plasma-ashing (× 40,000). (*e*) OM phase-contrast micrograph of thick (1 μ) section of embedded TIV pellet mounted on silica cover-slip. The pellet is seen as the crescent-shaped object in the center of the embedding methacrylate plastic. The dark lines across the section are EXM scan tracks. (*f*) Same section as (*e*) after plasma-ashing. The nearly invisible ash of the pellet is seen outlined by the small arrow heads. The light lines represent the EXM scan tracks which were precisely rescanned after ashing (× 31). (*g*) Normalized relative x-ray intensities (10 kV beam) for carbon (C), sulfur (S), and phosphorus (P), along lower track of (*e*) and (*f*). Upper plots (left ordinates) are from before ashing, lower plots (right ordinates) are from after. Abcissas are same scale as for (*e*) and (*f*). (Reproduced with permission from Thomas and Corlett, 1973.)

278

but see Thomas and Corlett, 1973). Complete ashing of sections such as in Figure 8.9c left a residue of the core which seemed to have a hollow center (Thomas, 1969), and this was dramatically confirmed when very thin sections (about 200 Å thick) were examined (Fig. 8.9d). The principal ash consists of well resolved rings about 600 Å in diameter. A very faint residue, surrounding the core ash and apparently representing the coat, also seen in Figure 8.9d, may represent a trace amount of phospholipid in the virus (Thomas, 1961).

To confirm the phosphorus nature of the ash and the completeness of ashing, and also show the relative absence of other mineral elements in the ash patterns, 1 μ thick sections of TIV pellet, either before or after ashing, were analyzed in the linear scan mode by EXM. The sensitivity of this technique and its applicability to thin specimens permitted microanalysis on samples which could be plasma-ashed under conditions comparable to those used on the ultrathin sections for TEM. The sensitivity of EXM analyses was further improved on ash patterns by the thermal stability of the latter, which permitted use of higher EXM-beam intensities. In some cases, the very same section and scan track were analyzed before and after ashing, as shown in Figures 8.9e–g. Carbon was quantitatively removed by the ashing and phosphorus quantitatively retained. A trace of the sulfur originally present in the virus was retained in the ash but its amount was slight relative to the phosphorus. The analytical results for numerous other elements are discussed in detail by Thomas and Corlett (1974), who concluded that the ash consisted of little else than polyphosphoric acid.

Because of the high local concentration of the phosphoric acid in the DNA of the virus it seems likely that the ash patterns seen in Figure 8.9d are not an artifact of fortuitous etching, as discussed above (*Comment on TEM techniques*), but rather indicate that the DNA within the intact virus is probably organized in a shell configuration about 600 Å in diameter inside the 900 Å diameter core. The shell interpretation has been discussed elsewhere (Thomas, 1969).

A further conclusion from this work is that combined use of TEM, EXM, and plasma-ashing on thin specimens mutually improves the performance or interpretation of all three techniques in applications to fine cytochemical analysis of mineral-containing biological materials and the combination should have wide usefulness on specimens which are not too complex (for further discussion see Thomas and Corlett, 1973, 1974).

TEM and SAED of isolated mitochondria. Work on isolated mitochondria (Thomas and Greenawalt, 1968) was undertaken primarily to study the possibility of identifying ash-pattern constituents by SAED. Most of the successful experiments involved high-temperature microincineration. At the same time, however, the work provided an early demonstration of the "membrane" ash patterns produced by plasma incineration which are typical of thin-sectioned whole cells and tissues (see work of Hohman and Schraer, and Frasier, below), and provided some interesting new information on the differences between the high-temperature and low-temperature plasma microincineration results.

The isolated mitochondria had been caused to incorporate large quantities of calcium phosphate in biochemical experiments on ion transport by these subcellular organelles, and as a result, they contained numerous noncrystalline dense granules (see Fig. 8.10a). When high-temperature (500–600°C) microincineration was applied to thin section preparations of this material, the mitochondrial structure disappeared completely, leaving behind, however, the noncrystalline granules. Still higher temperatures (700–800°C) applied to this residue caused the dense granules to crystallize (see Figs. 8.10b, 8.10c), and from fields of about 20 μ^2 containing numerous granules from many mitochondria, SAED powder patterns were obtained which allowed chemical identification of the granules (Fig. 8.10d).

When the same experiments were done with ash patterns produced by the low-temperature plasma treatment, SAED results were disappointing. The plasma ash patterns contained a noncrystalline residue of the mitochondrial structure in addition to the noncrystalline dense granules, and when the patterns were heated to induce crystallization, the two components apparently interfered with each other and crystallization was poor.

It was an interesting and curious fact from these experiments, however, that plasma-ashing produced a structural ash which was stable at high temperature, even though initial ashing at high temperature produced no such ash at all. This was studied further in control mitochondria (containing no dense granules). Figure 8.11a shows a typical thin section of the material before ashing. After high-temperature ashing, such a section is volatilized without a trace (Fig. 8.11b). After plasma-ashing, however, a pattern probably derived from structure-bound mineral of the internal membranes of the organelles remains, as seen in Figure 8.11c. When such a pattern is subsequently heated to a very high temperature—e.g., 900°C—it does not volatilize appreciably, it simply melts, as seen in Figure 8.11d. The ash in Figure 8.11d, previously noncrystalline, also crystallized to some extent, but not sufficiently to produce a chemically identifiable SAED pattern. The chemical nature of the ash thus remains unknown.

With respect to SAED, the current conclusion from these experiments is that the technique may be applicable in identifying ash pattern constituents in especially favorable, simple specimens, but probably will not generally work on chemically and structurally complex ash patterns which do not crystallize readily. The noncrystalline ash typically produced by plasma incineration may be especially unfavorable for SAED experiments because of

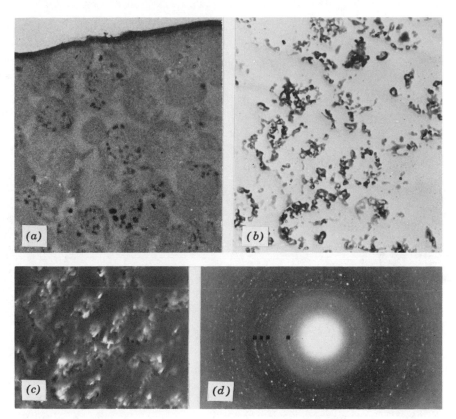

FIGURE 8.10. High-temperature microincineration, TEM, and SAED of calcium phosphate-loaded mitochondria: (a) Thin-sectioned (about 700 Å) pellet of loaded, formalin-fixed mitochondria isolated from rat liver. The unstained mitochondria appear only slightly darker than the embedding plastic (Epon) but the dense granules they contain stand out clearly. Edge of the section and the silicon monoxide support film seen at top (× 12,000). (b) Section similar to (a) but thicker (about 4000 Å) after incineration at 600°C and brief subsequent treatment at 700°C. Only the mineral residues of the granules remain and they have become largely crystalline (× 17,000). (c) Darkfield view of part of the field (lower left) shown in (b). Bright objects are crystalline particles which diffracted through darkfield aperture. Other settings of aperture show that nearly all particles are crystalline (× 17,000). (d) SAED pattern of 20 μ^2 area including the field shown in (b). The four strongest inner rings, which are marked, correspond to crystal spacings of 5.15, 3.18, 2.86, and 2.57 Å and serve, together with fainter arcs, to identify the crystals as β-tricalcium phosphate. (Reproduced with permission from Thomas and Greenawalt.)

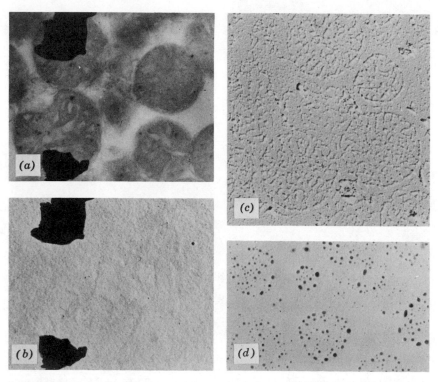

FIGURE 8.11. High-temperature furnace- and low-temperature plasma-micro-incineration and TEM of control (not-loaded) mitochondria: (*a*) Thin-sectioned pellet of osmium-fixed, Epon-embedded control mitochondria. The very dense, irregular objects are particles of silica deposited on the back side of the silicon monoxide support film to serve as location markers. (*b*) Same area as (*a*) after incineration at 600°C, and shadowing. Nothing remains but a small fortuitous con-taminant particle, seen upper right (× 24,000). (*c*) Preparation similar to (*a*) (except formalin-fixed) after ashing by low-temperature plasma and shadowing. A definite ash now remains, and appears to correspond to external and internal membranes of the mitochondria (faintly visible in (*a*) (× 24,000). (*d*) Unshadowed preparation similar to (*c*), after brief heating to 960°C. The granular ash has fused into droplet-like particles, some of which are crystalline and appear especially dense owing to diffraction contrast (× 17,000). (Reproduced with permission from Thomas and Greenawalt, 1968).

its better preserved chemical complexity hindering high-temperature crystallization.

TEM of rice hulls, human hair, plant cells. In the exploration and development of the TEM thin-section microincineration technique, we have preferred to study relatively simple homogeneous preparations of small objects such as spores, virus particles, and isolated mitochondria which are technically simpler to work with and do not pose the sampling problems of more structurally and chemically complex cells and tissues. We have begun in a preliminary way, however, to examine some of the simpler of these more complex objects.

Thin-section plasma microincineration work on rice hulls has been undertaken in connection with chemical-kinetic studies on the synthesis of silicon tetrachloride from the silica in this material (Thomas, et al. 1972, 1974). Although the material as a whole is structurally and chemically complex, in terms of its mineral content it is nevertheless simple, and amenable to easy interpretation. The 20% silica content (see Figs. 8.4b, 8.4c) is accompanied by only trace amounts of other mineral substances, and since the silica is largely insoluble, there has been no major problem of discounting for mineral losses in thin-section preparation. The silica has been found to exist in a variety of dispersed states within the carbonaceous structure of the hulls, ranging from solid silica in an outer crust of the hull to 100 Å colloidal micelles of silica interspersed in the cellulose within cell walls. TEM micrographs of the ash patterns may be seen in Thomas, et al. (1972). Important features of the chemical kinetics correlate with these variations in dispersion.

Preliminary studies on plasma-ashing of thin cross-sections of human hair have been initiated with a view to studying possible variations in the pattern of structure-bound ash correlating with nutritional or pathological states of the donor. Figure 8.12b from Thomas (1969a) shows the ash pattern near the edge of a cross-section of hair from a blond subject. The dense deposits represent the ash of melanin granules and the much finer background ash comes from macrofibrils, and at the right of the field, from the outer cuticle layers. Interpretation of the ash has been greatly aided by examination of partially ashed, or etched preparations such as seen in Figure 8.12a. The nonkeratinous materials of the hair are the first to be etched away, probably owing to their greater chemical lability relative to the highly cross-linked keratin. The final ash pattern, exclusive of the melanin ash, seems to reflect this variation in etching and may be to some extent artifactual in terms of the original fine distribution of mineral. The nonmelanin mineral content of the hair is quite low—probably less than 1%—and so this, at least, dictates caution in interpretation. The most interesting part of the ash pattern might prove to be the melanin residues, and their chemical natures in situ could be determined by EXM analyses.

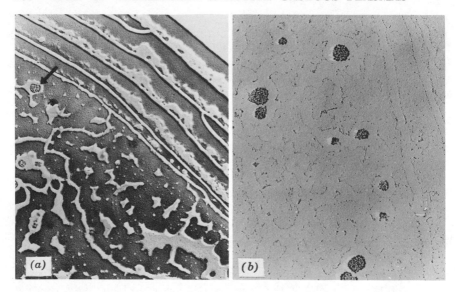

FIGURE 8.12. TEM of plasma-etched and -ashed sections of human hair: (*a*) Cross-section, about 0.3 μ thick, of human hair (no fixation or staining; embedding methacrylate removed) after brief treatment in oxygen plasma, and shadowing. The nonkeratinous material has been etched away, but keratin largely remains. Cuticle layers and the exterior boundary of the hair (and silicon monoxide support film) are seen upper right. The arrow points to the mineral residue of a melanin granule which has already been completely ashed (× 10,000). (*b*) Preparation similar to (*a*) but completely ashed (× 16,000). (Reproduced with permission from Thomas, 1969*a*.)

Preliminary plasma-ashing studies on isolated plant cells were performed simply to explore the character of ash patterns which would result from these more complex specimens. Using conventional fixing and embedding procedures, the ash patterns were confined to the ill-defined, "structure-bound" fraction of mineral which resisted extractions. An example TEM micrograph has been published by Thomas (1969*a*), showing the well-defined ash residue which results from the cell wall middle lamella, and also showing recognizable ash patterns of mitochondria within the cells. Since the patterns were basically similar to those which will be described below (Frazier, 1971; Hohman and Schraer, 1972) except on different tissues, they are not further discussed here.

TEM and EXM of avian egg shell gland. Other workers have begun to explore in some breadth the application of oxygen plasma microincineration to TEM thin sections of complex tissues, and have simply accepted the difficulties of interpretation resulting from uncontrolled mineral losses during section preparation and the unknown chemical identity of the ash.

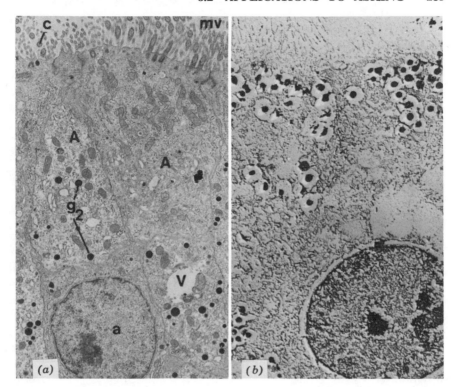

FIGURE 8.13. TEM micrographs of sectioned mucosal tissue of the shell gland: (*a*) Conventionally prepared thin-section (glutaraldehyde—osmium fixation, Epon embedding, sectioned at about 600 Å, and stained with uranyl acetate and lead citrate); A, apical cell; g_2, apical cell granules; a, apical cell nucleus containing dense nucleolus; V, vacuole; mv, microvilli; c, cilia (× 4000). (*b*) Section similar to (*a*), but thicker (0.5 μ), and fixed with only glutaraldehyde and not stained, after oxygen plasma-ashing and platinum-shadowing. Several dense nucleolar residues may be seen in the nucleus, lower right. Vacuolar spaces lie just above the nucleus, and ash-remains of microvilli or cilia are seen at the top of the field. Dense ash residues of the granules are scattered throughout. The mesh-work texture of the cytoplasmic ash represents largely the endoplasmic reticulum (× 7000). (Reproduced with permission from Hohman and Schraer, 1972.)

The first of these was Hohman (1967; see also Hohman and Schraer, 1972) who used the technique in connection with studies of calcium transport in the shell gland of the hen. The shell gland delivers large amounts of calcium from the blood stream to the developing egg shell but nevertheless the concentration of this element remains low within the gland. The route of the calcium through the gland has consequently been difficult to trace. Microincineration was applied to the thin-sectioned tissue in the hope of discovering previously undetected mineral deposits which could implicate specific ultrastructures in the transport process.

Figure 8.13a shows a typical, unashed thin section of the gland prepared by conventional procedures, including staining with heavy metals. The luminal border, with cilia (c) and microvilli (mv) attached to the cell surfaces is seen at the top of the field. A conspicuous feature of the cells is the population of densely stained granules (g₂), which have been thought to be involved in protein or mucopolysaccharide synthesis in the gland. Figure 8.13b shows a similar section, unstained, after plasma-ashing. It is now apparent that the opaque granules are quite rich in mineral material. This has necessitated a reevaluation of the role of these granules, and suggests that they may actually be involved in the calcium transport.

An equal objective of Hohman's study was simply to explore the method for its ability to preserve ultrastructural detail, and determine some of the technical factors involved. As may be seen in Figure 8.13b, nuclei, nucleoli vacuoles, endoplasmic reticulum, microvilli, or cilia can all be recognized in addition to the opaque granules. Mitochondria could not be recognized owing to a disorganization of their ash. Most of Hohman's sections were about 0.5 μ thick and this allowed some artifactual migration of the ash, as he has discussed. Hohman has compared his results with those of Boothroyd (1968) who explored high-temperature microincineration applied to TEM ultrathin sections of tissue, and he points out the much greater retention of cytoplasmic ash which the low-temperature plasma-ashing provides. This recalls our work (above) on isolated mitochondria. He has also compared his own results on both glutaraldehyde-fixed and glutaraldehyde–osium-fixed tissue and finds very little difference. Evidently the metallic osmium is reoxidized by the plasma to volatile osmium tetroxide which leaves no trace. Similar results have been obtained with osmium-fixed preparations after high-temperature incineration (see Figures 8.11a,b). The reader may consult Hohman's papers for discussion of a number of technical points.

The chemical nature of Hohman's ash patterns remain largely unknown. In preliminary experiments with EXM, however, he was able to show qualitatively that they contain calcium and phosphorus.

TEM and EXM of bones, teeth, and tendon. Frazier (1971) has used plasma microincineration and also plasma-etching in an extensive TEM study of early mineralization in developing bones and teeth. The technical problem of distinguishing electron opacities due to intrinsic mineral deposits from other opaque ultrastructures made visible only by their avidity for introduced heavy metal stains has been especially troublesome with this material. By use of plasma-etching on thin sections of the material it has been possible to develop sufficient contrast in translucent organic structures so that heavy metal stains were no longer necessary. At the same time, etching to completion—ashing—has served to confirm or refute the intrinsic mineral nature of electron opacities observed initially in unetched material. It may be mentioned in passing that the experiments were done in a custom-made device (Frazier, et al., 1967) but its design and performance were essentially similar to commercial instruments.

Figures 8.14a–c show shadowed sections of odontoblasts (tooth dentin matrix-secreting cells) which were etched in the plasma for varying lengths of time at a low-power level. After only 30 sec treatment (Fig. 8.14a), enough of the embedding Epon is preferentially removed to reveal a wealth of structural detail emerging from the previously smooth surface of the section. Plasma membranes, mitochondria, and endoplasmic reticulum are distinctly visible. After 2 min treatment (Fig. 8.14b), the structures themselves have been partially etched (ashed) and there is some indication of collapse of the residue (ash?) into the deeply etched surface. Nevertheless, the preservation of fine detail remains excellent. After 30 min treatment (Fig. 8.14c), ashing is complete, and mitochondria, plasma membranes, and other structures are seen as finely granular deposits on the support film. The fine definition of the ash pattern, better than most of those shown by Hohman and Schraer (1972, see above), is probably due principally to the initial thinness of the sections (about 600–1000 Å).

Figure 8.14d shows a thin section of developing enamel tissue after complete ashing. Masses of newly deposited enamel platelets are seen surrounding on three sides an ameloblast (enamel matrix-secreting cell). The arrows indicate ash residues of "dense" granules within the cell which in nonincinerated, conventionally stained sections have an electron opacity similar to that of the enamel. Their opacity as usually seen is evidently due to the stain rather than their intrinsic mineral content.

Similar graded etch-ash experiments were done on bone cells and also the collagen material of predentin and tendon (from rat tail). A particularly interesting result on the collagen was demonstration by etch pattern of the known 680 Å molecular periodicity in these polymeric fibrils. Four unknown subspacings within this periodicity could also be shown under proper etching conditions. The final ash of the collagen also displayed the 680 Å periodicity, as Hohman (1967) had seen earlier (although not mentioned above). Frazier was able to show this periodicity in the ash of fresh fibrils of collagen which had suffered no artifactual extraction of mineral, and has tentatively identified it as largely phosphoric in nature, using EXM. Extraction of the collagen with EDTA† removed traces of calcium present but did not appreciably affect the phosphorus signal in EXM or the banding pattern in TEM.

---

† Ethylenediaminetetraacetic acid.

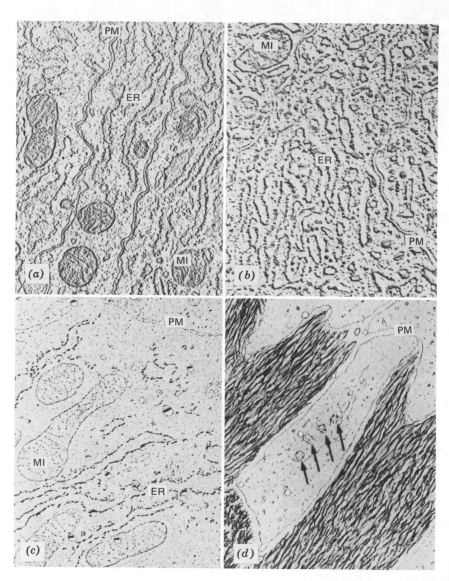

FIGURE 8.14. TEM micrographs of plasma-etched and -ashed thin sections of dental tissue: (*a*) Thin (about 700 Å) section of unstained, osmium-fixed, Epon-embedded odontoblast tissue after surface etching for 30 sec in oxygen plasma, and shadowing; MI, mitochondrion; ER, endoplasmic reticulum; PM, plasma membrane (× 18,000). (*b*) Similar to (*a*), but more deeply etched by 2 min treatment in plasma (× 18,000). (*c*) Similar to (*a*) but completely ashed after 30 min in plasma. The ER appears to yield the densest ash deposits (× 19,000). (*d*) Thin section (technically similar to (*a*) except unshadowed) of enamel tissue after complete ashing in plasma. Arrows indicate the ash residues of "dense" granules within an ameloblast (× 22,000). (Reproduced with permission from Frazier, 1971.)

Frazier has discussed the possible role of collagen phosphorus in nucleating deposition of calcium phosphate during initial stages of mineralization. His paper may also be consulted for discussion of numerous technical points of the microincineration procedure.

TEM of wood.   Zicherman (1970; see also Zicherman and Thomas, 1971, 1972) has recently reinvestigated early OM high-temperature microincineration work on mineral substances in wood (e.g., Uber and Goodspeed, 1935) using TEM and oxygen plasma microincineration. Pine wood yields only about 0.3% ash, but this causes difficulties in paper pulp manufacture and so the nature and distribution of mineral substances in wood continues to be of practical interest.

As a baseline for the studies, Zicherman and Thomas (1972) have presented data on the mineral composition of bulk samples of wood ashed both by the oxygen plasma procedure and by conventional furnace high temperatures. Calcium, magnesium, and potassium prove to be the major elements present, and the retention of most of the elements, but particularly potassium, is materially improved by use of the plasma-ashing method.

Ash patterns of thin (900–1300 Å) cross-sections show structure-bound mineral substance distributed through the wood cell walls but with appreciably different textures in the different layers, S_1, S_2, S_3. Whether these differences reflect fine variations in the mineral distribution is open to question, however, because of the very low concentration of the mineral. They may rather reflect differences in the organic fine structure of the wood and its effect on differential etching (similar considerations as discussed for human hair, Figure 8.12). Spit and Jutte (1965; see Etching of Bio-Organic Materials) have shown such textural differences in the oxygen plasma etch patterns of thin-sectioned wood.

In preliminary experiments, Zicherman (1970) also used microincineration to aid in the fine localization of lead salts introduced into the wood, and he suggests that the technique may have general usefulness in demonstrating impregnation with metal-containing wood preservatives, etc.

TEM of cotton.   Quite recently, Goynes, Muller, and Boyleston (1972; Goynes, et al., 1973) have used oxygen plasma microincineration on thin cross-sections of cotton to demonstrate the distribution of an infiltrated and in situ polymerized phosphonium compound, THPOH–NH₃,† introduced into the fiber to impart flame resistance. The nearly complete absence of ash in incinerated cross-sections of untreated cotton has made the distribution patterns from treated samples relatively easy to interpret. Residue, consisting either of polyphosphoric acid or perhaps the partially degraded polymer itself, is found as a lacey skeleton throughout the cross-section. In comparison

---

† Tetrakis(hydroxymethyl)phosphonium hydroxide-ammonia.

to an OM-staining method and a TEM chemical extraction technique, also used on the treated cotton, the microincineration procedure gives the most reliable demonstration of polymer penetration, according to the authors.

Two-Dimensional Ash Patterns from Thin Physical or Technological Specimens

Turkevich and Streznewski (1958) (see Particle Analysis) were perhaps the first to use a technique which is essentially plasma microincineration of physical objects. They studied carbon black particles, mounted on silicon monoxide membranes, under the TEM as they were sequentially burned away by atomic oxygen, and noted that the particle diameters decreased uniformly with time during treatment. Strictly speaking, however, this work should be classed with etching applications (see Etching of Carbon) since the preparation contained no mineral materials.

Later, Moscou (1962) used a technique which he, himself, considered to be etching, but which really qualifies as plasma microincineration. The specimens were samples of rubber, containing dispersed zinc oxide and carbon black particles, and the object of the work was to characterize the dispersion of the particles by TEM. Ultrathin sectioning of the rubber was difficult, and also not permissible because this caused physical displacement of the filler particles. To overcome the difficulty, thick $(1–5\ \mu)$ sections were cut on a freezing microtome, the sections mounted on silicon-monoxide-filmed TEM grids, and the rubber burned away at low temperature in an oxygen plasma, leaving the zinc oxide and [partially etched(?)] carbon particles deposited in place. Figure 8.15a shows a thick section, which had a fortuitous thin spot, before incineration. After incineration, the same section, same field, is seen in Figure 8.15b. The distribution of the filler particles is now easily seen over the entire field. The apparatus used in this work had previously been developed by Spit (1960) (see Etching of Polymers) and employs a 650 V dc discharge in low-pressure oxygen with the specimen mounted on either the cathode or anode. Removal of the rubber was evidently dependent on the chemical action of the oxygen plasma since a similar discharge in nitrogen was without effect. With oxygen, the best etching rate was obtained with specimen on the cathode.

Also qualifying as microincineration were experiments by Doberenz and Wyckoff (1967) on fossil teeth. In the hollow tubules of the dentine of a *Brontotherium* tooth from the Oligocene age, there were discovered masses of small cylindrical particles which resembled bacteria. Removal of the calcium phosphate from the dentine by a suitable decalcifying reagent did not affect the "bacteria," and so their possible organic nature, of recent origin, had to be considered. To test this, TEM thin sections mounted on silicon monoxide membranes were chemically decalcified and then subjected to oxygen plasma

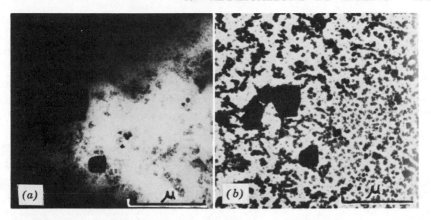

FIGURE 8.15. TEM micrographs of thick-sectioned SBR rubber filled with carbon black and zinc oxide particles, before and after oxygen plasma-ashing: (*a*) Before ashing; a few of the filler particles can be indistinctly seen in a thin area of the section. (*b*) Same area as (*a*) after ashing; the rubber component has been completely removed and all of the filler particles, retained in place, are distinctly visible (× 19,000). (Reproduced with permission from Moscou, 1962.)

microincineration (custom-made apparatus, similar to commercial rf instruments). The "bacteria" came through unscathed. They were, perhaps, truly bacteria, but of ancient age and fossilized.

Recently, Den Besten and Mancuso (1970) have extolled the usefulness of low-temperature plasma treatments in many analytical applications in geochemistry, including an application to microscopy similar to that of Doberenz and Wyckoff. They point out that oxygen plasma incineration of sections of rock specimens observed under the light microscope can be a general tool for distinguishing minute, organic carbon or graphite inclusions from finely divided, metallic minerals. Vacuum stages are available for the light microscope, such that observations can be made continuously during the plasma oxidation. They also consider that plasmas of oxygen or hydrogen may be useful metallographic etchants for suitable specimens (see Etching of Carbon and Inorganic Materials).

Irving (1971) has reported the use of low-temperature plasmas for removal of photoresist films from silicon surfaces in the manufacture of solid-state microelectronic devices (see Chapter 9). Although not so regarded by the author, the technique described (involving plasma incineration of thin, organic polymer films containing inorganic particles, with subsequent darkfield OM observation of the deposited particles) is essentially microincineration as described previously in this review. In this case, the inorganic particles were submicron grains of tin oxide occurring as a contaminant in the photoresist solution used to form the film. Oxygen plasma removed the polymer film but had no apparent effect on the unwanted contaminant particles, which were left behind. Interestingly enough, however, the polymer film could be removed also by active hydrogen generated in hydrogen–nitrogen, water-vapor–nitrogen, and ammonia plasmas. In these cases, the tin oxide particles were also removed, owing to the formation of volatile tin hydride (see Etching of Inorganic Materials). Alternatively, the polymer film could be removed by oxygen plasma and the tin particles subsequently removed by the hydrogen-based plasmas. The use of special plasmas for the specific removal of certain mineral constituents might offer interesting possibilities for the topochemical differentiation of biological ash patterns.

8.3. APPLICATIONS TO SURFACE-ETCHING

Surface-etching by chemically reactive plasmas has been used in the microscopic study of a wide range of materials, including organic polymers, bio-organic materials of various sorts, carbon and graphite, and metals and other inorganic materials. Unlike ashing applications which have mostly utilized a single type of apparatus and gas—oxygen flowing through a rf electrodeless discharge at low temperature—the various etching applications have employed not only this sort of plasma but also dc glow discharges with specimen on either anode or cathode, 50–60 cycle glow discharges, corona discharge, various types of rf sputtering apparatus, ion beam devices, and gas activation by radiolysis. Specimens have sometimes been treated at elevated temperatures. Gases have included not only O_2 but also CO_2, H_2, H_2O, halogens, noble gases, and others. In considering and selecting the literature on this broad range of materials, apparatus, operating conditions, and gases, it has been necessary to continually bear in mind that the present review deals with *chemically reactive* plasmas, and to judge in each case whether the interaction of plasma and surface qualified the application for inclusion.

Inert ion sputter-etching has long been used in the microscopic study of metals and other inorganics and is regarded as a physical process, largely or entirely. Energetic ions, of species which by their nature cannot react chemically with the specimen surface, do impart their kinetic energy and momentum to the surface atoms on which they impinge. In a momentum transfer, cascade impact effect, many of the surface layer atoms acquire a high-velocity component which causes them to eject from the surface. The theoretical basis of the effect, and numerous applications to inorganic specimen etching have been reviewed by Behrisch (1964), Haymann, Waldburger, and Trillat (1965), Kaminsky (1965), and Carter and Colligon (1968), among others. Microscopic applications involving this purely physical type of plasma-etching have been excluded from the review. Included in the review, on the other hand, have been those examples in which the interaction of plasma with the surface has created an *intrinsically volatile product* which then sublimed. This is what some authors (e.g., Kaminsky, 1965; Weiss, Held, and Moore, 1958) have called *chemical sputtering*.

Depending on the specimen and the gas, both chemical and physical sputtering can occur simultaneously, and the relative importance of the two effects has often been difficult to assess. Examples have not been excluded simply because physical sputtering might potentially contribute to the etching effect. A case in point is the etching of polymers and bio-organic materials by noble gases such as argon. Here, sputtering action might occur but is unnecessary since the impinging ion can work its etching effect simply by breaking a chemical bond, thereby creating a low molecular weight fragment of the polymer which is intrinsically volatile. This effect has been called "plasma pyrolysis" by Fu and Blaustein (1968, 1969; see also Fu, 1971). Their experiments on coal have illustrated very nicely the distinction between plasma pyrolysis and inert ion physical sputtering. When the coal is subjected to a rf electrodeless discharge in a noble gas, the hydrocarbon polymer components of the material are gasified to methane, ethane, etc., by the impinging low-energy ions but the elemental carbon and graphite components, removable only by sputtering, remain in situ, unaltered. On the basis of the plasma-pyrolysis effect, examples of noble gas plasma-etching of polymers and bio-organics have been included in the review. In actual fact, the exclusion of such examples would be difficult since many experimenters with noble gas-etching of polymers have also used reactive gases such as O_2 and H_2 and etching results have not infrequently been rather similar.

The examples of plasma-etching gathered together in the present review, although numerous, must be considered a sampling of the literature and not an exhaustive assemblage. Further searching would have undoubtedly turned up still more. The number acquired and their diversity, however, already pose a problem in reviewing them in detail. Because of space limitations and also because many applications range outside the author's experience, it seemed best to present the literature in annotated tabular form which would allow condensed presentation of technical details, if not experimental results, for all of the papers, and at the same time allow some space for general remarks and to present in more detail a few selected examples. A few comments on the tables: (*a*) To minimize the number of footnotes, a certain imprecision has been introduced in technical details—e.g., the precise voltage of the discharge in a given example. (*b*) It seemed valuable to indicate the affiliation of the authors, especially when this was the common link between several authors and their work; the listed affiliation of a group of authors and papers is not necessarily that of the senior author on each paper, but rather is the common link. (c) Authors and groups of authors are listed chronologically by the date of their first paper in the present bibliographical collection; this is not necessarily the earliest date of their first work on the subject. (d) One or more principal source references for apparatus design or etching technique are shown for the various authors or groups to indicate their relation to predecessors.

Etching of Organic Polymers

Among the many examples found for microetching applications of plasmas, the greatest number have been devoted to electron microscopic studies of organic polymer structure. These are summarized in Table 8.1. As seen from the principal source references in the table, most of the work has stemmed either directly or indirectly from the early publications of Kassenbeck, Spit, Anderson and Holland, or Grasenick and Jakopic. In many cases, the etching has been performed in apparatus built specifically for microscopic applications. Figure 8.16 shows a particularly convenient rf electrodeless discharge device described by Evko (1968). It consists of a fairly simple modification of a standard vacuum evaporator, and permits heating (to 1000°C), cooling (to −150°C) or deformation of the specimen during etching, and also shadowing and replication of the etched surface without transferring the specimen or breaking vacuum.

The relative popularity of plasma-etching on polymer materials is probably a reflection of the difficulties and limitations which other methods have encountered. Ultrathin sectioning to produce specimens thin enough for direct examination by TEM can lead to mechanical distortion of soft polymer structures, and in the case of filled polymers, to the rearrangement of hard, inorganic filler particles which may be of the same dimensions as the section thickness and hence pushed by the knife. Even if thin sectioning has been permissible, there has remained the problem of obtaining differential image contrast in thin sections when staining procedures were unsuccessful. Differential surface etching can provide structure-related contrast, and when applied to bulk specimens observed either indirectly by TEM of shadowed surface replicas, or directly in SEM, it also overcomes the problem of mechanical disruption of bulk structure. The reliability of structural information obtained in this way is only so good as the etching technique, however. In the past, surface-etching of polymers has been accomplished by solvent vapors or solutions and by chemical attack of suitable corrosive solutions and these methods are still much in use (e.g., Bucknall, Drinkwater, and Keast, 1972; Palmer and Cobbold, 1964; Reding and Walter, 1959). These methods can, however, lead to artifacts of swelling or disintegration of the

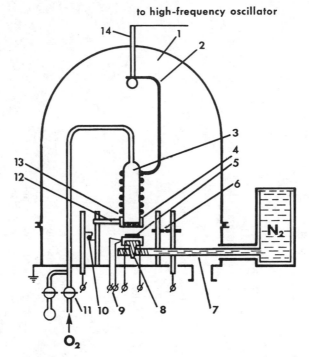

FIGURE 8.16. Diagram of device for the etching of organic objects with active oxygen (after Evko): (1) Bell jar, (2) high-frequency end of the coil, (3) reactor, (4) frame of earthed grid, (5) object or specimen, (6) electrodes for arc evaporation of carbon, etc., (7) cooling stage, (8) heating stage for the specimen, (9) thermocouple, (10) filament for evaporation of metals, etc., (11) needle valve, (12) earthed lead, (13) high-frequency coil, (14) high-frequency lead. (Reproduced with permission from Evko, 1968.)

bulk material, recrystallization of surface structures, reprecipitation of polymer material or other contaminants onto the surface, etc. (e.g., see Kubota, 1965). Plasma-etching has, and is, being explored as a means of overcoming these difficulties.

As indicated in Table 8.1, plasma-etching has been used in any of four general types of application: (*a*) To demonstrate the distribution and morphology of mineral or other highly etch-resistant filler particles (or conversely, voids) incorporated in the polymer; (*b*) to demonstrate the distribution of two or more chemically distinct polymer components which differ in their etch resistance; (*c*) to demonstrate the nature and distribution of structural

TABLE 8.1. Etching of Organic Polymers

Author(s) by affiliation	Type plasma[1] and source reference[2]	Gas	Specimen	Type application[3]	Type microscopy[4]	Special comments[5]
Technische Hochschule, Graz, Austria						
Grasenick, 1957	c(?)	O₂	Filled natural rubber	a	a, c	
Jakopic, 1960	c	O₂	Plastic, filled rubber	a, c, d	a, c, d	d
Blaha, et al., 1960	c, 4	O₂	Polyethylene, natural rubber	a, c, d	a, d	d
Grasenick, 1960, 1961	c, 4	O₂	Filled natural rubber	a	a, c	d
Aldrian, et al., 1965; Geymayer, 1965	c, 4	O₂	Indigo, PVC, Teflon, plastic in rubber, paint, etc.	a, b, c, d	a, c, d	d, e
Geymayer, 1966	c, 4	O₂	ABS-polymer, polymethyl-methacrylate, rubber, etc.	c	a. b. c	
Aldrian, et al., 1967	c, 4	O₂	Nylon, gelatin, rubber, etc.	a, c, d	a, c, d	d, e
Institut Textile de France, Paris						
Kassenbeck, 1958	a	Air	Nylon spherulites	c	a	a
Chemstrand Corp., Decater, Ala						
Anderson and Holland, 1960	a, 2	Air, A	Nylon-66, polyethylene-terephthalate	c	a	
Anderson, 1963	b', 3	A, He	Polyethylene, Nylon-610	c	a	
TNO-TH, Delft, Netherlands						
Spit, 1960	b, b', 1, 2	O₂, N₂, I₂, A	Regenerated cellulose	c	b	b, d
Spit, 1963	b, 3	O₂	Regenerated cellulose, polystyrene	c	b	
Spit, 1967	b, 3	O₂, N₂, Ne, Hg-vapor	Regenerated cellulose, poly-ethylene–polypropylene	b, c	b	c
Fibres Research, Manchester, England						
Dlugosz, 1962	b	O₂	Resin-rubber tire cord adhesives	b	a, c	
University of Bristol, England						
Keller, 1962	(?)	"Ionic etch"	Polyethylene terephthalate	d	d	

Reference	Code	Gas	Material			
British Nylon Spin., Ltd., Pontypool or Mellon Institute, Pittsburgh, Pa. Magill and Harris, 1962	b, 3	O_2	Nylon-56, -66	c	d	d
Magill, 1971	b, 3	O_2	Polyamides	c	d	d
Royal Sulphuric, Ltd., Amsterdam Moscou, 1962	b', 3	O_2, N_2	Filled natural rubber	a	d	f
German Academy of Science, East Berlin Casperson, et al., 1964, 1967	d, 4	He, A, Kr, N_2, O_2	Regenerated cellulose fibers	c	a	d, g
Fukai University, Japan Kurokawa, et al., 1964, 1964a	a	Air	Drawn polyethylene film	c	a, c	h
University of Mainz, West Germany Fischer and Goddar, 1965	d, e, 2, 3, 4, 8	O_2, A	Stretched polyethylene	c	a	d, i, j, k
Frank, et al., 1967	d, 7	O_2	Amorphous polycarbonate	c	a	
Fischer, et al., 1968	d, 7	O_2	Stretched polyethylene, polyethylene terephthalate, etc.	c	a	i
USSR Academy of Science, Moscow Luk'yanovich, 1965	c, 4	O_2	(?)	(?)	(?)	o
Kiselev, et al., 1966	c, 4, 5	O_2	Filled polyester resin, butadiene-styrene rubber	a, c	a, b	d
Gumargalieva, et al., 1966	c, 6	O_2	Polyethylene terephthalate films	c	a	i
Eliseeva, et al., 1967	c, 4, 6	O_2	Acrylic latex particles and films	c	a	
Sukhareva, et al., 1967	c, 5	O_2	Various synthetic rubber latex particles and films	c	a	
Zubov, et al., 1967	c, 6	O_2	Polyester resin coatings	c	a, b	
Evko, 1968	c, 4, 6	O_2	—	—	—	
Mikhaylova, et al., 1968	c, 6	O_2	Filled polyvinyl chloride	a, c	a	d
Pavlova, et al., 1968	c, 4, 6	O_2	Stretched natural and polyisoprene rubbers	c	a, c	i

(continued)

TABLE 8.1. (*Continued*)

Author(s) by affiliation	Type plasma[1] and source reference[2]	Gas	Specimen	Type application[3]	Type microscopy[4]	Special comments[5]
Zubov, et al., 1968 (Not indicated)	c, 6	O_2	Polyester resin coatings	c	a	
Gul and Rogovaya, 1966	b, 3	O_2	Polyisoprene rubber film	c	a	
Vladychina, et al., 1966	c, 4	O_2	Paint and varnish films	a	a	d
Cotton Cellulose Institute, Tashkent, USSR						
Nikonovich, et al., 1967	a, c, 2, 3, 4, 5	O_2, air	Regenerated cellulose fibers	c	a, b	
Nikonovich, et al., 1968	c, 6	O_2	HCHO-modified cellulose fibers	c	b	
Nikonovich, et al., 1968a	b, 3	O_2	Regenerated cellulose films	c	b	
Nikonovich, et al., 1968b	b, c, 3, 4	O_2	Regenerated cellulose films	c	a, b	
Ukrainian Academy of Science, Kiev, USSR						
Ochkivskii and Bezruk, 1967	c, 10	O_2	Kapron	c	a, b	
Bezruk, 1968	c, 5, 6	O_2	Regenerated cellulose and polyamide fibers, fiberglass	a, c, d	a, d	d
Bezruk, et al., 1968	c, 10	O_2	Various polyamides	c	a	d, i, l
Bezruk, et al., 1968a	c, 10	O_2	Regenerated cellulose and polyamide fibers	c	a	i
Bezruk, et al., 1969	c, 10	O_2	Polyamide fibers	c	a	i
Lipatov, et al., 1969; Bezruk, et al., 1970	c, 10	O_2	Amorphous polyurethanes and polymethacrylate	c	a	i
Ochkivskii, et al., 1970	c, 10	O_2	Stretched Kapron fibers	c	a	i
Musachi Institute of Technology, Tokyo						
Toriyama, et al., 1967	f, 9	Air, N_2, O_2	Polyethylene	c	a	d
Eastman Kodak, Kingsport, Tenn.						
Peck and Carter, 1968	c, 11(?)	O_2	Polyethylene terephthalate	c	a, b	l

Source			Application			
University of Stuttgart, West Germany Dietl, 1969	d, 4, 12	O_2	Polyamide-6, polyethylene, polypropylene, polystyrene–butadiene	b, c	a	d
Synthetic Resins Institute, Vladimir, USSR Kurbanova and Kafengauz, 1969	c, 4	O_2	Polyurethane latex film	c	a	
University of Wisconsin, Madison Hien, 1970; Hien, et al., 1972	a, 3, 8	A	Twisted fibers of Nylon, polypropylene, polyethylene, polyethylene terephthalate	c	a	d, i
International Plasma Corp., Hayward, Ca. Irving, 1970	c, 11(?)	O_2	Magnetic recorder tape (filled polymer)	a	a, e	
Farbenfabrik Bayer AG Krefeld-Uerdinger, West Germany Kämpf, 1970, 1970a	c, 4, 12	O_2	Magnetic tape, paint films, filled plastics	a	e	d
Synthetic Fibers Institute, Mytishchi, USSR Mikheleva and Vlasov, 1970	b', 3	A	Polypropylene fibers graft-coated with acrylonitrile, etc.	b	b	
Farbenfabrik Bayer AG, Dormagen, West Germany Orth, 1970	d, 12	O_2	Polyesters, polyamide-6, filled polyacrylonitrile	a, c	f	d
Hinrichsen and Orth, 1971	d, 12	O_2	Drawn polyacrylonitrile film	c	a	i

(continued)

TABLE 8.1. (*Continued*)

Author(s) by affiliation	Type plasma[1] and source reference[2]	Gas	Specimen	Type application[3]	Type microscopy[4]	Special comments[5]
Lenin Electrotechnical Institute, Moscow, USSR Basin, et al., 1971	c, 6	O_2	Various amorphous polyesters	c	a	m
C. Dreyfus Laboratory, Research Triangle Park, N.C. Yasuda, 1972	c, 13	N_2, He, styrene and mixes	Polyoxymethylene	b, c	a	n

[1] Letters designate the following types of plasma: (a) 50–60 cycle, 700–6000 V glow discharge; (b) 500–1600 V dc discharge with specimen on anode: (b') same as (b) but specimen on cathode; (c) rf electrodeless discharge; (d) combined rf discharge and dc discharge, variable 0–3000 V, with specimen on cathode; (e) positive ion beam, 3000–5000 V; (f) corona discharge at atmospheric pressure, 6000 V, 50–60 cycles.

[2] Numbers indicate principal source references for apparatus design or etching technique: (1) König and Helwig, 1951; (2) Kassenbeck, 1958; (3) Spit, 1960 and/or 1963; (4) Jakopic, 1960; (5) Luk'yanovich, 1965; (6) Kiselev, Evko, and Luk'yanovich, 1966; (7) Fischer and Goddar, 1965; (8) Anderson and Holland, 1960, and/or Anderson, 1963; (9) Ross and Curdts, 1956; (10) Bezruk, 1968; (11) Gleit and Holland, 1962; (12) Aldrian, et al., 1967; (13) Smolinsky and Heiss, 1968.

[3] Letters designate the following general types of application: (a) demonstration of mineral or other highly etch-resistant filler particles; (b) demonstration of two or more chemically distinct polymer components differing in their etch resistance; (c) demonstration of structure in a single polymer component with two or more phases (e.g., crystalline and amorphous) which have differing etch resistances; (d) thinning, without structural differentiation, of single-component, single-phase polymer material for examination by TEM and/or SAED.

[4] Letters designate the following types of microscopical technique: (a) TEM examination of shadowed surface replicas; (b) TEM examination of ultrathin sections, films, or particles, usually shadowed after etching; (c) TEM and/or SAED examination of ultrathin polymer surface film, removed from bulk sample by pseudo-replica technique after etching; (d) TEM and/or SAED examination of bulk polymer sample thinned by plasma-etching; (e) TEM examination of particles embedded in extraction replica of surface; (f) direct examination of solid surface in SEM.

[5] Letters designate the following special comments: (a) early TEM demonstration of polymer fine (100 Å) structure; (b) argon did not etch. Author concludes that O_2, N_2, or I_2 etch effects were chemical, not sputtering; (c) determined the nature of the etch products, other than CO_2 and H_2O—e.g., methane, ethane, ethylene, hydrogen; (d) provides detailed description of plasma apparatus; (e) reviews literature on microscopic applications of plasmas; (f) discharge with N_2 effected no etching, in contrast to O_2; (g) most experiments done with gases other than O_2. Concluded that etch patterns were too critically dependent on discharge parameters to give valid indication of structure; (h) conclude that etch patterns were artifactual and resulted from relaxation of surface structures to more stable state; (i) etching parameters are critical, but use of fixed, optimized parameters gives reproducible results. Etch patterns correlate with x-ray diffraction data, etc.; (j) etch patterns correlate with chemical etch results and with EM selected-diffraction darkfield observations on ultrathin films; (k) similar etch results obtained by argon ion beam and rf-activated oxygen; (l) correlated etching rate with polymer molecular weight and degree of crystallinity in specially prepared specimens; (m) globular etch pattern was correlated with turbidimetric measurements on dissolved polymers; (n) studied simultaneous processes of polymer-etching and plasma deposition of polymer; (o) this key reference was not available for detailed review.

differentiation in a single-component polymer with two or more phases (e.g., crystalline, or stretch-oriented, or amorphous) which differ in their etch resistance; (*d*) to produce thin, representative foils for TEM and/or SAED examination of single-component, single-phase polymer material, by uniform etch-thinning of the bulk material. The work of Moscou (1962) which was described above could be considered not only as an example of microincineration, but also as a special technical example of a type (*a*) application applied to thin sections. More commonly in this type of application, the filler particles would be seen emerging from the surface of an etched bulk specimen observed either by TEM replica technique or by SEM.

Applications of the general types (*a*), (*b*), (*d*) have produced readily interpreted results generally regarded as reliable by the investigators using plasma-etching for these purposes. Hansen, et al., (1965) provided a firm basis for type (*b*) applications with a quantitative study of the oxygen plasma-etching rates of various, chemically different polymers. Etching rates differed widely—for example, polypropylene etched at a 16-fold greater rate than sulfur-vulcanized natural rubber.

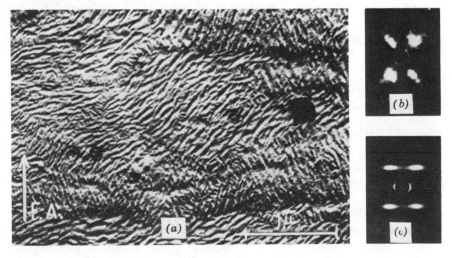

FIGURE 8.17. TEM micrograph of oxygen plasma-etched polyethylene specimen, and corresponding optical diffraction and small-angle x-ray diffraction patterns: (*a*) TEM micrograph of platinum–carbon surface replica of a polyethylene filament which was cold-drawn, tempered 1 hr at 125°C, and then surface-etched by oxygen plasma. The arrow lies parallel to the fiber axis (× 24,000). (*b*) Optical diffraction pattern obtained from (*a*). The optical diffraction mask consisted of a photographically reduced (to approximately 1 × 1 cm) very high contrast negative transparency from (*a*). (*c*) Small angle x-ray pattern of fiber from which (*a*) was obtained. The layer line spacing in both patterns indicates an axial periodicity of about 250 Å. (Reproduced with permission from Fischer, et al., 1968.)

With type (c) etching, variations in the accessibility of active plasma species to the different phases, crystalline, amorphous, etc., have been thought responsible for the differential etching effects (e.g., see Hansen, 1964; Hansen, Martin, and De Benedictus, 1963). There has been some disagreement on the reliable interpretation of this rather subtle type of etching, however. Kurokawa, Ban, and Motoji, (1964), for example, concluded that surface structures observed on drawn polyethylene film after plasma-etching were artifacts created by ion-bombardment-induced relaxation to stable states of metastable surface structures. Casperson, et al. (1967) decided that plasma etch patterns observed on regenerated cellulose specimens were too dependent on the gas-discharge parameters and the gases used to allow reliable interpretation in terms of underlying structure. Other workers, however, have accepted the fact that valid results by differential etching do require very careful adjustment and standardization of the etching conditions, and they have obtained useful results which could be correlated with other determinations of fine structure, such as small-angle x-ray scattering (see Table 8.1). Fischer, Goddar, and Smith, (1968) have provided a particularly elegant example of this. They made optical diffraction masks from the electron micrographs of the oxygen plasma surface etch pattern and showed that the resultant optical diffraction patterns (see Klug and Berger, 1964) were very similar to small-angle x-ray patterns obtained on the bulk material. See Figure 8.17.

Etching of Bio-Organic Materials

Collected examples of plasma-etching of organic biological materials are summarized in Table 8.2. Most of the work can be conveniently classified into one or the other of three groups. For one, there is work on textile or other fibrous materials of biological origin by some of the same researchers who conducted parallel studies on textile synthetic polymers. Just as with synthetic polymers, etching has been used in lieu of staining to develop TEM contrast in the specimens.

Apparently the first of these textile researchers was Kassenbeck (1958) who used plasma-etching and surface replication of cut and polished ends of embedded cotton and wool fibers to reveal cross-section, TEM ultrastructure of the fibers at a time when adequate thin-sectioning techniques had not yet been developed for these materials. His replicated etch patterns from wool fiber ends showed ultrastructural details similar to those seen in Figure 8.12a. More recently, Fabergé (see Table 8.2) has uniquely explored a similar approach on fixed and embedded plant tissue, and has obtained results similar to those which Frazier (1971) showed on thin-sectioned odontoblast tissue (Fig. 14a,b). Now that it is possible to cut thin sections of these materials for etching experiments, as Frazier, Spit, Nikonovich, and others have done (see Table 8.2), the Kassenbeck–Fabergé approach using replicas seems unnecessarily indirect. However, for materials which remain difficult to thin-section, such as highly mineralized tissue, the Kassenbeck–Fabergé approach might be very useful and replication could be circumvented in many cases by use of SEM.

The second class of examples consists of etching in connection with microincineration experiments. The etched specimens have served to aid in the interpretation of the final ash patterns, but as pointed out by Frazier (1971) and utilized earlier by Berkley, et al. (1965), the partially ashed or etched preparations are useful in their own right for enhancing contrast without the use of stains. This class of examples has already been discussed above (Two-Dimensional Ash Patterns).

A third class consists of recent attempts to use plasma-etching (i.e., ion-etching) to dissect three-dimensional specimens for viewing by SEM. Since SEM operated in the usual secondary electron mode sees only the surface of the specimen, there have oftentimes been needs to carry out delicate dissiec tions so that interior structures could be studied. The possibility of using on--etching for such dissections has recently become particularly appealing with

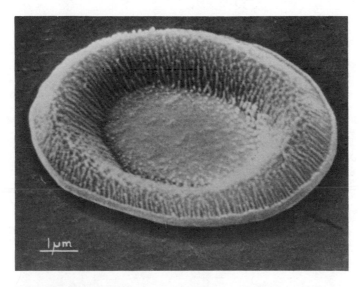

FIGURE 8.18. SEM micrograph of normal red blood cell after ion-etching in hydrogen for 35 min at 1 μtorr. The membrane has been completely removed to show an underlying internal structure within the cell (×10,000). (Reproduced with permission from Stuart, et al., 1969.)

TABLE 8.2. Etching of Bio-Organic Materials

Author(s) by affiliation	Type plasma[1] and source reference[2]	Gas	Specimen	Type Application[3]	Type microscopy[4]	Special comments[5]
Institut Textile de France, Paris Kassenbeck, 1958	a-	Air	Cotton, wool	a, b	a	c
TNO-TH, Delft, Netherlands Spit, 1960	b, 1, 2	O_2	Cellulose fibrils from bacterium, rye straw	b	b	a, c
Isings and Spit, 1964	b, 3	O_2	Cotton, native and resin impregnated	b	b	
Spit and Jutte, 1965; Spit, 1967	b, 3	O_2	Wood, fresh or dried	b	b	
University of Texas, Austin Fabergé, 1964, 1965, 1966, 1967, 1968	b', c, d, 2, 4, 8	A, H_2, He, N_2, O_2, Cl_2, Br_2, I_2, Hg, H_2O, NO, CO, SO_2, NH_3, N_2H_4	Root tips of rye, chromosomes from rye, grasshopper testis	b	a	a, b, c
Mt. Sinai Hospital, New York Berkley, et al., 1965	a, d, 5	O_2	Lung tissue containing mineral bodies	b, c	c	
Cotton Cellulose Institute Tashkent, USSR Nikonovich, et al., 1967	d, 2, 3, 11, 12	O_2	Cotton	b	b	
Nikonovich, et al., 1968 U.S. Department of Agriculture, Albany, Ca.	d, 6	O_2	HCHO-modified cotton	b	b	
Thomas and Greenawalt, 1968	d, 5	O_2	Isolated mitochondria from rat liver	c	b	c
Thomas, 1969a	d, 5	O_2	Human hair	c	b	c
Thomas and Corlett, 1973	d, 5	O_2	Crystallized virus (TIV)	c	b	c

Royal Postgraduate Medical School, London, and/or National Physics Laboratory, Teddington, England Lewis, et al., 1968; Osborn, et al., 1969; Stuart, et al., 1969, 1969a	e, 4, 7, 8	A, H$_2$, O$_2$, H$_2$O	Human red blood cells, normal and sickle, etc.	d	d	a, d
Lewis, et al., 1969	e, 9	A, H$_2$, O$_2$, H$_2$O	Malaria-infected red blood cells	d	d	d
Lewis and Stuart, 1970	e, 9	A, H$_2$, O$_2$, H$_2$O	Various normal and abnormal red blood cells	d	d	d
Ambrose, et al., 1970	e, 9	H$_2$	Cultured mammalian cells, normal and malignant	d	d	d
University of Southern California, Los Angeles Baker, 1969	a, c, 9	Air, A	Red blood cells and isolated cell membranes	d	b, d	a, e
University College, England London, Boyde and Wood, 1969; Boyde, 1971	e, 9	Various	Various	d	d	e
University of Sao Paulo, Brazil Silveira and Arruda, 1969	c, 10	Air	Planarian (worm) tissue	b	b	c
University of Leeds, England Fulker and Holland, 1970	c, e, f, 9	A, H$_2$	Red blood cells, rabbit epithelial tissue	d	d	e
University of Washington, Seattle Frazier, 1971	d, 5	O$_2$	Tooth and bone tissue	b, c	b	c
Imperial Cancer Research Fund, London, England Hodges, et al., 1972	e, f, 9	A, O$_2$, He, H$_2$	Chick red blood cells, cultured mammalian cells	d	d	e
University of Pittsburgh, Pa. Rizk and Bendet, 1972	c	N$_2$	Isolated virus (TMV)	b	b	a

(continued)

TABLE 8.2. (*Continued*)

[1] Letters designate the following types of plasma: (a) 50–60 cycle, 600–3000 V glow discharge; (b) 300–600 V dc discharge with specimen on anode; (b') same as (b) but specimen on cathode; (c) positive ion beam, 300–5000 V; (d) rf electrodeless discharge; (e) rf electrode sputtering apparatus; (f) positive ion beam, 10,000–15,000 V, operated inside SEM specimen chamber, to permit alternate etching and viewing of specimen.

[2] Numbers indicate principal source references for apparatus design or etching technique: (1) König and Helwig, 1951; (2) Kassenbeck, 1958; (3) Spit, 1960; (4) Anderson and Holland, 1960, and/or Anderson, 1963; (5) Gleit and Holland, 1962; Gleit, 1963; (6) Kiselev, et al., 1966; (7) Boyde and Stewart, 1962; (8) Anderson, et al., 1962; and Davidse, 1967; (9) Lewis, et al., 1968; (10) Fabergé, 1966; (11) Luk'yanovich, 1965; (12) Jakopic, 1960.

[3] Letters designate the following types of application: (a) substitute method for thin-sectioning technique which was either unsatisfactory or too difficult; (b) creation of specimen contrast without use of stains; (c) etch pattern obtained as an intermediate in the process of ashing specimen, used as an aid in interpreting the final ash pattern; (d) removal of superficial structures to reveal interior detail.

[4] Letters designate the following types of microscopical technique: (a) TEM examination of surface replica from etched block face (previously cut and polished) of polymer-embedded specimen; (b) TEM examination of ultrathin sections (polymer-embedded), films, fibrils or particles, usually shadowed after etching; (c) OM examination of thick-sectioned material; (d) SEM examination of whole-mount preparations.

[5] Letters designate the following special comments: (a) provides detailed description of plasma apparatus; (b) the three different types of apparatus produced similar results. Of the gases tested, O_2 and H_2O gave the finest etch patterns, H_2 and Hg produced very coarse patterns, and Cl_2, Br_2, I_2, NH_3, and N_2H_2 were all relatively inactive etchants; (c) etch patterns showed ultrastructural details similar to those seen in stained ultrathin sections; (d) etch patterns vary with discharge parameters, including gas used. Consistent differences between normal and abnormal cells suggest that patterns are related to real structure, however; (e) discusses problems of artifact in ion-etching unmineralized, unembedded, soft tissue, and concludes that interior details demonstrated by this technique are uninterpretable, and probably not real structures.

the advent of SEM instruments with ion beam devices built into the specimen chambers. This would permit convenient sequential etching and viewing of specimens without their removal from the instrument. Lewis and co-workers (see Table 8.2) published a series of papers showing remarkable, previously unseen structures within red blood cells and tissue culture cells whose surfaces had been dissected off by plasma-etching (Fig. 8.18). More recently, other workers (Baker, 1969; Boyde, 1971; Boyde and Wood, 1969; Fulker and Holland, 1970; Hodges, et al., 1972) concluded that these structures are largely artifacts of the very deep etching required to gain access to the cell interior. Electrical charging, geometrical shielding, and other factors are more important determinants of the etch pattern than intrinsic differences in plasma resistance. Boyde and Wood (1969), as mentioned above (Three-Dimensional Ash Structures), have considered the possibility that intrinsic differences in etch resistance might be heightened by specific introduction of inorganic materials into the structure, thus allowing the plasma etch dissection to proceed primarily on the basis of structure rather than artifact.

Etching of Carbon and Inorganic Materials

Microscopic studies on carbon and graphite have constituted another major area of application for reactive plasma-etching, and collected examples are cataloged in Table 8.3. To a much lesser extent, reactive plasma-etching has also been used in microscopic studies of inorganic materials. These latter, only few in number, are included as a subsection in the table.

Carbon

Applications to carbon have been largely of two sorts, or a combination of the two. The greatest number have used plasma-microetching to study the chemical kinetics of the plasma oxidation of carbon and graphite, and the role which structural anisotropy, microheterogeneity, and surface impurities may play in these kinetics. This sort of work has usually been done on single crystals of graphite of very well-known structure, trace element composition, etc. Occasionally, other plasma reactions have been studied and compared to oxidation. A second group of applications has used reactive plasma-etching as a structure differentiating tool, usually applied to the polished surface of a specimen, similar to microetching applications to polymers and in metallography. Specimens here have typically been polycrystalline materials consisting of graphite particles in amorphous carbon binders, or in some cases carbon deposits in or on inorganic materials. The number of examples in this second group has been limited by the exclusion of plasma-etching applications utilizing chemically inert ions and involving physical sputtering (e.g., Hales

TABLE 8.3. Etching of Carbon and Inorganic Materials

Author(s) by affiliation	Type plasma[1] and source reference[2]	Gas	Specimen	Type application[3]	Type microscopy[4]	Special comments[5]
		CARBON				
Argonne National Laboratory, Argonne, Ill.						
Hennig, et al., 1958	a, b	O_2	Madagascar graphite crystals	a	a, b	a
Hennig, 1965	b	O_2	Graphite single crystals	a	b, c	
Hennig, 1965a, 1966	b	O_2	Ticonderoga graphite crystals	a	c	
Princeton University, Princeton, N.J.						
Turkevich and Streznewski, 1958; Streznewski and Turkevich, 1959	c, 1	O_2	Carbon-black particles, thin evaporated-carbon films	a, c	d	a, b
Technische Hochschule, Graz, Austria						
Blaha, et al., 1960	d, 2	O_2	Spherical graphite particles within steel, or isolated; thin evaporated-carbon films	b, c	b, d	b
Geymayer, 1965	d, 2	O_2	Thin evaporated-carbon films	c	d	
Horn and Warbichler, 1965	d, 2	O_2	Carbonaceous "decoration" deposit on silicon crystal	b	h	
King's College, Newcastle-upon-Tyne, England and/or University College, Bangor, Wales						
Marsh, et al., 1963, 1964, 1965	a, 3, 4, 5	O_2, CO_2	Ticonderoga or pyrolytic graphite crystals, amorphous carbons	a	a, b, d	c
Hughes, et al., 1964	a	O_2	Ticonderoga graphite crystals	a	b	

Thomas, J. M., 1965	a	O_2, CO_2	Graphite single crystals	a	a, b	d
Marsh and O'Hair, 1969	a, 6, 7	O_2	Ticonderoga graphite crystals	a	a	
Marsh, et al., 1969	a, 6, 7	O_2	Ticonderoga graphite crystals; porous, paracrystalline carbon	a	b, e	
Atomic Energy Research Establishment, Harwell, England and/or Argonne National Laboratory, Argonne, Ill.						
Adamson, et al., 1966	e, 8	CO_2, CO_2–CH_4 mixture, etc.	Powdered graphite	a	d	g
Montet, et al., 1967; Feates, 1968	e	O_2, CO_2, CH_4, CO_2–CH_4	Single-crystal graphite	a	c	g
Feates, 1969	e	O_2, CO_2	Single-crystal graphite	a	c	
École Superieure, Mulhouse, France						
Papirer, et al., 1967	b, 9	O_2	Carbon-black particles	a	d	e
Los Alamos Scientific Laboratory N.M.						
Reiswig, 1967, 1970	f, 11	H_2	Various embedded carbons	b	a	f
Reiswig, et al., 1967	f, 10	O_2, N_2, CO_2, A, He, H_2O, Xe, Kr, air	Polymer-embedded porous carbons (polished block)	b	a, b, f	b, f
Green, et al., 1967	f, 11	H_2, Kr	Polycrystalline, porous graphite block	b	a, b	
Chard, et al., 1968	f, 11	H_2, Kr	Graphitized, compacted glassy carbon	b	a, b	
Reiswig, et al., 1968	f, 11	H_2, Kr	Graphitized polyfurfural alcohol, embedded block	b	a, b	
Levinson, et al., 1970	f, 11	H_2, Kr	Pyrographite, embedded	b	b	
Reiswig, et al., 1970	f, 11	H_2, Kr	Embedded petroleum coke	b	a, b	
Commissariat à l'Energie Atomique, Gif-sur-Yvette, France						
Lang, et al., 1968	a, 6	O_2	Graphite crystals	a	b	

(continued)

TABLE 8.3. (*Continued*)

Author(s) by affiliation	Type plasma[1] and source Reference[2]	Gas	Specimen	Type application[3]	Type microscopy[4]	Special comments[5]
State University, Bowling Green, O. Den Besten and Mancuso, 1970	d, 15	O_2	Carbon particles in sectioned minerals	b	a	j
Batelle Memorial Institute, Richland, Wash. Jones, 1970, 1970*a*	d, 12	O_2, CO_2, H_2O, H_2	Various polycrystalline nuclear graphites	a, b	e, g	b
Jones, 1970*b*	d', 13	O_2	Nuclear graphites	a, b	g	g
Jones and Woodruff, 1971	d, 13	O_2	Graphitized coal-tar pitch, nuclear graphites	a, b	g	g
General Electric Co., Schenectady, N.Y. McCarrol and McKee, 1970, 1971	a, 6, 12	H_2, N_2, O_2–A mixture	Ticonderoga graphite crystals	a	a, e	b, h
INORGANIC MATERIALS						
German Academy of Science, Berlin Meyer and Berger, 1964	c, 14	H_2	Antimony single crystals	b	a, b	b, i
Los Alamos Scientific Laboratory, N.M. Reiswig, 1967, 1970	f, 11	H_2	Metal surfaces, Cu, Ag, Au	b	a	f, i
Technishe Hochschule, Graz, Austria Golob, et al., 1968	d, 2	O_2	Zinc sulfide single crystals	b	b	
Signetics Corp., Sunnyvale, Ca. Blish, et al., 1970	d	CF_4	Silicon surface	d	a	
Irving, 1971	d	H_2, NH_3, H_2–N_2 mixture	Tin oxide particles	d	a	i
State University, Bowling Green, O. Den Besten and Mancuso, 1970	d, 15	H_2, O_2	Unspecified minerals	b	a, b, e	

[1] Letters designate the following types of plasma: (a) atomic oxygen (or atomic hydrogen or nitrogen) generated in a rf (or microwave) electrodeless discharge, flowing gas system with specimen downstream; (b) ozone produced by silent discharge; (c) atomic oxygen (or atomic hydrogen) generated in a 3000–5000 V dc discharge, flowing-gas system with specimen downstream; (d) rf (or microwave) electrodeless discharge, flowing-gas system, with specimen in glow region; (d') same as (d), but stopped-flow; (e) radiolytic decomposition of gas to active species by vacuum ultraviolet or gamma irradiation; (f) 3000 V dc glow discharge with specimen on cathode.

[2] Numbers indicate principal source references for design of apparatus, etching technique, or rationale of experiment: (1) Harteck and Kopsch, 1931; (2) Jakopic, 1960, and/or Aldrian, et al., 1967; (3) Streznewski and Turkevich, 1959; (4) Blackwood and McTaggart, 1959; (5) Hennig, et al., 1958; (6) Marsh, et al., 1963 and/or 1965; (7) Feates, 1968, 1969; (8) Feates and Sach, 1965; (9) Kinney and Friedman, 1952; (10) Hales and Woodruff, 1962; (11) Reiswig, et al., 1967; (12) Thomas, J. M., 1965; (13) Jones, 1970; (14) Wood, 1922; (15) Gleit, 1963.

[3] Letters designate the following types of application: (a) experiments to determine the effect of known heterogeneity or anisotropy of structure and/or the effect of localized impurities on chemical reactivity; (b) determination of unknown distribution of structural heterogeneity, as in metallography; (c) thinning of thick film or bulk specimen for examination by TEM or SAED; (d) microscopy incidental to development of practical product.

[4] Letters designate the following types of microscopical technique: (a) direct examination by OM; (b) TEM examination of shadowed surface replica, showing etch pits, etc.; (c) TEM examination of gold-decorated surface of ultrathin, cleaved lamella, showing monoatomic step edges of etched vacancies, dislocations, etc.; (d) TEM and/or SAED examination of thin films, flakes or particles, sometimes shadowed after etching; (e) direct examination of surface by SEM; (f) TEM and SAED examination of thin lamellae removed from etched surface by pseudo-replication technique; (g) TEM and SAED examination of ultrathin wafer produced, before etching, by mechanical grinding; (h) direct examination of surface in EEM (electronemission microscope).

[5] Letters designate the following special comments: (a) reported that atomic oxygen burned carbon at room temperature with practically no activation energy; (b) provides detailed description of plasma apparatus; (c) rigorous precautions to exclude plasma species other than atomic oxygen showed that atomic oxygen burning of carbon does have an appreciable activation energy which is provided by the action of other species; (d) extensive review of microscopic studies of graphite oxidation; (e) plasma oxidation produces hydrophilic surfaces, which are of technological interest; (f) for OM examination, hydrogen gives best etch; success is due to impurity-promoted (H_2O, O_2, air) presence of atomic hydrogen in discharge: (g) deposition of carbonaceous material onto specimen counteracted etching effects when plasma contained organic gases; (h) specimens were maintained at elevated temperatures (700–1100°C) during exposure to atomic gases; (i) etching is, or is presumed to be, due to formation of volatile hydride; (j) plasma-etching can be observed continuously in special vacuum chamber with OM.

311

and Woodruff, 1962) rather than chemical sputtering with the formation of volatile products from the carbon.

A third type of application, somewhat trivial in terms of microscopy, has been the use of reactive plasmas to produce uniform thinning of amorphous or graphitic carbon films. The object here has been simply to show that the plasma could, in fact, thin uniformly without altering the structure of the film.

Microscopic studies on the oxidation kinetics of reactive plasmas on carbon have been part of the much larger study of carbon oxidation in general, on which there is vast literature. Some feel for the field may be obtained by consulting the Series of Advances edited by Walker (1965, and later). J. M. Thomas (1965) has written a very useful review on the microscopy of graphite oxidation, and the review by Amelinckx, Delavignette, and Heerschap (1965) on the crystallography of graphite is also useful. Interest in oxidation by plasmas, specifically, has stemmed to a great extent from problems arising in the use of carbon and graphite as structural materials in gas-cooled nuclear reactors. The inert cooling gas inevitably becomes contaminated with traces of O_2, CO_2, CO, H_2O, etc., and radiolysis of these gases by intense gamma irradiation produces ozone, atomic oxygen, and other highly reactive oxidative species which attack the graphite. To understand this attack it has been desirable to study the action of individual active species, and so most of the oxidation kinetic studies have tried to use atomic oxygen or other definite species as the etching agent, rather than a heterogeneous, ill-defined plasma. It has not always been easy to achieve a single species for experimentation, however. In pioneer microetching experiments on graphite by Hennig, Dienes, and Kosiba (1958), and independently on carbon particles by Turkevich and Streznewski (1958), both groups of workers reported that the kinetics of atomic oxygen attack on carbon were quite independent of temperature—i.e., the reaction had zero activation energy. Later studies by Marsh, et al. (1963–1965, see Table 8.3) showed, however, that the reaction did have an appreciable activation energy and this energy was provided by surface bombardment of other species in the plasma accompanying the atomic oxygen.

A variety of microscopic techniques has been used to study the etched surfaces of carbon specimens as may be seen in Table 8.3. A particularly interesting technique, and one of great sensitivity, was developed and used by Hennig (1966) and later adopted by Feates and co-workers. Slight etching of a cleaved graphite surface by the active oxidative species causes initially the removal of single atoms from the basal plane surface, and further etching enlarges the single atom vacancies into flat, circular depressions in the surface which are uniformly one atom deep. Evaporation of gold onto the surface under conditions which cause it to migrate produces a decoration pattern,

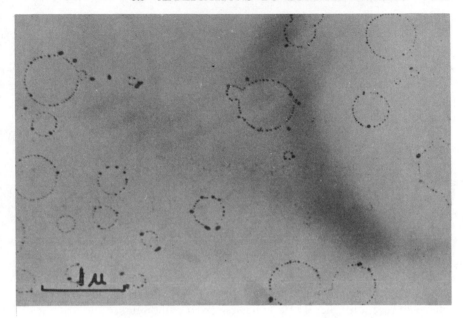

FIGURE 8.19. TEM micrograph of ultrathin, cleaved lamella from Ticonderoga graphite crystal after surface-etching with radiolyzed (Co^{60} gamma source) carbon dioxide, and gold decoration ($\times$ 22,000) (reproduced with permission from Feates, 1968).

visible in TEM, with the gold collected exclusively at the monoatomic step edges at the boundaries of the flat depressions. An example of this is shown in Figure 8.19. It has been possible to determine quantitatively the rate of initial vacancy creation and also the rate of monatomic oxidation in the basal planes by counting and measuring these decorated loops after appropriate experimental manipulation.

Inorganic Materials

A wide variety of etching techniques have long been used with adequate success in preparing inorganic materials for microscopic examination—e.g., corrosion by chemical solutions, electrolytic etching in solution, chemical etching by intrinsically reactive gases forming volatile products, high-temperature thermal etching, inert ion physical sputter etching. The success of these "tried-and-true" methods probably accounts for the relative scarcity of examples of chemically reactive plasma-etching, which is a relatively new, not well-validated technique in this field. The examples entered in the table have mostly been of an exploratory sort.

Meyer and Berger (1964), for example, tried atomic hydrogen-etching of antimony crystals in preference to the usual corrosive solution etch method because they expected it to be cleaner. They found, however, that the volatile product of the etching, SbH_3, is relatively unstable and decomposed to redeposit antimony onto the etched surfaces. The etch patterns were also highly dependent on the working parameters of their apparatus, and so they did not finally recommend the procedure for routine work. Reiswig (1967, 1970), who has found hydrogen discharge (presumed atomic hydrogen) etching of carbon to be a useful routine technique (see Table 8.3), has casually applied the technique to specimens of copper, silver, and gold, and finds a remarkably efficient etching action which he ascribes to the formation of volatile hydrides of these metals. The etching mechanism requires further study, however. It is interesting to recall in this regard that Weiss et al. (1958) doing hydrogen ion-beam-sputtering experiments on silver targets, found that most of the sputtering was due to neutral species, and they also invoked volatile hydride formation as an explanation.

Quite apart from microetching applications, there exists a considerable amount of literature on volatilization of inorganic materials by plasma-chemical reactions. Siegel (1961) has reviewed and discussed the reactions of atomic hydrogen to form volatile hydrides of many solid elements and compounds, including B, Ga, TlCl, Si, Ge, Sn, Pb, $PbCl_2$, P, As, Sb, Bi, S, Se, Te, halides, etc. Bergh (1965) has reviewed atomic hydrogen as a reducing agent for various metal oxides but also discusses volatile hydride formation. Rosner and Allendorf (1969) have reviewed kinetic studies on the volatilization of solid refractory materials by the attack of dissociated gases at high temperatures (400–2900°C). At sufficiently high temperatures, atomic oxygen forms volatile oxides of Ge, Si, SiC, Pt, Mo, Re, and Cr, and atomic halogens volatilize Mo, B, and W. These have all been studied, as has also the high-temperature volatilization of SiC by atomic nitrogen. The undissociated gases will also form volatile products at these high temperatures in most cases, but the reactions are greatly accelerated by the atomic species. The authors also cite work on the low-temperature formation of volatile sulfur and selenium oxides from atomic oxygen attack, and germanium hydride from atomic hydrogen.

Many of these studies have practical importance in space technology. Thus it seems likely that with macroscopic chemical and kinetic studies as the starting point, the finer details of the reactions, as they are affected by the surface structures of the reacting solids, will eventually be studied by microscopy in a manner analogous to the kinetic structural studies of carbon oxidation. More examples of reactive plasma-etching for microscopy of inorganics will then be available for review.

8.4. APPLICATIONS INVOLVING SURFACE MODIFICATIONS (OTHER THAN ETCHING) AND THIN FILMS

The examples rather arbitrarily gathered together in the present section form a heterogeneous group of applications which share in common only that they involve surfaces in one way or another, other than etching. It is however, possible and convenient to divide them into three main categories: (*a*) Plasma-chemical treatments of specimen surfaces to create a modified surface layer from the original material; (*b*) plasma-chemical deposition of thin films onto surfaces; (*c*) cleaning and/or hydrophilization of surfaces and thin films by plasma-chemical treatment. The examples, again a sampling rather than an exhaustive collection, have been selected from large bodies of related literature on the basis of the involvement of microscopy and chemical action in the plasma treatment; just as for examples of etching, treatments involving only or largely physical action of plasmas have been excluded. Although emphasis has remained on applications using plasma treatment as a technique *for* microscopy, it seemed desirable in showing the scope of potential applications to also include to some extent microscopy *of* applications for other purposes. A few such were also included in the section on etching.

Just as with etching applications, the examples presented here have involved a diversity of experimental approaches, types of plasma, types of specimen, and types of microscopical technique, so again a tabular form has been used to present the details.

Creation in situ of Modified Surface Layers or Thin Films

Most of the examples here have been concerned with metals or other inorganics, and most commonly the modified surface layer has been an oxide. See Table 8.4. Modified surface layers of organic specimens are listed separately as a subsection.

Inorganic Specimens

The first use of plasma to chemically modify surface layers as a technique in preparing specimens for microscopy was probably by Mahl (1942, 1945) who was a pioneer in developing surface replica methods for TEM. He found that faithful replicas of many metal surfaces could be produced by forming a chemically resistant oxide layer on the surface and then chemically dissolving the metal itself. The remaining thin oxide film was quite suitable

TABLE 8.4. In situ Creation of Modified Surface Layers or Thin Films

Author(s) by affiliation	Type plasma[1]	Gas	Specimen	Type application[2]	Type microscopy[3]	Special comments[4]
		INORGANIC MATERIALS				
Research Institute AEG, Berlin						
Mahl, 1942, 1945	a	O_2	Various etched metal surfaces	a	a	
Max-Planck Institute, Düsseldorf, West Germany						
Hempel, et al., 1962	c'	N_2, air, H_2, O_2, A	Iron and steel surfaces	a	d, f, g	a, f
Central Institute for Nuclear Physics, Rossendorf, West Germany						
Hilbert, 1962, 1962a	b	A–air mixture	U, Fe, Cu metal surfaces	a	b	a, b
Czarnecki and Hilbert, 1962	b	A–N_2–O_2 mixture	Etched U metal surface	a	b	c
Czarnecki and Hilbert, 1952a	b	N_2–O_2 mixture	Carbon–steel surface	a	b	
Hilbert and Lorenz, 1963, 1964	b	N_2–O_2 mixture	U, sintered-U_3Si_2, Cu, Ni, Zr, steel, various alloys	a	b	
Hilbert, 1965	b	A–O_2, A—air mixtures	Various	a	b	c
Indiana University, Bloomington						
Moore, et al., 1962	c	Argon	Polycrystalline Ag_2S surface	a	e	g
C.N.R.S., Bellevue, France						
Trillat, 1962	c	Air, O_2	FeS_2 and AgBr single crystal thin films	a	c	a, b, g
Haymann, et al., 1963	c	A–O_2 mixture	Fe surfaces at 800–1000°C	a	d, e	
Meyer and Haymann, 1964; Meyer, et al., 1965	c	O_2	Cu single-crystal surface at 350–800°C	a	d, e	c
Meyer, et al., 1968, 1969	c	O_2	Cu single-crystal surface at 20–350°C	a	a, d, e	d
Institute of Nuclear Research, Warsaw, Poland						
Bochenek, et al., 1964	b	O_2, A–O_2 mixture	Niobium metal surface	a	b, d, g	a

Institution / Reference		Reagent	Material			
Technische Hochschule, Graz, Austria Aldrian, et al., 1965, 1967	d	O_2	Cu metal surface	a	b	a, b
Shevchenko University, Tiraspol, USSR Kozlovskii, 1966	e	S or Se vapor	Ag single-crystal thin film on NaCl substrate	a	a	a, e
Illinois Institute of Technology, Chicago Dzoanh, 1967	f	O_2	Cu and Al thin films	c	c, d	a
Physicotechnical Research Institute, Gor'kii, USSR Pavlov and Shitova, 1967	c′	O_2	Silicon single crystal	c	a, e	
University of Bologna, Italy De Maria, et al., 1968	c′	Air, O_2, $A-O_2$, $He-O_2$ mixtures	Zn single-crystal surface	a	f	b, c
General Electric Co., Schenectady, N.Y. Phillips and Seybolt, 1968; Seybolt, 1969	a	N_2-H_2 mixture	Various iron alloys and steels at 550–600°C	c	g, h	f
Atomic Energy Research Establishment, Harwell, England Matthews, 1969	c′	Antimony vapor	Silicon surface	b	h	h
Plessey, Co., Caswell, England Bicknell and Allen, 1970	c′	Boron vapor	Silicon surface	b	h	h
International Plasma Corp., Hayward, Ca. Irving, 1971	g	O_2	Silicon surface	c	e	
ORGANIC MATERIALS						
Du Pont Co., Wilmington, Del. Scott and Ferguson, 1956	h	(Electrons)	Dacron polyester, cellulose acetate, rayon	d	a	
Technische Hochschule, Graz, Austria Grasenick, 1957	h	(Electrons)	Natural rubber with in-organic filler	d	a	

(continued)

317

TABLE 8.4. (*Continued*)

Author(s) by affiliation	Type plasma[1]	Gas	Specimen	Type application[2]	Type microscopy[3]	Special comments[4]
Jakopic, 1960; Grasenick, 1960, 1961; Aldrian, et al., 1965, 1967	g	O_2	Natural rubber, as above	d	i	a
Geymayer, 1966	g	O_2 (Electrons)	Unspecified polymers	d	i	
Grasenick and Jakopic, 1969	h		Millipore filter	d	a	
Fukai University, Japan Kurokawa, et al., 1964a	i	Air	Drawn polyethylene film	d	i	
USSR Academy of Science, Moscow Pavlova, et al., 1968	g	O_2	Polyisoprene, natural rubber	d	i	
Indiana State University, Terre Haute Alväger and Swez, 1970	c′	Lead vapor	Whole bacterial cells	b	c	i

[1] Letters designate the following types of plasma: (a) dc glow discharge, 500 V, with specimen on cathode; (b) dc discharge, 2000–5000 V, in magnetic field, specimen on cathode; (c) positive ion beam, 8000–15000 V; (c′) similar to (c) but 20,000–100,000 V; (d) combined rf electrodeless discharge and 5000 V dc discharge with specimen on cathode; (e) vacuum (thermal) evaporation of substance into 50-cycle, 500–2000 V glow discharge with evaporant source and condensation substrate at common ground; (f) "self-sustaining dipole discharge," 20,000 V at atmospheric pressure; (g) rf electrodeless discharge; (h) irradiation of surface with 40–100 kV electron beam; (i) 50–60 cycle, 6000 V glow discharge.

[2] Letters designate the following types of application: (a) production of epitaxial surface layer (usually oxide) for metallographic demonstration of crystal boundaries, precipitates, etc., and/or study of variations in chemical reactivity of different crystal faces or phases; (b) study of chemical changes, phase separations, precipitates, etc., produced by ion implantation at some depth below the immediate surface; (c) practically oriented study of plasma-surface treatment or corrosion; (d) production of cross-linked surface layer with greater chemical or mechanical resistance than bulk material.

[3] Letters designate the following types of microscopical technique:

resulting from varying optical interference effects of differentiated oxide surface film; (c) direct TEM and SAED examination of thin film or particle specimen prepared before chemical modification by ion bombardment; (d) direct examination of solid surface by OM, and/or TEM (and SAED) examination of surface by replica (or extraction replica) technique; (e) glancing angle SAED of solid surface; (f) direct examination of surface by EEM during ion bombardment; (g) OM examination of section of specimen oriented perpendicular to ion-bombarded surface; (h) TEM and SAED examination of thin film produced by electrolytic (or similar) thinning of thick specimen after ion bombardment; (i) TEM and/or SAED examination of thin film mechanically stripped from surface by pseudo-replica technique.

[4] Letters designate the following special comments: (a) provides detailed description of plasma apparatus; (b) reviews other literature pertaining to method, and/or other examples; (c) physical sputter-etching and oxidation occur simultaneously and final result is the net effect of both processes; (d) character of epitaxial layer depends on angle of incidence of bombarding ions; (e) crystal character of product films (Ag_2Se or Ag_2S) depends on presence or absence of discharge; (f) N ions produced precipitate of Fe_4N. (g) silver compound was reduced to metallic silver; (h) ions implanted at high doses precipitate

for direct examination in TEM. The films could be produced by high-temperature oxidation in air or by electrolytic oxidation, or quite conveniently by oxidation in a gas discharge. After inert-ion, physical sputter-etching of a metal surface, air was simply admitted to the discharge to oxidize the surface. The oxides were found to grow epitaxially on the etched metal surface and thus the replicas could be used to study crystallite boundaries, orientation, etc.

The oxidation of metal surfaces in gas discharges containing trace amounts of air is now well known and has most often been regarded as an unwanted complication in experiments aimed at producing clean, chemically unaltered surfaces by physical sputter-etching in inert gases (e.g., see Gribi, 1962). Hilbert and co-workers (see Table 8.4), however, recognized the value of the phenomenon in preparing specimens for OM metallography. Owing to differences in rate of growth and final thickness of the oxide on different crystal faces of a polycrystalline surface, and also due to differences in

FIGURE 8.20. OM micrograph by reflected light of the surface of a technical-grade copper specimen after inert etching by argon ions and subsequent surface oxidation by oxygen plasma. The various crystal faces exposed at the surface are differentiated by varying interference colors produced by the epitaxial oxide layer. Different shades of gray in the present reproduction represent various shades of red, yellow, and blue in the original micrograph (× 300) (reproduced with permission from Aldrian et al., 1967).

refractive index of the different epitaxial oxides, the oxidized metal surface film viewed in reflected light by OM can show varying interference colors which display the grain boundaries, etc. The technique has been used by Bochenek, Fiett, and Mizera (1964) and by Aldrian, et al. (1965, 1967). Figure 8.20, from the latter authors, provides an example. The original micrograph, in color, conveys more information than is possible in the black and white reproduction shown here.

Trillat and co-workers and also a number of others (see Table 8.4) have used microscopic techniques to conduct basic studies on oxidation or other chemical alterations produced on surfaces and in thin films by interactions with plasmas. In most cases, the plasma has consisted of a carefully controlled ion bean of a chemically reactive species, and in some cases the specimen has been at elevated temperature. Also included in the table are a couple of examples (Moore, et al. 1962; Trillat, 1962) in which the bombarding species might have been considered chemically inert but also produced a chemical change in the specimen: Silver compounds were reduced to metallic silver.

Other microscopic studies of plasma-chemically modified surface layers have resulted from practical problems—e.g., surface oxidation of semiconductor silicon devices (Irving, 1971; Pavlov and Shitova, 1967; see Chapter 9) or ion-nitriding of steel (Phillips and Seybolt, 1968; Seybolt, 1969; see Chapter 3), or surface corrosion by intense plasma (Dzoanh, 1967). The table also includes a couple of practically motivated studies from the field of high-energy ion implantation in semiconductor devices. Most work in this field (e.g., see Eisen and Chadderton, 1970) has been excluded from the present review since it involves the minor introduction of a dispersed, trace impurity in the crystal lattice rather than a major chemical alteration in the specimen. At high dosages, however, the implanted ions may precipitate as separate chemical phases, and such examples (Bicknell and Allen, 1970; Matthews, 1969) qualify for inclusion.

Organic Specimens

Plasma treatments of organic polymers can cause cross-linking of a surface layer, with a resultant increase in the mechanical and chemical stability of the layer, and can also cause increased chemical reactivity of the immediate surface. These effects have been used in practical applications to increase the strength of adhesive bonds to various polymers (e.g., see Hansen and Schonhorn, 1966; and Chapter 3) and they have also found use in preparing polymer specimens for TEM. In TEM applications, the polymer surface, after exposure to the plasma, is coated with a suitable adhesive material—e.g., gelatin on a rubber surface (Grasenick, 1960) and the cross-linked surface layer is then stripped from the bulk polymer by mechanical removal of the adhesive coating, i.e., the adhesive joint breaks at the boundary of cross-linked layer with bulk polymer rather than at the true surface. The adhesive coating is finally dissolved in a suitable solvent, leaving an ultrathin lamella of polymer for direct observation by TEM. Alternatively, the chemical stability of the cross-linked surface layer can be exploited; the bulk polymer is simply dissolved in a suitable reagent which does not attack the stabilized surface layer and the latter is liberated for direct examination in TEM. The table lists several examples of these techniques. In addition to examples involving plasma exposure, preparations involving exposure to electron beams are also included since the net result is quite similar. As discussed elsewhere (Chapter 3) surface cross-linking of polymers is thought to be mediated in many cases by the surface bombardment of high-energy electrons from the plasma.

The last entry in Table 8.4, the work of Alväger and Swez (1970) is a quite different application of chemically reactive plasma to an organic specimen surface. They have explored the possibility of high-energy ion implantation as a means of introducing TEM-contrast enhancing, heavy metal stains into biological specimens.

Deposition of Thin Films by Plasma-Chemical Reactions

Deposition of thin films by plasma-chemical reactions, and their numerous practical uses, has been discussed at length in Chapters 5 and 9. Plasma-chemical techniques have also been used to make thin films for several purposes in preparing specimens for electron microscopy: (a) Production of thin-film surface replicas for TEM, particularly of highly convoluted surfaces; (b) production of thin specimen-support films for TEM; (c) oblique deposition of thin coatings to shadow specimens for TEM; (d) deposition of thin, electrically conducting coatings on specimens for SEM. Examples of these uses are listed in Table 8.5 together with a few special, one-of-a-kind applications.

In choosing examples for inclusion in the table, a somewhat broader definition of plasma-chemical reactions has been used than in Chapter 5. Thus, in addition to film deposition by reactions involving polymerization of gaseous reagents, the phenomenon of deposition by reactive sputtering is also included. In the latter, the reagent is a condensible solid vaporized by ion-bombardment, physical sputtering from a cathode, but the vapor is chemically altered (usually oxidized) in the plasma before condensation into a film. Numerous examples of film deposition by non-reactive sputtering (e.g., see Davidse and Maissel, 1965) have been excluded, however, as have also examples of plasma ion-plating (e.g., see Mattox, 1964), since these involve a purely physical, not chemical, action of plasmas. A somewhat unique

TABLE 8.5. Deposition of Thin Films by Plasma Reactions (Including Reactive Sputtering or Evaporation)

Author(s) by affiliation	Type application[1]	Deposited film	Type plasma[2]	Gas	Sputter target or evaporation source[3]	Special comments[4]
Research Institute AEG, Berlin						
Mahl, 1942	a	Nickel oxide	a	O_2	Nickel	
University of Göttingen, West Germany						
Helwig and König, 1950	a, b	PtO_2	a	O_2	Platinum	a, b
	a	Carbonaceous polymer	a	Hydrocarbons	—	
König and Helwig, 1951	a, c	Carbonaceous polymer	a	C_6H_6 and other hydrocarbon vapors	—	a, c
König, 1953	a, b	PtO_2	a	O_2	Platinum	a, b, c, d
	a	Carbonaceous polymer	a	Hydrocarbons	—	
Technische Hochschule, Graz, Austria						
Grasenick and Haefer, 1952	a	Carbonaceous polymer	a	C_6H_6, C_6H_6–CCl_4	—	a, c
	a	Carbonaceous polymer	a	A, H_2	Carbon	
Haefer, 1954, 1954a	a, c	Carbon	b	CH_4, C_2H_2, C_6H_6	—	a, e
	a, c	Si–SiO	b	SiH_4	—	
	a, c	Boron	b	B_2H_5	—	
Grasenick, 1956, 1957	a, b, c	Various, e.g., C, SiO_2, SiC, PtO_2	a, b	Various	Various	d
Grasenick and Reiter, 1958	a, c	Carbon	c	Hydrocarbons	—	
Haefer, 1958	a, c	Carbonaceous polymer	a	Benzene	—	a
Grasenick, 1961	a	Carbon	b	Hydrocarbons	—	
Aldrian, et al., 1965; Ziegelbecker, 1965	a	Carbon	d	Hydrocarbons	—	a, d, e
Aldrian, et al., 1967	a	SiO	d	Unspecified	—	
University of Munich, West Germany	a	Carbon	d	Benzene–alcohol	—	a, d
	e	Carbonaceous polymer	a, f	Oil vapor in H_2 or O_2	—	a, h
	f	Carbonaceous polymer	f	Oil vapor in air	—	a, h
Bauer, et al., 1958	c'	Carbonaceous polymer	a	Benzene	—	i

Reference		Material		Gas/Vapor	Product	
University of Tübingen, West Germany Speidel, 1959 Mollenstedt and Speidel, 1961 Mollenstedt, et al., 1965	a, b	Carbonaceous polymer–platinum	a	A plus C_6H_6 or CHCH	Platinum	a, b
TNO-TH, Delft, Netherlands Spit, 1959	a	Carbonaceous polymer	a	Hydrocarbon	—	f
(Not indicated) Gritsayenko and Gorshkov, 1961	a	Carbon	a	Benzene	—	c
USSR (not further indicated) Pravdyuk and Golyanov, 1961; Golyanov, 1964	b, c'	Various metal oxides	b	Xenon (plus residual air)	Various metals	a, b, j
Golyanov, et al., 1970	d	Tin oxide	b	Kr plus air	Tin	b
Weston Instruments, Newark, N.J. Lakshmanan, et al., 1964	d	Titanium oxide	a	A plus O_2	Titanium	a
Thin Flims Laboratory, Mulhous, France Perny and Laville-Saint-Martin, 1964	d	Copper oxides	a	A plus O_2	Copper	
University of Glasgow, Scotland Baird, et al., 1965	d	Carbonaceous particles	g	Carbon monoxide	—	a
Shevchenko University, Tiraspol, USSR Kozlovskii, 1965	d	Sulfur	e	Sulfur vapor	Solid sulfur	g
Kozlovskii, 1966	d	Lead sulfide	e	PbS vapor	Solid PbS	a, g
Kozlovskii, 1967	a, c	Carbon	e	C vapor	Solid C	a
University of California, La Jolla Reimann, et al., 1965	c	Carbonaceous polymer	a	Hydrocarbon	—	f
Sanyo Electric, Osaka, Japan Kuwano, 1969	d	Silicon nitride	c	SiH_4 plus N_2	—	a

(continued)

TABLE 8.5. (Continued)

Author(s) by affiliation	Type application[1]	Deposited film	Type plasma[2]	Gas	Sputter target or evaporated source[3]	Special comments[4]
IBM Corp., Yorktown Heights, N.Y.						
Pennebaker, 1969	d	Strontium titanate	h	A plus O_2	$SrTiO_3$	a, k
University of Cambridge, England Echlin and Hyde, 1972	g	Carbon	a	Methane	—	
University Muenster, West Germany Pfefferkorn, et al., 1972	g	Carbonaceous polymer	j	Residual organic vapors	—	

[1] Letters designate the following types of application: (a) production of three-dimensional surface replicas for TEM; (b) production of specimen-shadowing films (including self-shadowing replicas) for TEM; (c) production of specimen support films for TEM; (c') production of film specimens for special TEM and/or SAED experiments; (d) TEM and/or SAED and/or OM examination of thin film (or particles, as indicated) produced for some practical or experimental purpose other than specimen preparation technique; (e) contact microionograph or ion beam microscopy (ion shadow image is recorded by variations in deposition of polymer film, and this microscopic image is viewed by OM using an optical interference technique); (f) similar to (e) but deposited film is used as mask in manufacture of TEM apertures and specimen grids by electroplating process; (g) production of electrically conductive coatings on SEM specimens.

[2] Letters designate the following types of plasma: (a) 800–5000 V dc glow discharge with sputter target (if any) as cathode, specimen or substrate on anode; (b) low-pressure, Penning-type, 300–8000 V dc discharge in magnetic field, specimen on anode, sputter target (if any) on cathode; (c) rf electrodeless discharge; (d) combined rf electrodeless discharge and 0–5000 V dc discharge at low pressure; (e) vacuum (thermal) evaporation of substance into 50-cycle, 500–3000 V glow discharge with evaporant source and condensation substrate at common ground; (f) 40–50 kV ion beam microscope; (g) alpharadiolysis

of gas in apparatus installed within TEM to allow undisturbed TEM examination of deposited particles; (h) rf-sputtering apparatus utilizing magnetic field; (i) 5000 V ion-beam sputtering apparatus; (j) "gas discharge," used for specimen etching, no further details provided.

[3] Absence of a listed target or source indicates that the process is polymerization of a gaseous reagent.

[4] Letters designate the following special comments: (a) provides detailed description of plasma apparatus; (b) unlike condensed metal film, the deposited metal oxide (or metal–carbon) showed no crystalline grain at high magnification; (c) polymer films are, or can be converted to more stable carbon films by heating under vacuum, or by intense electron bombardment; (d) extensive review of TEM specimen preparation methods; (e) films deposited at low pressure are stable and require no heat or electron treatment; (f) no details of plasma film deposition other than citation of König and Helwig (1951); (g) passage of evaporant vapor through glow discharge affects crystallinity of condensed film; (h) C-polymer film serves as grainless, high-resolution substitute for conventional silver bromide photographic film; (i) in experiment on effect of electron irradiation in TEM, electrons caused chemical reaction between C-polymer film and overlying thin film of vacuum-evaporated beryllium; (j) films were used to record radiation damage tracks of nuclear particles; (k) Oxygen in plasma is necessary to maintain stoichiometric content of oxygen in film.

procedure related to ion-plating, namely, vacuum evaporation into a glow discharge (Kozlovskii, 1965, 1966, 1967), has been included for its interest, even though uncertainly classed as a chemically reactive process. Another unusual variant included is a combination of nonreactive physical sputtering (of platinum) with plasma polymerization from the gas phase (of hydrocarbon) to produce an integral carbonaceous–metallic film (Bauer, Fritz, and Kinder, 1958).

Mahl (1942) may have provided the first example of reactive sputter deposition for preparation of TEM specimens—a nickel oxide replica of inorganic crystals. Condensed metal films used for the same purpose were found to be unsatisfactory because of coarse, crystalline grain. The oxide, on the other hand, was almost grainless. Other authors (e.g., König, 1953; Pravdyuk and Golyanov, 1961; also see review by Grasenick, 1956) have capitalized on this advantage of reactive sputtering to produce fine grain self-shadowing platinum oxide replicas or deposit thin, shadow coatings of various oxides on TEM specimens. A similar advantage of fine grain was found by Bauer, et al. (1960) for their self-shadowing platinum–carbon polymer replicas. In a rather unusual application Pravdyuk and Golyanov (1961) used reactively sputtered thin films as ultrafine grain recording media for nuclear particle tracks to be viewed in TEM. Reactively sputtered oxide films produced for various practical applications have been studied by TEM and SAED (e.g., Golyanov et al., 1970; Lakshmanan, Wysocki, and Slegesky, 1964; Pennebaker, 1969; Perny and Laville-Saint-Martin, 1965) to determine the factors which control crystallite grain size.

König and Helwig (1951; Helwig and König, 1950) first described what has probably become the best known application of plasma-polymerized film deposition in electron microscopy technique. They showed that excellent surface replicas of difficult, three-dimensional specimens could be produced by exposing them to various hydrocarbon vapors in a glow discharge. The specimens, surrounded by the plasma, became uniformly coated with a hydrocarbon polymer film which faithfully reproduced the surface features and maintained its shape when the specimen was dissolved away in an appropriate solvent. Specimen support films were also made this way. The stability of the films and replicas was found to be improved by heating them or exposing them to an intense electron beam before dissolving the specimen. In initial replica applications the specimens were inorganic materials so such treatments were permissible. The technique was further developed by Grasenick and Haefer (1952 and later; see Table 8.5) who found that by proper choice of organic vapor and discharge conditions, replicas could be produced which required no further stabilizing treatment. Subsequent to this early development, the technique has been used by various authors (e.g., Gritsayenko and Gorskov, 1961; Spit, 1959) to replicate especially difficult three-dimensional specimens and also to make specimen support films (Riemann, Lewin, and Volcani, 1965). This method of carbon film deposition has never become widely popular, however, because another method developed at about the same time, vacuum deposition from a carbon arc (e.g., Bradley, 1956) proves to be easier and requires less special equipment. The plasma method, nevertheless, continues to have an advantage of more readily forming uniform film thicknesses on highly irregular or reentrant surfaces. Recently, it has been explored as a possible method for coating SEM specimens to provide electrical conductivity (Echlin and Hyde, 1972; Pfefferkorn, Gruter, and Pfautsch, 1972).

Grasenick and Haeffer in their early work also showed that replicas of silica, silicon carbide, boron, and other substances could be produced by use of appropriate gases in the discharge. An example of a silica replica is seen in Figure 8.21. Recently Kuwano (1969) has provided TEM and SAED characterization of glow-discharge–deposited silicon nitride films, which would provide another possibility.

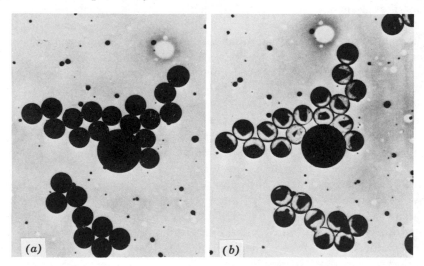

FIGURE 8.21. TEM micrographs of plasma-deposited silicon oxide replica film on polystyrene latex particles, with the latter subsequently removed by oxygen plasma: (*a*) Polystyrene latex particles on thin, organic-polymer support film, uniformly coated with silicon oxide film deposited by plasma reaction. (*b*) Same preparation, same field as (*a*), after brief treatment in oxygen plasma. The polystyrene has been partially removed by the plasma but the silicon oxide replica film retains perfectly the original three-dimensional shape of the latex spheres. Continued treatment in the oxygen plasma removes all of the polystyrene (× 11,000). (Reproduced with permission from Aldrian et al., 1967.)

Several unique microscopic applications of plasma-deposited carbonaceous polymer have been reported by Mollenstedt and Speidel (1961; see Table 8.5). These include use of polymer deposition to record ion microscope images and to provide electroplating masks for the production of TEM specimen grids and apertures. And Baird, Dawson, and Feates, 1965, have described a rather unique TEM experiment in plasma deposition of polymer. In connection with studies of carbon deposition inside nuclear reactors, they examined carbon polymer particles formed by plasma reactions which were conducted entirely within the TEM.

Cleaning and/or Improving the Wettability of Surfaces and Thin Films

Treatment in glow discharges has long been used to clean glass and other surfaces (e.g., Holland 1958) and also to make hydrophobic surfaces of polymers and other materials wettable by aqueous reagents (e.g., Hansen, et al., 1965; Mantell, 1963). Such treatment has also been used to good effect on TEM specimen support films, particularly carbon films which are notoriously hydrophobic, to improve the spreading of small particles and reagents applied to the films as aqueous suspensions or solutions. Some examples are listed in Table 8.6. The plasma treatment has been especially valuable for thin-film cleaning because the delicacy of the films to mechanical disruption and the possibility of recontamination has often argued against other treatments such as by aqueous cleaning agents. Also the ability of the plasma techniques to produce a very short duration, controlled treatment has been an advantage, since overtreatment with a cleaning agent might destroy the film.

The workers at the Technische Hochschule, Graz (see Table 8.6) have described or mentioned a number of other applications of plasma-cleaning and hydrophilization, including improved, wrinkle-free mounting of ultrathin sections retrieved onto plasma-treated TEM grids from water surfaces, hydrophilization of the surfaces of bulk specimens to permit use of aqueous or polar replication or embedding media, and also the removal of adhering organic material from thin film surface replicas. In the last mentioned application, a prolonged cleaning treatment is typically required, and so this procedure has been restricted to replica films which are highly resistant to the plasma treatment, such as silica films. An example of a partially cleaned silica replica is shown in Figure 8.21. The plasma-cleaning treatment has been especially valuable for highly convoluted or three-dimensional replicas which would be easily collapsed or disrupted by the surface tension of liquid reagents and also for specimen materials such as polymers which swelled excessively in the course of their solvent dissolution.

TABLE 8.6. Cleaning and/or Improving the Wettability of Surfaces and Thin Films

Author(s) by affiliation	Type application,[1] specimen and/or technique	Type plasma[2]	Special comments[3]
Technische Hochschule, Graz, Austria			
Grasenick, 1957	(a) SiO_2 and SiC replicas of graphite particles	a	b
Blaha, et al., 1960	(a) SiO replicas of graphite and polystyrene latex particles	a	b
Grasenick, 1960	(a) SiO replicas of rubber surface	a	a
Jakopic, 1960	(a) SiO replicas of rubber latex particles	a	a
Horn and Jakopic, 1961	(b), (b') Unspecified	a	b
Aldrian, et al., 1965	(b) Aragonite crystals dispersed on collodion or formvar film; (b'), (d) unspecified	a	a
Geymayer, 1965	(a) SiO replica of freeze-dried plastic emulsion particles; (c) thylose replica of Teflon surface	a	b
Aldrian, et al., 1967	(a) SiO replica of polystyrene latex particles; (b) kaolinite and aragonite crystals dispersed on collodion or formvar film; (b'), (d) unspecified	a	a
Baylor University, Houston, Texas			
Mayor, 1963	(b) Negative-stain preparation of virus particles on C-film	b	b
California Institute of Technology, Pasadena			
Griffith, 1969	(b) Adsorption of DNA molecules to C-film	c	a
Johns Hopkins University, Baltimore, Md.			
Reissig and Orrell, 1970	(b) Negative-stain preparation of glycogen molecules on C-film	d	a
National Institute of Medical Research, London, England			
Nermut, 1972	(b) Negative stain preparations of virus particles on C-film	b	b
University of Pittsburgh, Pa.			
Rizk and Bendet, 1972	(b) Preparations of proteins or nucleic acids	e	b
University of California, Berkeley			
Richards, et al., 1973	(b) Negative stain preparation of bacteriophage particles on C-film	d	b

[1] Letters designate the following types of application: (a) removal of organic or carbonaceous specimen material from TEM surface replica; (b) improved wettability and/or adsorptive surface characteristics of TEM specimen support film to promote better spreading and dispersion of aqueous particle suspensions, etc.; (b') same as (b) but for better, wrinkle-free mounting of ultrathin sections; (c) improved wettability of solid specimen surface to permit use of aqueous or polar replication medium; (d) improved wettability of solid specimen surface to permit use of aqueous- or polar-embedding media.

[2] Letters designate the following types of plasma: (a) rf electrodeless discharge using o_2; (b) "glow discharge," not otherwise specified; (c) Tessler coil applied to electrical feed-through terminal of vacuum evaporator maintained under poor vacuum (air, 50 μ); (d) 60 cycle, 5000–10,000 V glow discharge in reduced-pressure air using conventional apparatus in vacuum evaporator; (e) "ac glow discharge in vapors of amylamine."

[3] Letters designate the following special comments: (a) provides detailed description of plasma device and/or treatment procedure; (b) few or no technical details.

8.5.   CONCLUSIONS

In gathering together the many diverse applications of reactive plasmas in widely different areas of microscopy, the isolation of activity in any one area from work in the others has frequently been quite evident. Workers using plasmas in biological ashing applications, for example, have been unaware of the considerable related literature on etching of organic polymers (and this included the present author until stimulated to write this review) and vice versa. Workers doing plasma-etching of polymers have apparently been unacquainted with possibly related work on etching of carbon, etc. This provincialism seems to been partly engendered by language barriers. Unless the present sampling of the literature has been highly biased by omissions, it shows that work on etching of carbon has been largely in English, etching of polymers has been predominantly in German and Russian, and nearly all ashing work has been in English. There has been relatively little citation between languages. One purpose of this review will have been served if it stimulates creation of new applications or understanding by the transfer of experience from one disciplinary area or school of activity to another. Some present or potential examples of this have been mentioned above—e.g., results on etching of organic polymers used to help interpret biological ash patterns; artificial petrifaction of organic specimens as used to create three-dimensional structures in ashing experiments for SEM, also used to impart differential etch resistance in SEM ion-etching experiments; use of hydrogen, fluorine, and other gas plasmas which can etch inorganic materials, also used to selectively remove and hence identify inorganic constituents in biological ash patterns; ion implantation as used on semiconductor devices used also to implant heavy metal stains in thin-sectioned biological specimens; potential use of approaches developed in chemical kinetic studies of the plasma oxidation of graphite, to better understand the etching attack of various plasmas on metals and other inorganic materials. Hopefully, there will be many more examples of this in the future.

Many of the applications here reviewed have become current routine techniques well validated by the accumulated experience of those who have used them; these include among others, ashing for the recovery of microscopic particles; etching of filled polymers for demonstration of filler particle distribution; etching of carbon and graphite for demonstration of either structure or chemical reactivity; differentiation or replication of metal surfaces by epitaxial oxide formation; plasma-cleaning and hydrophilization of TEM specimen support films. As these current methods and their advantages over other techniques become more widely known, they will undoubtedly receive greater use, in spite of the special plasma apparatus their performance requires. Other methods—e.g., production of TEM surface replicas by plasma deposition of thin films—which were used in the past, have been replaced by alternative techniques requiring less special equipment. These also may return to use as plasma devices become more common in microscopy laboratories.

More exciting than routine methods are plasma techniques for microscopy still in the exploratory and development stage. Reactive etching of inorganic materials, etching of organic polymers for demonstration of crystalline or multiphase fine structure, artificial petrifaction for development of three-dimensional ash structures, and microincineration of thin-sectioned biological materials for high-resolution demonstration of mineral distributions, are among examples which should probably be placed in this category. A second purpose of the present review will have been served if, in calling attention to these promising new techniques, their further exploration is spurred.

REFERENCES

ADAMSON, I. Y. R., I. M. DAWSON, F. S. FEATES, and R. S. SACH (1966), *Carbon* 3, 393–396.

ALDRAIN, A., E. JAKOPIC, O. REITER, and R. ZIEGELBECKER, (1965), *in* "Eine Apparatur zur Anwendung von Hochfrequenz und Gleichspannungsgasentladungen. Vorabdruk einen ausführlichen Arbeit in der *Radex-Rundschau*," Radenthein, Kärnten, Austria, pp. 2–6.

ALDRIAN, A., E. JAKOPIC, O. REITER, and R. ZIEGELBECKER (1967), *Radex-Rundschau* **1967** (2), 510–521.

ALVÄGER, T., and J. SWEZ (1970), *J. Appl. Phys.* **41**, 5030–5031.

AMBROSE, E. J., U. BATZDORF, J. S. OSBORN, and P. R. STUART (1970), *Nature* **227**, 397–398.

AMELINCKX, S., P. DELAVIGNETTE, and M. HEERSCHAP (1965). *in* P. L. WALKER, JR. (ed.), *Chemistry and Physics of Carbon*, Vol. 1, Marcel Dekker, New York, pp. 1–71.

ANDERSON, F. R. (1963), *J. Appl. Phys.* **34**, 2371–2373.

ANDERSON, F. R., and V. F. HOLLAND (1960), *J. Appl. Phys.* **31**, 1516–1518.

ANDERSON, G. S., W. N. MAYER, and G. K. WEHNER (1962), *J. Appl. Phys.* **33**, 2991–2992.

APPLETON, T. C. (1972), *Micron* **3**, 101–105.

BAIRD, T., I. M. DAWSON, and F. S. FEATES (1965), *in* "Proceedings 2nd Conference on Industrial Carbon and Graphite, London, 1965," Soc. Chem. Ind., London, 1966, pp. 225–231.

BAKER, R. F. (1969), *in* G. A. JAMIESON and T. J. GREENWALT (eds.), *Red Cell Membrane Structure and Function*, Lippincott, Philadelphia, pp. 13–35.

BASIN, V., T. N. GANINA, A. G. GROZDOV, and L. M. KORSUNSKII (1971), *Polymer Sci. USSR* (trans *Vysokomol. Soedin. A*) **13**, 1931–1936.

BAUER, E., B. FRITZ, and E. KINDER (1958), *in* "Proceedings 4th International Congress for Electron Microscopy, Berlin, 1958," Springer, Berlin, 1960, pp. 430–433.

BEHRISCH, R. (1964), *Ergeb. Exakt. Naturwiss.* **35**, 295–443.

BERGH, A. A. (1965), *Bell System Tech. J.* **44**, 261–271.

BERKLEY, C., J. CHURG, I. J. SELIKOFF, and W. E. SMITH (1965), *Ann. N.Y. Acad. Sci.* **132**, 48–63.

BERKLEY, C., A. M. LANGER, and V. BADEN (1967), *Trans. N.Y. Acad. Sci., II* **30**, 331–350.

BERSIN, R. L., J. R. HOLLAHAN, and W. D. HOLLAND (1966), *in* "Proceedings Symposium on Trace Characterization—Chemical and Physical," Nat. Bur. Stds., Washington, D.C., pp. 514–518.

BEZRUK, L. I. (1968), *Fiz.-Khim. Mekh. Mater.* **4**, 105–106.†

BEZRUK, L. I., G. A. GOROKHOVSKII, and YU. S. LIPATOV (1968), *Polymer Sci. USSR* (trans. *Vysokomol. Soedin. A*) **10**, 1665–1668.

BEZRUK, L. I., YU. S. LIPATOV, V. I. GRABOSHNIKOVA, and A. P. OCHKIVSKII (1968a), *Vysokomol. Soedin. B* **10**, 237–238.†

BEZRUK, L. I., YU. S. LIPATOV, V. K. IVASHCHENKO, T. E. LIPATOVA, and YU. V. PASECHNIK (1970), *Vysokomol. Soedin. B* **12**, 35–37.†

BEZRUK, L. I., A. P. OCHKIVSKII, YU. V. PASECHNIK, and YU. S. LIPATOV (1969), *Vysokomol. Soedin. B*, **11**, 180–181.†

BICKNELL, R. W., and R. M. ALLEN (1970), *in* F. H. EISEN and L. T. CHADDERTON (eds.), "Ion Implantation, A Conference, Thousand Oaks, Ca. 1970," Gordon and Breach, New York, 1972, pp. 63–67.

BLACKWOOD, J. D. and F. K. MCTAGGART (1959), *Australian J. Chem.* **12**, 114–121.

BLAHA, J., F. GRASENICK, H. HORN, and E. JAKOPIC (1960), *Radex-Rundschau* **1960** (6), 429–446.

BLISH, R. C., II, S. M. IRVING, and K. E. LEMONS (1970). "Low Temperature Plasma Etching of Silicon Substrates and Dielectric Films," technical bulletin, 13 pp., Signetics Corp., Sunnyvale, Ca.

BOCHENEK, B., M. FIETT, and E. MIZERA (1964), Institute of Nuclear Research (Warsaw, Poland) Report No. 555/XIV, 29 pp, Nuclear Energy Information Center, Warsaw.

BOOTHROYD, B. (1968), *J. Roy. Microsc. Soc.* **88**, 529–544.

BOYDE, A. (1971), *in* O. JOHARI and I. CORVIN, (eds.), "Proceedings 4th Annual Scanning Electron Microscope Symposium," Part I, IIT Research Inst., Chicago, Ill, pp. 1–8.

† An English translation is available from The National Translations Center, The John Crerar Library, Chicago, Ill.

BOYDE, A., and A. D. G. STEWART (1962), *in* S. S. BREESE, Jr. (ed.), "Proceedings 5th International Congress for Electron Microscopy," Academic Press, New York, p. QQ–9.

BOYDE, A., and C. WOOD (1969), *J. Microscopy* **90**, 221–249.

BRADLEY, D. E. (1956), *J. Appl. Phys.* **27**, 1399–1412.

BRADLEY, D. E. (1961), *in* D. KAY, (ed.), *Techniques for Electron Microscopy*, Thomas, Springfield, Ill., pp. 63–81.

BUCKNALL, C. B., I. C. DRINKWATER, and W. E. KEAST (1972), *Polymer* **13**, 115–118.

CARTER, G., and J. S. COLLIGON (1968), *Ion Bombardment of Solids*, American Elsevier, New York, 442 pp.

CASPERSON, G., H. HÄNSEL, and G. HOFFMANN (1967), *Faserforsch. Textiltech.* **18**, 455–460.

CASPERSON, G., H. R. LEHMAN, and K. MEHNERT (1964), *Wiss. Z. Päd. Hochschule Potsdam* **8**, 45–50.

CASSELMAN, W. G. B. (1959), *Histochemical Technique*, Wiley, New York, pp. 148–161.

CHARD, W. C., R. D. REISWIG, L. S. LEVINSON, and T. D. BAKER, (1968), *Carbon* **6**, 950–951.

CHRISTENSEN, A. K. (1971), *J. Cell Biol.* **51**, 772–804.

CRABLE, J. C., R. G. KEENAN, F. R. WOLOWICZ, M. J. KNOTT, J. L. HOLTZ, and C. H. GORSKI (1967), *Am. Ind. Hyg. Ass. J.* **28**, 8–12.

CZARNECKI, R., and F. HILBERT (1962). *Kernenergie* **5**, 566–570.

CZARNECKI, R., and F. HILBERT (1962a), *Neue Hütte* **7**, 765.

DAVIDSE, P. D. (1967), *Vacuum* **17**, 139–145.

DAVIDSE, P. D., and L. I. MAISSEL (1965), *in* H. ADAM, (ed.), "Transactions of 3rd International Vacuum Congress, Stuttgart, 1965," Pergamon Press, Oxford, pp. 651–655.

DE MARIA, R., P. SPINEDI, and E. SUSI-DE MARIA (1968), *Corrosion* **24**, 237–242.

DEN BESTEN, I. E., and J. J. MANCUSO (1970), *Chem. Geol.* **6**, 245–253.

DIETL, J. J. (1969), *Kunststoffe* **59**, 792–798.

DLUGOSZ, J. (1962), *in* S. S. BREESE, (ed.), "Proceedings 5th International Congress for Electron Microscopy," Vol. 1, Academic Press, New York, p. BB–11.

DOBERENZ, A. R., and R. W. G. WYCKOFF (1967), *J. Ultrastruct. Res.* **18**, 166–175.

DRUM, R. W. (1968), *Nature* **218**, 784–785.

DRUM, R. W. (1968a), *Science* **161**, 175–176.

DZOANH, N. T. (1967), *Surface Sci.* **6**, 422–439.

ECHLIN, P., and P. W. J. HYDE (1972), *in* O. JOHARI and I. CORVIN (eds.), "Proceedings 5th Annual Scanning Electron Microscope Symposium," Part I, IIT Research Institute, Chicago, Ill., pp. 137–146.

EISEN, F. H., and L. T. CHADDERTON (eds.) (1970), "Ion Implantation, A Conference, Thousand Oaks, Ca., 1970," Gordon and Breach, New York, 1972, 468 pp.

ELISEEVA, V. I., N. G. ZHARKOVA, E. I. EVKO, and V. M. LUK'YANOVICH (1967), *Polymer Sci. USSR* (trans. *Vysokomol. Soedin. A*) **9**, 2803–2808.

ESAU, K. (1965), *Plant Anatomy*, 2nd ed., Wiley, New York, 767 pp.

ESTEP, P. A., J. J. KOVACH, and C. KARR, JR. (1968), *Anal. Chem.* **40**, 358–363.

ESTEP, P. A., J. J. KOVACH, C. KARR, JR., E. E. CHILDERS, and A. L. HISER (1969). Preprints, *Am. Chem. Soc., Div. Fuel Chem.* **13** (1), 18–34.

EVKO, E. I. (1968), *Russ. J. Phys. Chem.* (trans. *Zh. fiz. Khim., SSSR*) **42**, 1682–1683.

FABERGÉ, A. C. (1964), *J. Appl. Phys.* **35**, 3077.

FABERGÉ, A. C. (1965), *J. Appl. Phys.* **36**, 2615.

FABERGÉ, A. C. (1966), *J. Appl. Phys.* **37**, 3920.

FABERGÉ, A. C. (1967), *Genetics* **56**, 558–559.

FABERGÉ, A. C. (1968), *in* "Study in Genetics, IV. Research Reports," Texas Univ. Austin, pp. 21–47.

FEATES, F. S. (1968), *Trans. Faraday Soc.* **64**, 3093–3099.

FEATES, F. S. (1969), *Trans. Faraday Soc.* **65**, 211–218.

FEATES, F. S., and R. S. SACH (1965), *in* "Proceedings 2nd Conference on Industrial Carbon and Graphite, London, 1965," Soc. Chem. Ind., London, 1966, pp. 329–336.

FISCHER, E. W., and H. GODDAR (1965), *J. Polymer Sci, C* (16) (Pt. 8), 4405–4427 (published 1969).

FISCHER, E. W., H. GODDAR, and G. F. SCHMIDT (1968), *Kolloid-Z. Z. Polym.* **226**, 30–40.

FLINCHBAUGH, D. A. (1969), *Anal. Chem.* **41**, 2017–2023.

FLINCHBAUGH, D. A. (1971), *Anal. Chem.* **43**, 178–182.

FRANK, W., H. GODDAR, and H. A. STUART (1967), *J. Polymer Sci., B* **5**, 711–713.

FRAZIER, P. D. (1971), "An Electron Microscopic Investigation of Mineralizing Tissues," Ph.D. Thesis, Washington Univ., Seattle, 1971, University Microfilms, Ann Arbor, Mich. 294 pp.

FRAZIER, P. D., F. J. BROWN, L. S. ROSE, and B. O. FOWLER (1967), *J. Dent. Res.* **46**, 1098–1101.

FU, Y. C. (1971), *Chem. Ind.* (31), 876–877.

FU, Y. C., and B. D. BLAUSTEIN (1968), *Fuel* **47**, 463–474.

FU, Y. C., and B. D. BLAUSTEIN (1969), *Ind. Eng. Chem. Process Des. Dev.* **8**, 257–262.

FULKER, M. J., and L. HOLLAND (1970), *Micron* **2**, 117–123.

GEYMAYER, W. (1965), *in* "Eine Apparatur zur Anwendung von Hochfrequenz- und Gleichspannungsgasentladungen. Vorabdruck einer ausführlichen Arbeit in der *Radex-Rundschau*," Radenthein, Kärnten, Austria, pp. 7–9.

GEYMAYER, W. (1966), *in* "Proceedings, 6th International Congress for Electron Microscopy, Kyoto, 1966," Maruzen Co., Tokyo, pp. 273–274.

GLEIT, C. E. (1963), *Am. J. Med. Elec.* **2**, 112–118.

GLEIT, C. E. (1964), *in* "Proceedings of Third Annual Technical Conference of the American Association for Contamination Control," Los Angeles, 1964, paper IX-2, 8 pp.

GLEIT, C. E. (1966), *Microchem. J.* **10**, 7–26.

GLEIT, C. E., and W. D. HOLLAND (1962), *Anal. Chem.* **34**, 1454–1457.

GLUSKOTER, H. J. (1965), *Fuel* **44**, 285–291.

GLUSKOTER, H. J. (1967), *J. Sediment. Petrol.* **37**, 205–214.

GLUSKOTER, H. J., and P. C. LINDAHL (1973), *Science* **181**, 264–265.

GOLOB, P., E. KRAUTZ, and U. KÜMMEL (1968), *Z. angew. Phys.* **25**, 222–227.

GOLYANOV, V. M. (1964), *in* "Issled. Ob'ektov, Izmenyayushchikhsya Protsesse Prep. Nablyudeniya Elektron, Mikrosk., Mater. Simp., Moscow, 1964," pp. 81–86, (publ. 1966).†

GOLYANOV, V. M., A. P. DEMIDOV, M. N. MIKHEEVA, and A. A. TEPLOV (1970), *Soviet Physics JETP* **31**, 283–286, (trans. *Zh. Eksp. Teor. Fiz.* **58**, 528–534).

GOYNES, W. R., E. K. BOYLSTON, L. L. MULLER, and B. J. TRASK (1974). *Text. Res. J.*, **44**, 197–203.

GOYNES, W. R., L. L. MULLER, and E. K. BOYLSTON (1972), *in* C. J. ARCENEAUX, (ed.), "Proceedings 30th Annual Meeting Electron Microscopy Society of America" Claitor's Publishing Div., Baton Rouge, La., pp. 210–211.

GRASENICK, F. (1956), *Radex-Rundschau* **1956** (4/5), 226–246.

GRASENICK, F. (1957), *Radex-Rundschau* **1957** (5/6), 843–856.

GRASENICK, F. (1960), *in* "Proceedings Natural Rubber Research Conference, Kuala Lumpur, 1960," pp. 619–625.

GRASENICK, F. (1961), *Technische Rundschau (Bern)* **1961** (21), 3–15.

GRASENICK, F., and R. HAEFER (1952), *Monatshefte Chemie* **83**, 1069–1082.

GRASENICK, F., and E. JAKOPIC (1969), *Naturwiss.* **56**, 413–414.

GRASENICK, F., and O. REITER (1958), *in* "Proceedings 4th International Congress for Electron Microscopy, Berlin, 1958," Springer, Berlin, 1960, p. 427.

GREEN, W. V., L. S. LEVINSON, R. D. REISWIG, and E. G. ZUKAS (1967), *Carbon* **5**, 583–586.

GRIBI, M. M. (1962), *in* J.-J. TRILLAT, (ed.), "Le bombardement Ionique. Theories et Applications," Colloq. Intern. Centre Nat. Rech. Sci. (Paris) No. 113, 77–82.

GRIFFITH, J. (1969), "Procedure for Making Surface-Active Carbon Films and Mounting Nucleic Acids," 1 p. Distributed at Biophysical Society Workshop on Electron Microscopy of Nucleic Acids, California Institute of Technology, Pasadena, Feb. 26, 1969.

GRITSAENKO, G. S., and A. I. GORSHKOV (1961), *Zapiski Vsesoyuz. Mineral. Obshchestva* **90**, 266–269.

GUL, V. E., and E. M. ROGOVAYA (1966), *Dokl. Akad. Nauk SSSR* **170**, 366–368.†

GUMARGALIEVA, K. Z., E. M. BELAVTSEVA, M. R. KISELEV, E. I. EVKO, and V. M. LUK'YANOVICH (1966), *Polymer Sci. USSR* (trans. *Vysokomol. Soedin.*) **8**, 1923–1926.

HAEFER, R. (1954), *in* "Proceedings 3rd International Conference for Electron Microscopy, London, 1954," pp. 466–473.

HAEFER, R. (1954a), *Acta Phys. Austr.* **9**, 1–17.

HAEFER, R. (1958), *in* E. THOMAS, (ed.), "Proceedings of the 1st International Congress on Vacuum Techniques, Namur, Belgium, 1958" Pergamon Press, New York, 1960, pp. 760–763.

HALES, R. L., and E. M. WOODRUFF (1962), *in* "Proceedings 5th Conference on Carbon, University Park, Penn., 1962," Pergamon Press, New York, pp. 456–465.

HALL, T. A. (1971), *in* G. OSTER, (ed.), *Physical Techniques in Biological Research*, 2nd ed., Vol. 1, Part A, Academic Press, New York, pp. 157–275.

HANSEN, R. H. (1964), *in* "Proceedings of the Symposium Polypropylene Fibers, Birmingham, Ala., 1964," pp. 137–182.

HANSEN, R. H., W. M. MARTIN, and T. DE BENEDICTIS (1963), *Trans. Inst. Rubber Ind.* **39**, T301–T313.

HANSEN, R. H., J. V. PASCALE, T. DE BENEDICTIS, and P. M. RENTZEPIS (1965), *J. Polymer Sci.*, *A* **3**, 2205–2214.

HANSEN, R. H., and H. SCHONHORN (1966), *J. Polymer Sci.*, *B* **4**, 203–209.

HARTECK, P., and U. KOPSCH (1931), *Z. phys. Chem.* **B12**, 327–347.

HAYMANN, P., C. WALDBURGER, and J.-J. TRILLAT (1963), *in* Colloq. Intern. Centre Nat. Rech. Sci. (Paris) No. 122, 135–139 (published 1965).

HAYMANN, P., C. WALDBURGER, and J.-J. TRILLAT (1965), *Mem. Sci. Rev. Met.* **62**, 515–525.

HELWIG, G., and H. KÖNIG (1950), *Optik* **7**, 294–302.

HEMPEL, M., A. KOCHENDÖRFER, and E. HILLNHAGEN (1962), *Arch. Eisenhütten-wesen* **6**, 405–416.

HENNIG, G. R. (1965), *Carbon* **3**, 107–114.

HENNIG, G. R. (1965a), *in* "Proceedings 2nd Conference on Industrial Carbon and Graphite, London, 1965," Soc. Chem. Ind., London, 1966, pp. 109–113.

HENNIG, G. R. (1966), *in* P. L. WALKER, (ed.), *Chemistry and Physics of Carbon* Vol. 2, Marcel Dekker, New York, pp. 1–49.

HENNIG, G. R., G. J. DIENES, and W. KOSIBA (1958), *in* "Proceedings 2nd United Nations International Conference on Peaceful Uses of Atomic Energy, Geneva, 1958," Vol. 7, pp. 301–306.

HIEN, N. V. (1970), "Morphology and Mechanical Properties of Twisted Homo-polymer and Polyblend Fibers," Ph.D. Thesis, Wisconsin Univ., Madison, 1970, University Microfilms, Ann Arbor, Mich. 133 pp.

HIEN, N. V., S. L. COOPER, and J. A. KOUTSKY (1972), *J. Macromol. Sci. Phys.*, *B* 6, 343–363.

HILBERT, F. (1962), *Neue Hütte* 7, 368–374.

HILBERT, F. (1962*a*), *Neue Hütte* 7, 416–420.

HILBERT, F. (1965), *Z. Metallkde.* 56, 461–464.

HILBERT, F., and H. LORENZ (1963), *Jena Rev.* (English trans.) 8, 218–223.

HILBERT F., and H. LORENZ (1964), *Alluminio* 33, 405–411.

HINRICHSEN, G., and H. ORTH (1971), *J. Polymer Sci.*, *B* 9, 529–531.

HINTZSCHE, E. (1956), *Das Aschenbild Tierischer Gewebe und Organe. Methodik, Ergebnisse und Bibliographie*, Springer, Berlin, 140 pp.

HODGES, G. M., M. D. MUIR, C. SELLA, and A. J. P. CARTEAUD (1972), *J. Microsc.* 95, 445–451.

HOHMAN, W. R. (1967), "A Study of Low Temperature Ultramicroincineration of Avian Shell Gland Mucosa by Electron Microscopy," Ph.D. Thesis, Pennsylvania State Univ., 1967, University Microfilms, Ann Arbor, Mich. 159 pp.

HOHMAN, W. R., and H. SCHRAER (1972), *J. Cell. Biol.* 55, 328–354.

HOLLAHAN, J. R. (1966), *J. Chem. Ed.* 43, A401–A416.

HOLLAND, L. (1958), *Brit. J. Appl. Phys.* 9, 410–415.

HOLLAND, W. D. (1970), "Use of Gaseous Plasmas for Small Particle, Fiber and Trace Material Analysis," Summary of presentation for course by Program Design, Inc., Trapelo/West, a Division of LFE Corporation, Richmond, Ca., 33 pp.

HOLLAND, W. D., and R. A. ABELMANN (1967), Presentation at Pacific Conference on Chemistry and Spectroscopy, October 31, 1967, Anaheim, Ca.

HOLT, S. C , and E. R. LEADBETTER (1969), *Bacteriol. Revs.* 33, 346–378.

HORN, H., and JAKOPIC, E. (1961), Presented at 10th Jahrestagung der Deut. Ges. f. Elektronmikroskopie, Kiel, 1961, *Mikroskopie* 17, 30 (1962).

HORN, H., and P. WARBICHLER (1965), *Naturwiss.* 52, 616–617.

HOZUMI, K. (1971), *Kagaku-no-Ryoiki* 25, 713–723.

HOZUMI, K., M. HUTOH., and K. UMEMOTO (1972), *Microchem. J.* 17, 173–185.

HOZUMI, K., and M. MATSUMOTO (1972), *Bunseki Kagaku* 21, 206–214.

HUGHES, G., J. M. THOMAS, H. MARSH, and R. REED (1964), *Carbon* 1, 339–343.

IRVING, S. M. (1970), The International Plasma Bulletin, Vol. 2, No. 4, 4 pp. Internat. Plasma Corp., Hayward, Ca.

IRVING, S. M. (1971), *Solid State Technol.* 14, 47–51.

ISINGS, J., and B. J. SPIT (1964), *in* "Proceedings 3rd European Regional Conference on Electron Microscopy," Vol. A, Publishing House of the Czechoslovak Academy of Science, Prague, pp. 407–408.

JAKOPIC, E. (1960), *in* A. L. HOUWINK and B. J. SPIT, (eds.), "Proceedings European Regional Conference on Electron Microscopy, Delft, 1960," Vol. 1, Nederlandse Vereniging voor Electronenmicroscopie, Delft, 1961, pp. 559–563.

Jones, S. S. (1970), *Carbon* **8**, 673–679.

Jones, S. S. (1970*a*), *Carbon* **8**, 681–683.

Jones, S. S. (1970*b*), *Carbon* **8**, 685–688.

Jones, S. S., and E. M. Woodruff (1971), *Carbon* **9**, 259–264.

Kaminsky, M. (1965), *Atomic and Ionic Impact Phenomena on Metal Surfaces*, Academic Press, New York, 432 pp.

Kämpf, G. (1970), *Farbe Lack* **76**, 25–34.

Kämpf, G. (1970*a*), *Optik* **31**, 113–115.

Karr, C., Jr., P. A. Estep, and J. J. Kovach (1968), Preprints, *Am. Chem. Soc., Div. Fuel Chem.* **12**(4), 1–12.

Kassenbeck, P. (1958), *Melliand Textilber.* **39**, 55–61.

Keenan, R. G. (1968), *Ind. Hyg. Highlights* **1**, 57–83.

Keenan, R. G., and R. E. Kupel (1968), Presentation at 9th Conference on Methods in Air Pollution and Industrial Hygiene Studies. California State Dept. of Public Health at Pasadena.

Keenan, R. G., and J. R. Lynch (1970), *Am. Ind. Hyg. Ass. J.* **31**, 587–597.

Keller, A. (1962), *in* S. S. Breese, Jr., (ed.), "Proceedings 5th International Congress for Electron Microscopy," Vol. 1, Academic Press, New York, p. BB–3.

Kemezys, M., and G. H. Taylor (1964), *J. Inst. Fuel* **37**, 389–397.

Kindig, O. (1969), Reported by M. E. Ebeling, J. A. Grach, and D. T. Jeter, Jr., "Applications of a Low Temperature RF Reactor," Technical Bulletin No. T-211, Coleman Instruments Div. Perkin Elmer Corp., Maywood, Ill.

Kinney, C. R., and L. D. Friedman (1952), *J. Am. Chem. Soc.* **74**, 57–61.

Kiselev, M. R., E. I. Evko, and V. M. Luk'yanovich (1966), *Zavodsk. Lab.* **32**, 201–202.†

Klug, A., and J. E. Berger (1964), *J. Mol. Biol.* **10**, 565–569.

Koehler, J. K. (1971), *in* O. Johari, (ed.), "Proceedings 4th Annual Scanning Electron Microscope Symposium," Part I, IIT Research Institute, Chicago, Ill., pp. 243–248.

Kondo, M., and J. W. Foster (1967), *J. Gen. Microbiol.* **47**, 257–271.

König, H. (1953), *Ergebnisse exakt. Naturwiss.* **27**, 188–247.

König, H., and G. Helwig (1951), *Z. Physik* **129**, 491–503.

Kozlovskii, M. I. (1965), *Soviet Physics- Crystallog.* (trans. *Kristallografiya*) **10**, 101–103.

Kozlovskii, M. I. (1966), *Applied Electrical Phenomena* (trans. *Elektran. Obrab. Mater., Akad. Nauk Mold. SSR*) **1966**(3), 203–207.

Kozlovskii, M. I. (1967), *Ind. Lab.* (trans. *Zavodsk. Laboratoriya*) **33**, 235–237.

Kruszynski, J. (1963), *Acta Histochem.* **15**, 58–77.

Kruszynski, J. (1966), *in* W. Graumann and K. Newmann, (eds.), *Handbuch der Histochemie*, Vol. 1, Part 2, Gustav Fischer, Stuttgart, pp. 96–187.

Kubota, K. (1965), *J. Polymer Sci., B* **3**, 545–547.

KURBANOVA, I. I., and A. P. KAFENGAUZ (1969), *Polymer Sci. USSR* (trans. *Vysokomol. Soedin.*, *A*) **11**, 1709–1714.

KUROKAWA, M., T. BAN, and N. MOTOJI (1964), *J. Electron Microsc.* (*Japan*) **13**, 51–52.

KUROKAWA, M., N. MOTOJI, and T. BAN (1964*a*), *J. Electron Microsc.* (*Japan*) **13**, 195–199.†

KUWANO, Y. (1969), *Jap. J. Appl. Phys.* **8**, 876–882.

LAKSHMANAN, T. K., C. A. WYSOCKI, and W. J. SLEGESKY (1964), *IEEE Trans. Component Parts* **CP-11**, 14–18.

LANG, F. M., P. GILLES, S. KERN, and P. MAIRE (1968), *J. Chim. Phys.* **65**, 580–582.

LANGER, A. M. (1970), *in* F. W. SUNDERMAN and F. W. SUNDERMAN, JR., (eds.), *Laboratory Diagnosis of Disease Caused by Toxic Agents*, Green, St. Louis, Mo, pp. 126–136.

LANGER, A. M., A. D. MACKLER, I. RUBIN, E. C. HAMMOND, and I. J. SELIKOFF (1971*a*), *Science* **174**, 585–587.

LANGER, A. M., I. RUBIN, and I. J. SELIKOFF (1970), *in* H. A. SHAPIRO, (ed.), "Pneumoconiosis. Proceedings of International Conference, Johannesburgh, 1969," Oxford Univ. Press, Cape Town, S. Africa, pp. 57–69.

LANGER, A. M., I. RUBIN, and I. J. SELIKOFF (1972), *J. Histochem. Cytochem.* **20**, 723–734.

LANGER, A. M., I. J. SELIKOFF, and A. SASTRE (1971), *Arch. Environ. Health* **22**, 348–361.

LÄUCHLI, A. (1972), *in Microautoradiography and Electron Probe Analysis. Their Application to Plant Physiology*, Springer, New York, pp. 191–236.

LEVINSON, L. S., R. D. REISWIG, and T. D. BAKER (1970), *Carbon* **8**, 100.

LEWIS, E. R. (1971), *in* O. JOHARI (ed.), "Proceedings 4th Annual Scanning Electron Microscope Symposium," Part I, IIT Research Institute, Chicago, Ill., pp. 281–288.

LEWIS, S. M., J. S. OSBORN, and P. R. STUART (1968), *Nature* **220**, 614–616.

LEWIS, S. M., J. S. OSBORN, P. R. STUART, and J. WILLIAMSON (1969), *Trans. Roy. Soc. Trop. Med. Hyg.* **63**, 418.

LEWIS, S. M., and P. R. STUART (1970), *Proc. Roy. Soc. Med.* **63**, 465–468.

LIPATOV, YU. S., L. I. BEZRUK, YU. V. PASECHNIK, V. K. IVASHCHENKO, and T. E. LIPATOVA (1969), *Dopov. Akad. Nauk Ukr. RSR*, *B* **31**, 525–528.†

LUK'YANOVICH, V. M. (1965), *in* V. M. LUK'YANOVICH (ed.), *Techniques for Electron Microscopy*, (Revised Russian edition of book in English edited by D. KAY.) MIR Publishing House, Moscow, p. 106.

MAGILL, H. J. (1971), *J. Polymer Sci.*, *A-2* **9**, 815–827.

MAGILL, H. J., and P. H. HARRIS (1962), *Polymer* **3**, 252–256.

MAHL, H. (1942), *Naturwiss.* **30**, 207–217.

MAHL, H. (1945), *Ergebn. exakt. Naturwiss.* **21**, 262–312.

MANTELL, R. M. (1963), U.S. Patent No. 3,309,299, application Aug. 22, 1963, awarded March 14, 1967.

MARSH, H., T. E. O'HAIR, R. REED, and W. F. K. WYNNE-JONES (1963), *Nature* **198**, 1195–1196.

MARSH H., T. E. O'HAIR, R. REED, and W. F. K. WYNNE-JONES (1964). *Carbon* **1**, 356.

MARSH, H., T. E. O'HAIR, and R. REED (1965). *Trans. Faraday Soc.* **61**, 285–293.

MARSH, H., and T. E. O'HAIR (1969). *Carbon* **7**, 702–703

MARSH, H., T. E. O'HAIR, and W. F. K. WYNNE-JONES (1969), *Carbon* **7**, 555–566.

MARTIN, J. H., and J. L. MATTHEWS (1970), *Clin. Orthopaedics Relat. Res.* (68), 273–278.

MATTHEWS, J. L., J. H. MARTIN, H. W. SAMPSON, A. S. KUNIN, and J. H. ROAN, (1970), *Calc. Tissue Res.* **5**, 91–99.

MATTHEWS, M. D. (1969), *J. Mater. Sci.* **4**, 997–1002.

MATTOX, D. M. (1964), *Electrochem. Tech.* **2**, 295–298.

MAYOR, H. D. (1963), *J. Appl. Phys.* **34**, 2522. (Glow-discharge technique described in presentation at meeting is not included in abstract.)

McCARROLL, B., and D. W. McKEE (1970), *Nature* **225**, 722–723.

McCARROLL, B., and D. W. McKEE (1971), *Carbon* **9**, 301–311.

McCRONE, W. C., and J. G. DELLY (1972), *The Particle Atlas*, 2nd ed., Vol. 1, Ann Arbor Science, Ann Arbor, Mich., pp. 221–222.

MEYER, K., and U. BERGER (1964), *Z. phys. Chem.* **225**, 145–160.

MEYER, M., and P. HAYMANN (1964), *Compt. Rend. Acad. Sci. Paris* **258**, 4690–4693.

MEYER, M., P. HAYMANN, and J.-J. TRILLAT (1965), *Compt. Rend. Acad. Sci. Paris* **261**, 4353–4358.

MEYER, M., P. HAYMANN, and J.-.J TRILLAT (1968), *Growth of Crystals* (trans. *Rost Kristallov, Akad. Nauk SSSR, Inst. Kristallogr.*) **1968** (8), 120–124.

MEYER, M., C. MARELLE, and P. HAYMANN (1969), *Compt. Rend. Acad. Sci. Paris*, B **268**, 1145–1148.

MIKHAILOVA, S. S., S. N. TOLSTAYA, V. M. LUK'YANOVICH, and E. I. EVKO (1968), *Vysokomol. Soedin.*, B **10**, 524–527.†

MIKHELEVA, G. A., and A. V. VLASOV (1970), *Vysokomol. Soedin.*, B **12**, 363–366.†

MOLLENSTEDT, G., and R. SPEIDEL (1961), *Z. angew. Phys.* **13**, 231–232.

MOLLENSTEDT, G., H. POLLAK, and H. SEILER (1965), *Z. Phys.* **182**, 445–450.

MONTET, G., F. FEATES, and G. MYERS (1967), "Electron-Microscopic Studies of Fundamental Physical and Chemical Processes in Single Crystals of Graphite," U.S. Atomic Energy Commission Research and Development Report ANL-7352, 49 pp.

MOORE, W. J., S. R. LOGAN, L. C. LUTHER, and S. N. BROWN (1962), in J.-J. TRILLAT (ed.), "Le Bombardement Ionique. Théories et Applications," Colloq. Intern. Centre Natl. Rech. Sci. (Paris) No. 113, 35–48.

MOSCOU, L. (1962), in S. S. BREESE, JR. (ed.), "Proceedings 5th International Congress for Electron Microscopy," Vol. 1, Academic Press, New York, p. BB–5.

MUELLER, P. K., G. R. SMITH, L. M. CARPENTER, and R. L. STANLEY (1972), in C. J. ARCENEAUX (ed.), "Proceedings 30th Annual Meeting, Electron Microscopy Society of America," Claitor's Publishing Div, Baton Rouge, La., pp. 356–357.

NATHANS, M. W., R. THEWS, and I. J. RUSSEL (1970), Advan. Chem. Ser. (93), 360–380.

NEFF, J. M. (1972), Tissue and Cell 4, 311–326.

NENADIC, C. M., and J. V. CRABLE (1970), Am. Ind. Hyg. Ass. J. 31, 81–86.

NERMUT, M. V. (1972), J. Microscopy 96, 351–362.

NIKONOVICH, G. V., N. D. BURKHANOVA, S. A. LEONT'EVA, and KH. U. USMANOV (1967), Cellul. Chem. Tech. 1, 171–177.†

NIKONOVICH, G. V., N. D. BURKHANOVA, S. A. LEONT'EVA, and KH. U. USMANOV (1968a), Khim. Volokna 1968 (4), 64–66.†

NIKONOVICH, G. V., N. D. BURKHANOVA, S. A. LEONT'EVA, and KH. U. USMANOV (1968b), Cell. Chem. Tech. 2, 231–257.

NIKONOVICH, G. V., S. A. LEONT'EVA, and KH. U. USMANOV (1968), Polymer Sci. USSR (trans. Vysokomol. Soedin., A) 10, 3112–3123.

NOWELL, J. A., J. PANGBORN, and W. S. TYLER (1970), in O. JOHARI (ed.), 'Proceedings 3rd Annual Scanning Electron Microscope Symposium," IIT Research Institute, Chicago, Ill., pp. 249–256.

NOWELL, J. A., J. PANGBORN, and W. S. TYLER (1972), in C. J. ARCENEAUX (ed.), "Proceedings 30th Annual Meeting Electron Microscopy Society of America," Claitor's Publishing Div., Baton Rouge, La., pp. 308–309.

OCHKIVSKII, A. P., and L. I. BEZRUK (1967), in "Tezisy Doklada na XVII Ukrainskoy Nauchnotekhnicheskoy Konferentsii, Posvyashchennoy Dnyu Radio," Izdatel'stvo Znaniye, Kiev, 1967, p. 215. (Described in OCHKIVSKII, et al., 1970.)

OCHKIVSKII, A. P., N. P. CHEMERIS, and M. P. NOSOV (1970), Khim. Volokna 1970 (4), 13–15.†

ONG, P., P. K. LUND, and W. M. CONRAD (1972), in "Proceedings of 7th National Conference on Electron Probe Analysis," Electron Probe Anal. Soc. Am. (G. CLEAVER, General Electric Co., P.O. Box 846, Pleasanton, Calif. 94556), pp. 46a–46c.

ORTH, H. (1970), Z. wiss. Mikr. mikr. Tech. 70, 179–188.

OSBORN, J. S., P. R. STUART, and S. M. LEWIS (1969), Sci. Res. 4(5), 14–15.

PALMER, R. P., and A. J. COBBOLD (1964), Makromolek. Chem. 74, 174–189.

PAPIRER, E., J.-B. DONNET, and A. SCHUTZ (1967), Carbon 5, 113–125.

PAVLOV, P. V., and E. V. SHITOVA (1967), *Sov. Phys. Dokl.* **12**, 11–13 (trans. *Dokl. Akad. Nauk SSSR* **172**, 588–590).

PAVLOVA, I. P., E. I. EVKO, M. R. KISELEV, O. A. SINYAEVA, A. L. ZAIDES, K. A. PECHKOVSKAYA, and V. M. LUK'YANOVICH (1968), *Vysokomol. Soedin.*, *B* **10**, 557–559.†

PECK, V., and W. L. CARTER (1968), *in* C. J. ARCENEAUX (ed.), "Proceedings 26th Annual Meeting Electron Microscopy Society of America" Claitor's Publishing Div., Baton Rouge, La., pp. 408–409.

PENNEBAKER, W. B. (1969), *IBM J. Res. Develop.* **13**, 686–695.

PERNY, G., and B. LAVILLE-SAINT-MARTIN (1964), *J. Phys. (Paris)* **25**, 993–998.

PFEFFERKORN, G. E., H. GRUTER, and M. PFAUTSCH (1972), *in* O. JOHARI and I. CORVIN (eds.), "Proceedings 5th Annual Scanning Electron Microscope Symposium" Part I, IIT Research Institute, Chicago, Ill., pp. 147–152.

PHILLIPS, V. A., and A. U. SEYBOLT (1968), *Trans. Met. Soc. AIME (Am. Inst. Mining, Met., Petrol. Eng.)* **242**, 2415–2422.

PRAVDYUK, N. F., and V. M. GOLYANOV (1961), *in* D. J. LITTLER (ed.), "Properties of Reactor Materials and the Effects of Radiation Damage. Proceedings of International Conference, Berkeley Castle, Gloucestershire, England, 1961," Butterworths, London, 1962, pp. 160–175.

RASPAIL, V. F. (1833), *Noveau Systèm de Chimie Organique Fondé sur des Méthods Nouvelles d'Observation*, Bailliere, Paris, p. 528.

REDING, F. P., and E. R. WALTER (1959), *J. Polymer Sci.* **38**, 141–155.

REIMANN, B. E. F., J. C. LEWIN, and B. E. VOLCANI (1965), *J. Cell Biol.* **24**, 39–55.

REISSIG, M., and S. A. ORRELL (1970), *J. Ultrastruct. Res.* **32**, 107–117.

REISWIG, R. D. (1967), "Cathodic Vacuum Etching with Hydrogen," U.S. Atomic Energy Commission Research and Development Report LA-DC-9603, 7 pp.

REISWIG, R. D. (1970), *Microstructures* **1**(2), 15–17.

REISWIG, R. D., L. S. LEVINSON, and T. D. BAKER (1967), *Carbon* **5**, 603–606.

REISWIG, R. D., L. S. LEVINSON, and J. A. O'ROURKE (1968), *Carbon* **6**, 124.

REISWIG, R. D., E. M. WEWERKA, L. S. LEVINSON, and J. A. O'ROURKE (1970), *Carbon* **8**, 241–242.

RICHARDS, K. E., R. C. WILLIAMS, and R. CALENDER (1973), *J. Mol. Biol.*, **78**, 255–259.

RIZK, N., and I. BENDET (1972), *in* C. J. ARCENEAUX (ed.), "Proceedings 30th Annual Meeting Electron Microscopy Society of America," Claitor's Publishing Div., Baton Rouge, La., pp. 692–693.

RODE, L. J., and M. G. WILLIAM (1966), *J. Bact.* **92**, 1772–1778.

ROSNER, D. E., and H. D. ALLENDORF (1969), *in* G. R. BELTON and W. L. WORRELL (eds.), "Heterogeneous Kinetics at Elevated Temperatures. Proceedings of International Conference in Metallurgy and Materials Science, University of Pennsylvania, Philadelphia, 1969," Plenum, New York, 1970, pp. 231–251.

Ross, C. W., and E. B. Curdts (1956), *Trans. AIEE (Power Apparatus Systems)* **75**, 63–67.

Sampson, H. W., J. L. Matthews, J. H. Martin, and A. S. Kunin (1970), *Calc. Tissue Res.* **5**, 305–316.

Scott, R. G., and A. W. Ferguson (1956), *Text. Res. J.* **26**, 284–296.

Seybolt, A. U. (1969), *Trans. Met. Soc. AIME (Am. Inst. Mining, Met., Petrol. Eng.)* **245**, 769–778.

Shen, A. C. (1972), *Am. J. Pathol.* **67**, 417–440.

Siegel, B. (1961), *J. Chem. Ed.* **38**, 496–501.

Silveira, M., and P. R. Arruda (1969), *Ciência Cultura (Brazil)* **21**, 484–485.

Smerko, R. G., and D. A. Flinchbaugh (1968). *J. Metals* **20**, 43–51.

Smolinsky, G., and J. H. Heiss (1968), Preprints, *Am. Chem. Soc., Div. Org. Coating Plast. Chem.* **28**(1), 537–544.

Speidel, R. (1959), *Z. Phys.* **154**, 238–263.

Spit, B. J. (1959), *J. Text. Inst., Trans.* **50**, T553–T557.

Spit, B. J. (1960), *in* A. L. Houwink and B. J. Spit (eds.), "Proceedings European Regional Conference on Electron Microscopy, Delft, 1960," Vol. 1, Nederlandse Vereniging voor Electronenmicroscopie, Delft, 1961, pp. 564–567.

Spit, B. J. (1963), *Polymer* **4**, 109–117.

Spit, B. J. (1967), *Faserforsch. Textiltech.* **18**, 161–169.

Spit, B. J., and S. M. Jutte (1965), *Acta Botan. Neerlandica* **14**, 403–408.

Stafford, B. B. (1969), *in* D. D. Hemphill (ed.), "Trace Substances in Environmental Health, II," Missouri Univ., Columbia, pp. 269–277.

Streznewski, T., and J. Turkevich (1969), *in* S. Mrozowski (ed.), "Proceeding of 3rd Conference on Carbon," Pergamon Press, New York, pp. 273–278.

Stuart, P. R., J. S. Osborn, and S. M. Lewis (1969), *in* O. Johari (ed.), "Proceedings 2nd Annual Scanning Electron Microscope Symposium," IIT Research Institute, Chicago, Ill., pp. 243–248.

Stuart, P. R., J. S. Osborn, and S. M. Lewis, (1969a), *Vacuum* **19**, 503–506.

Sukhareva, L. A., M. R. Kiselev, and P. I. Zubov (1967), *Colloid J. USSR* (trans. *Kolloid. Zh.*) **29**, 206–208.

Thomas, J. M. (1965), *in* P. L. Walker (ed.), *Chemistry and Physics of Carbon*, Vol. 1, Marcel Dekker, New York, pp. 121–202.

Thomas, R. S. (1961), *Virology* **14**, 240–252.

Thomas, R. S. (1962), *in* S. S. Breese, Jr. (ed.), "Proceedings 5th International Congress for Electron Microscopy," Vol. 2, Academic Press, New York, p. RR–11. (Preliminary results of plasma microincineration presented at meeting were not included in abstract.)

Thomas, R. S. (1964), *J. Cell Biol.* **23**, 113–133.

Thomas, R. S. (1965), *J. Cell Biol.* **27**, 106A.

Thomas, R. S. (1967), *J. Ultrastruct. Res.* **21**, 159.

Thomas, R. S. (1969), *Adv. Opt. Electr. Microsc.* **3**, 99–154.

Thomas, R. S. (1969a), *in* "Proceedings 22nd Annual Reciprocal Meat Conference of the American Meat Science Association," National Live Stock and Meat Board, Chicago, Ill., pp. 249–272.

THOMAS, R. S., and M. I. CORLETT (1969), *in* "Abstracts 3rd International Biophysics Congress, Cambridge, Mass., 1969," p. 211.

THOMAS, R. S., and M. I. CORLETT (1973), *in* C. J. ARCENEAUX (ed.), "Proceedings 31st Annual Meeting Electron Microscopy Society of America," Claitor's Publishing Div., Baton Rouge, La., pp. 334–335.

THOMAS, R. S., and M. I. CORLETT (1974), Manuscript in preparation.

THOMAS, R. S., and J. W. GREENAWALT (1968), *J. Cell Biol.* **39**, 55–76.

THOMAS, R. S., and F. T. JONES (1970). *in* C. J. ARCENEAUX (ed.), "Proceedings 28th Annual Meeting Electron Microscopy Society of America" Claitor's Publishing Div., Baton Rouge, La., pp. 276–277.

THOMAS, R. S., and R. C. WILLIAMS (1961), *J. Biophys. Biochem. Cytol.* **11**, 15–29.

THOMAS, R. S., P. K. BASU, and F. T. JONES (1972), *in* C. J. ARCENEAUX (ed.), "Proceedings 30th Annual Meeting Electron Microscopy Society of America" Claitor's Publishing Div., Baton Rouge, La., pp. 236–237.

THOMAS, R. S., P. K. BASU, and F. T. JONES (1974), Manuscript in preparation.

TIPPER, D. J., and J. J. GAUTHIER (1971), *in* H. O. HALVORSON, R. HANSON, and L. L. CAMPBELL (eds.), "Spores V, Papers Presented at 5th Internat. Spore Conference, Fontana, Wis., Oct. 1971," American Society for Microbiology, Bethesda, Md., 1972, pp. 3–12.

TOLGYESI, W. S., and E. M. COTTINGTON (1970), *in* L. REBENFELD and W. H. WARD (eds.), "Proceedings 4th International Wool Textile Research Conference, Berkeley, Cal., 1970" Part I, Appl. Polymer Symp. No. 18. Wiley-Interscience, New York, 1971, pp. 735–742. (SEM results described at meeting are not included in publication.)

TORIYAMA, Y., H. OKAMOTO, M. KANAZASHI, and K. HORII (1967), *Trans. Elec. Insul.* **2**, 83–91.

TRACERLAB ADVERTISEMENT (1966), see for example, *Anal. Chem.* **38**(9), 108A.

TRILLAT, J.-J. (1962), *in* J.-J. TRILLAT, (ed.), "Le Bombardement Ionique. Théories et Applications," Colloq. Intern. Centre Nat. Rech. Sci. (Paris) No. 113, 13–33.

TURKEVICH, J. (1959), *Am. Sci.* **47**, 97–119.

TURKEVICH, J., and T. STREZNEWSKI (1958), *Rev. Inst. Pétrole* **13**, 686–691.

UBER, F. M., and T. H. GOODSPEED (1935), *Proc. Nat. Acad. Sci.* **21**, 428–433.

UMEMOTO, K., and K. HOZUMI (1971), *Yakugaku Zasshi* **91**, 828–833.

UMEMOTO, K., and K. HOZUMI (1971*a*), *Yakugaku Zasshi* **91**, 845–849.

UMEMOTO, K., and K. HOZUMI (1971*b*), *Yakugaku Zasshi* **91**, 850–854.

UMEMOTO, K., and K. HOZUMI (1971*c*), *Yakugaku Zasshi* **91**, 890–895.

UMEMOTO, K., and K. HOZUMI (1971*d*), *Yakugaku Zasshi* **91**, 907–908.

UMEMOTO, K., and K. HOZUMI (1971*e*), *Yakugaku Zasshi* **91**, 1047–1057.

UMEMOTO, K., and K. HOZUMI (1971*f*), *Chem. Pharm. Bull. (Tokyo)* **19**, 217–219.

UMEMOTO, K., and K. HOZUMI (1972), *Microchem. J.* **17**, 689–702.

UMEMOTO, K., and K. HOZUMI (1972*a*). *Mikrochim. Acta* **1972**(5), 748–757.

UMEMOTO, K., M. HUTOH, and K. HOZUMI (1973), *Mikrochim. Acta* **1973**(2), 301–313.

VILKS, G. (1970), *Mar. Sediments* **6**, 72–73.

VLADYCHINA, E. N., O. R. D'YACHENKO, V. I. TROENKO, and T. P. KALININA (1966), *Mater. Ikh Primen.* **1966**(5), 59–60.†

WALKER, P. L., JR. (ed.) (1965), *Chemistry and Physics of Carbon. A Series of Advances*, Vol. 1, Marcel Dekker, New York, 390 pp.

WEISS, A., L. HELD, and W. J. MOORE (1958), *J. Chem. Phys.* **29**, 7–8.

WOOD, R. W. (1922), *Proc. Roy. Soc.*, *A* **102**, 1–9.

YASUDA, H. (1972), "A Study of Vapor Phase Polymerization and Crosslinking to Prepare Reverse Osmosis Membranes," Final Report to Office of Saline Water, U.S. Dept. of Interior, Contract No. 14-30-2658, Research Triangle Institute, Research Triangle Park, N.C., 104 pp.

ZICHERMAN, J. B. (1970), "Characterization of Loblolly Pine Ash Residues Prepared by Low Temperature Plasma Oxidation," M.S. Thesis, North Carolina State University, Raleigh, 1970, 79 pp.

ZICHERMAN, J. B., and R. J. THOMAS (1971), *Tappi* **54**, 1727–1730.

ZICHERMAN, J. B., and R. J. THOMAS (1972), *Holzforsch.* **26**, 150–152.

ZIEGELBECKER, R. (1965), *in* "Eine Apparatur zur Anwendung von Hochfrequenz- und Gleichspannungsgasentladungen. Vorabdruck einer ausführlichen Arbeit in der *Radex-Rundschau*," Radenthein, Kärnten, Austria, pp. 10–11.

ZUBOV, P. I., M. R. KISELEV, and L. A. SUKHAREVA (1967), *Dokl. Chem.* (trans. *Dokl. Akad. Nauk SSSR*) **176**, 802–804.

ZUBOV, P. I., M. R. KISELEV, and L. A. SUKHAREVA (1968), *Colloid J. USSR* (trans. *Kolloid Zh.*) **30**, 623–625.

Note Added in Proof

Following are some additional references.

General

THOMAS, R. S., and J. R. HOLLAHAN (1974), "Use of Chemically Reactive Gas Plasmas in Preparing Specimens for Scanning Electron Microscopy and Electron Probe Microanalysis, *in* O. JOHARI and I. CORVIN (eds.), "Proceedings 7th Annual Scanning Electron Microscope Symposium," Part I, IIT Research Institute, Chicago, Ill., pp. 83–92, 323–326.

Ashing for Particle Analysis

BADEN, V., J. A. SCHWARTZ, J. CHURG, and I. J. SELIKOFF (1968), "Demonstration of Asbestos Bodies in Tissue: Comparison of Available Techniques," *in* M. ANSPACH (ed.), "Internationale Konferenz über die Biologischen Wirkungen des Asbestes, Dresden, 1968," Deutsches Zentralinstitut für Arbeitsmedizin, Berlin, DDR, 1973, pp. 22–29.

BERKLEY, C., A. M. LANGER, A. SASTRE, and A. ARNESEN (1968), "Electron Microprobe Analysis of Asbestos Bodies," *in* M. ANSPACH (ed.), "Internationale Konferenz über die Biologischen Wirkungen des Asbestes, Dresden, 1968," Deutsches Zentralinstitut für Arbeitsmedizin, Berlin, DDR, 1973, pp. 12–22.

NICHOLSON, W. J., C. J. MAGGIORE, and I. J. SELIKOFF (1972), "Asbestos Contamination of Parenteral Drugs," *Science* **177**, 171–173.

SCHWARTZ, J. A., and A. M. LANGER (1968), "Technique of Removal and Analysis of Single Fibrous Particles from Human Lung Tissue," *in* M. ANSPACH (ed.), "Internationale Konferenz über die Biologischen Wirkungen des Asbestes, Dresden, 1968," Deutsches Zentralinstitut für Arbeitsmedizin, Berlin, DDR, 1973, pp. 8–12.

Microincineration Applications

HOHMAN, W. R. (1974), "Ultramicroincineration of Thin-Sectioned Tissue," *in* M. A. HAYAT (ed.), *Principals and Techniques of Electron Microscopy*, Vol. 4, Van Nostrand Reinhold, Cincinnati, O., Chapter 7, in press.

MITSCHE, R., G. HAENSEL, W. GEYMAYER, and P. WARBICHLER (1970), "Research on Graphite in Cast Iron, in Metallurgical Slags and in Minerals," (in German, with abstract in English), *in* "Proceedings 37th International Foundry Congress, Brighton, England, 1970," Foundry Trades Equipment and Supplies Association, John Adam House, WC-2 London, Official Exchange Paper No. 7, 25 pp.

UMEMOTO, K., and K. HOZUMI (1973), "Mode of Crystal Growth of Calcium Oxalate in Life Cycle of Willow Leaf," (in Japanese, with abstract in English), *Yakugaku Zasshi* **93**, 1069–1083.

WARBICHLER, P., and W. GEYMAYER (1973), "Micromorphological Investigations on Graphite Precipitates in Cast Iron," (in German), *Giessereiforschung* **25**, 29–38.

Etching of Organic Polymers

BLAHA, J., H. BRAUNEGG, and W. GEYMAYER (1972), "The Potential of Electron Microscopy in the Evaluation of Natural and Synthetic Materials, Demonstrated in the Case of Research on Domestic and Foreign Printing Media," (in German), *in* "Neuheiten und Verbesserungen—Ergebnisse geförderter Projekte aus dem Bereich der Roh- und Werkstoff-Forschung der kooperativen Forschunginstitute Östereichs," Forschungsförderfonds der Gewerblichen Wirtschaft, Vienna, 1972, 19 pp.

GODDAR, H. (1970), "Physical and Chemical Etching Techniques for Preparing Bulk Polymers for Electron Microscopy," (in German with abstract in English), *in* G. SCHIMMEL and W. VOGELL (eds.), *Methodensammlung der Elektronenmikroskopie*, Wissenschaftliche Verlagsgesellschaft, Stuttgart, 1970, 6 pp.

HINRICHSEN, G., and H. ORTH (1971), "On the Structure of Drawn Films and Fibers, and also Solution-Grown Single Crystals of Polyacrylonitrile," (in German with abstract in English), *Kolloid-Z. Z. Polymere* **247**, 844–850.

MAYOUX, C., and B. AI (1973), "Scanning Electron Microscope Observations of the Effects of Discharges on Polyethylene," *J. Appl. Phys.* **44**, 3423–3424.

Etching of Bio-Organic Materials

AMBROSE, E. J., U. BATSDORF, and D. M. EASTY (1972), "Morphology of Astrocytomas in Tissue Culture: Optical and Stereoscan Microscopy," *J. Neuropath. Exp. Neurol.* **31**, 596–610.

BENDET, I., and N. RIZK (1973), "Quantitative Measurements on the Ion Etching of TMV," *in* C. J. ARCENEAUX (ed.), "Proceedings 31st Annual Meeting Electron Microscopy Society of America," Claitor's Publishing Div., Baton Rouge, La., pp. 336–337.

HOLT, L., S. G. PAL, R. BENNETT, C. MAUNDER, J. EDWARDS, and S. LEWIS (1971), "The Ultrastructure of Cartilage," *Clin. Sci.* **40**, 27pp.

Etching of Carbon and Inorganic Materials

BARNET, F. R., and M. K. NORR (1973), "Carbon Fiber Etching in an Oxygen Plasma," *Carbon* **11**, 281–288.

BAUMAN, A. J., and J. R. DEVANEY (1973), "Allende Meteorite Carbonaceous Phase: Intractable Nature and Scanning Electron Morphology," *Nature* **241**, 264–267.

DEVANEY, J. R., and A. M. SHEBLE (1974), "Plasma Etching for PROM Failure Analysis and Other Problem Solutions," *in* "Proceedings 1974 IEEE Physics Reliability Symposium, Las Vegas, 1974," Institute of Electrical and Electronic Engineers, Inc., New York, in press.

FRANKLIN, P., and D. BURGESS (1974), "Reliability Aspects of Nichrome Fusable Link PROM's," *in* "Proceedings 1974 IEEE Physics Reliability Symposium, Las Vegas, 1974," Institute of Electrical and Electronic Engineers, Inc., New York, in press.

JONES, K. (1974), "Applications of Plasma Etching to Failure Analysis," *in* "Proceedings 1974 IEEE Physics Reliability Symposium, Las Vegas, 1974," Institute of Electrical and Electronics Engineers, Inc., New York, in press.

JONES, S. S. (1971), "Conical Microstructures Formed by Oxidation of Polycrystalline Graphite," *in* "Proceedings 3rd Conference on Industrial Carbons and Graphite," Society of Chemical Industry, London, 1971, pp. 238–245.

Deposition of Thin Films by Plasma-Chemical Reactions

REIMANN, B. (1973), personal communication. Dr. Reimann has provided a considerable amount of additional information on the techniques and applications of plasma-deposited, carbonaceous replica films. His address is: Electron Microscopy Laboratory, Department of Pathology, William Beaumont Army Medical Center, Department of the Army, El Paso, Texas 79920.

Chapter 9

Applications of Plasma Technology to the Fabrication of Semiconductor Devices

Ralph W. Kirk

9.1. INTRODUCTION

In a plasma the electric field interacts with the gas molecules to generate a wide variety of highly reactive species, including electrons, positive ions, neutral atoms, and molecules in both ground and excited states. Consequently, the reactions which occur in a plasma are often complex and differ markedly from those occurring under normal conditions. Gases which combine only at elevated temperatures react near room temperature in a glow discharge. Solid surfaces can be sputtered away by the bombardment of the energetic species or can react with them to form new substances, some of which are unique to plasma processing.

Plasma reactions have been extensively studied in the laboratory but have found only limited use in industry. A notable exception has been in the microelectronics industry where plasma techniques are replacing standard chemical processes in several important areas. This chapter will give a brief description of general semiconductor device fabrication and will then concentrate on the areas in which plasma techniques are now, or are likely to become, important.

9.2. SEMICONDUCTOR DEVICE FABRICATION

The device shown in Figure 9.1 is an integrated circuit 0.15 in. long and 0.125 in. wide. It contains approximately 3000 electrical components,

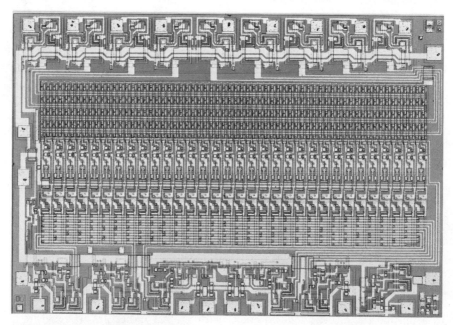

FIGURE 9.1.   Integrated circuit.

including transistors, diodes, and resistors. The electrical characteristics of the device are determined by the concentration of electrically active "dopant" atoms in specific areas of the silicon. The dopant concentrations are strictly controlled, often at the 10^{14} atoms/cm^3 level and never more than 10^{21} atoms/cm^3. The performance of device can be completely degraded by quantities of ionic impurities on the surface as small as 10^{-9} g.

The device shown in Figure 9.1 is not a specially made, one-of-a-kind curiosity. It is a standard product, commercially available. The semiconductor industry produces more than a million devices of comparable complexity per week. The fabrication of such small, complex, and electrically precise circuits has been developed into a highly specialized technology involving the interaction of photolithography, diffusion, metallization, and final passivation.

Photolithography

The starting point in all device fabrication is a uniform substrate or wafer of silicon, germanium, or a III-V compound, e.g., gallium arsenide. The substrate is subjected to a series of chemical and thermal treatments which modify the electrical properties of certain well-defined areas. The definition is achieved by photolithography.

A layer of photosensitive polymer, called a photoresist, is applied to the surface of the substrate. The photoresist is exposed to optical radiation through a very precisely made mask. The light initiates a chemical reaction in the polymer forming a pattern which is exactly defined by the mask. The pattern is developed by rinsing away the parts of the film which are soluble after exposure. The wafer is then immersed in an etchant which selectively attacks the unprotected areas of the substrate surface. After etching, the photoresist must be completely stripped off to prevent interference in any of the subsequent processing steps. The device in Figure 9.1 required nine separate photolithography applications.

Diffusion

Figure 9.2 is a cutaway view of a simple *npn* transistor. The areas labeled *n* have a net concentration of electron donor impurity atoms. The *p* regions contain excess electron acceptors. The electrical characteristics of the transistor are a function of the impurity concentrations as well as the concentration profiles. The profiles are produced by diffusing the electrically active atoms into selected areas of the substrate. The wafer is first covered by a diffusion-masking material which is either grown or deposited on the surface.

The masking material is patterned by photolithography and subsequent

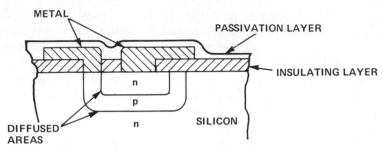

FIGURE 9.2. Side view of *npn* transistor.

etching. After etching away the appropriate areas of the diffusion-masking material, the wafer is exposed to a known concentration of dopant atoms, either from a solid source applied to the surface or from a gas in a controlled atmosphere furnace. The wafer is then heated to increase the diffusion rate of dopant atoms into the silicon. The concentration of dopant in the source, as well as the diffusion temperature and time, are strictly controlled to give exactly reproducible concentration profiles. Two or three separate diffusion steps are normally required.

The diffusion-masking material is extremely important. It must be impervious to penetration by the dopant atoms at temperatures as high as 1250°C. It must form continuous films which adhere strongly to the substrate and can be etched into the fine, complex patterns used in devices.

Metallization and Passivation

After the desired number of diffusions have been completed, the wafer is further etched to open areas in the diffusion-masking material for electrical contact. A layer of metal is then deposited on the surface (approximately 7000 Å of Al for a typical device). The metal is patterned and covered with a passivation layer. The passivation layer is patterned to expose areas of the metal either for wire-bonding to the leads of the device package or for evaporation of a second layer of metallization for complex multilayer devices. The passivation layers must be compatible with the metal system used, must be good electrical insulators, and must protect the device from contamination which could cause electrical failure.

Areas of Plasma Application

Plasma processes have several features which make them attractive for use in semiconductor device fabrication; they are clean, fast, and lend themselves readily to automation. Plasma technology is now being applied in production for photoresist removal, etching of silicon compounds, and sequential processing. An extensive amount of work has also been done on the application of plasmas to the deposition and growth of dielectric films for use on semiconductor devices as diffusion masks, passivation, and insulation films. Each of these areas will now be discussed in detail.

9.3. PHOTORESIST REMOVAL

Nature of Photoresists

The photoresist materials used in the semiconductor industry are supplied as proprietary mixtures by several chemical companies. Each company keeps the actual formulations a secret but practical experience has shown that the principal differences lie mainly in the area of solvent choice, solid content, and impurity level. There are two basic types of resist material available. The negative resist which is typically a solution of polyisoprene and photoinitiator in xylene. Upon exposure to uv light, the initiator catalyzes cross-linking of the polyisoprene into an insoluble polymer film. Development consists of rinsing away the unexposed and therefore unpolymerized material, leaving a perfect negative reproduction of the mask. Proprietary developers are available but electronic grade xylene works equally well.

Positive photoresists are solutions of fairly low molecular weight polymers which are insoluable in aqueous solutions. Photosensitive sites in the chains react forming organic acid groups when exposed to uv light. The polymer units containing the acid sites can be washed away in dilute aqueous alkaline solutions. The polymer which remains is an exact positive copy of the mask.

After development, both types of resist are postbaked to drive out the remaining solvent and increase the chemical stability and adhesion of the polymer. The postbake time and temperature cycles are determined by the nature of the succeeding etch step.

Conventional Removal Techniques

Photoresist strippers are supplied in a confusing array of proprietary mixtures but are of two basic types. The first is a mixture of halocarbon solvents, phenols, and surface active agents. These liquids dissolve the polymer from the surface of the wafer without extensively degrading it. The actual removal procedure varies from one company's mixture to another and is also dependent upon the thermal history of the resist. A typical stripping process can be summarized:

1. Soak several minutes in hot stripping solution
2. Rinse in trichloroethylene

3. Rinse in isopropyl alcohol
4. Rinse in running deionized water
5. Dry

Some production areas add a supplementary chromic acid dip prior to step 4 to ensure complete removal of all organic residue.

The stripping solutions rapidly become saturated with polymeric material and must be changed often to prevent deposition of residue back on the wafer.

The second type of resist stripper is a solution of an oxidizing agent such as hydrogen peroxide or chromic oxide in sulfuric acid. All of the reagents must be specially prepared electronic grade to prevent contamination of the wafers. The solutions are used hot, with time and temperatures determined by the thickness and thermal history of the resist. The mixtures oxidize the polymer to CO_2 and H_2O. Therefore, only prolonged deionized water rinsing is needed after stripping. Control of removal time is rather more critical than with the organic strippers due to the reactive nature of the acid bath. The actual removal rate is determined by the concentration of oxidizing agent in the solution and is, therefore, a function of the total amount of photoresist removed.

Both methods require the use of expensive chemicals which must be kept free from contaminants. The solutions must be used in hoods by operators using face shields, safety aprons, and rubber gloves. The resist removal rate depends on the age of the reagents and the condition of the polymer; so no prescribed removal time can be specified. The operator must watch the reaction to determine the end point in removal. Both types of removal require extensive rinsing to remove all traces of contamination from the wafers.

Plasma Reactions with Photoresist

Photoresists are highly branched saturated hydrocarbon polymers. They do not react completely with molecular oxygen at temperatures below 600°C. These materials do, however, react with atomic oxygen very rapidly at temperatures as low as 40°C. Hansen, et al. [1], summarize the plasma oxidation of polymers:

$$O_2 \rightarrow O\cdot + O\cdot$$

$$RH + O\cdot \rightarrow R'\cdot + R''O\cdot \quad \text{or} \quad R\cdot + OH\cdot$$

$$R\cdot + O\cdot \rightarrow RO\cdot$$

$$R\cdot + O_2 \rightarrow ROO\cdot$$

The final products of these chain reactions are CO_2 and H_2O.

The oxidation of the polymers has no induction period and proceeds at a linear rate with time. The reaction rate is greatly affected by the structure of the polymer with highly cross-linked hydrocarbons reacting as fast or faster than other materials. Cross-linking of polyethylene by electron bombardment actually increases the removal rate of the polymer.

The application of plasma oxidation to photoresist removal was first suggested in 1968 by Irving [2]. He showed that conventional photoresists could be removed at rates as high at 2000 Å/min. The rate was a function of the oxygen flow through the system, the pressure in the reactor, and the rf power level. Plasma photoresist removal has become widely used in the microelectronic industry with the advent of commercially available automatic plasma strippers or "ashers."

As the design of the ashers has become more sophisticated, the difference between the commercially available units has increased. The basic ashing machine is shown in Figure 9.3a. An rf generator is inductively or capacitively coupled to a quartz reaction chamber. The chamber has an oxygen inlet and is connected to a mechanical vacuum pump. There is no need for expensive high-vacuum equipment since the presence of small amounts of N_2 or H_2O from leaks actually increases the concentration of atomic oxygen in the plasma [3]. The size of the mechanical pump is important since the oxidation rate is dependent on the oxygen pressure and flow rate in the chamber. The higher the flow rate for a given pressure, the faster the resist removal rate.

The actual stripping of photoresists is extremely simple. After etching and rinsing, the wafers are loaded vertically in quartz boats (Fig. 9.3b). The boat is placed in the reaction chamber which is then pumped down to less than 10^{-2} torr. Oxygen is then introduced into the system to give the desired flow

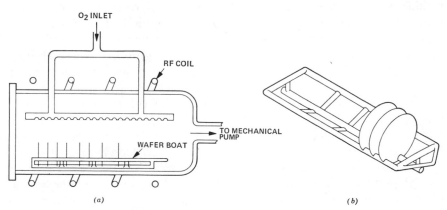

FIGURE 9.3. (a) Plasma asher, (b) wafer boat.

rate. This flow rate is often determined by opening the oxygen valve enough to establish and maintain the reactor pressure at 1 torr. The oxygen flow and rf power are turned off after ashing has proceeded for a time which has been empirically determined to be sufficient for complete oxidation of the resist. No further rinsing of the wafers is necessary. The wafers just proceed on to the next fabrication step.

Bersin [4] claims that a three-step process minimizes the time required for resist removal. The sequence for a typical run on his equipment is:

1. 2.3 min O_2 flow 700 cm³/min   power level 340 W
2. 3.3 min · O_2 flow 500 cm³/min   power level 280 W
3. 5.3 min O_2 flow 250 cm³/min   power level 260 W

This complicated procedure can best be handled by an automatic programming system. The programmer can be set up to turn on the O_2 and rf power at the preset levels after the initial pump down has been completed, cycle through all three steps, turn off the O_2 and rf power, vent the system to the atmosphere and signal that the run is over. All the operator has to do is insert the boat, close the reactor door, and push the start button. Several manufacturers use only one set of flow-rate pressure-power settings throughout the ashing cycle. They too include automatic programmers with their machines. An automated system can remove a 1 μ layer of photoresist from 300 wafers/hr. It is not uncommon for such a unit to process 20,000 wafers/week.

Limitations of Plasma-Ashing

Early work on plasma-ashing [2,5] indicated that a large amount of inorganic residue was left on the surface of the wafer after removal of photoresist layers. The residue was identified as being mostly tin oxide which could be removed by rinsing in hot 20% aqueous KOH or in buffered HF solution [5]. The tin was present in several photoresists having served as a cyclization catalyst, but had never caused problems in conventional removal methods [6]. Once the source of this problem was identified, new purer forms of photoresist were developed. With these new resists the amount of inorganic residue left after ashing is negligible. Any inorganic material which does remain can be removed in a nitrogen–hydrogen plasma which can be carried out in the same chamber after all of the organic material has been removed [7].

Metal-oxide–silicon (MOS) devices are extremely sensitive to surface charges or to electrically active surface sites. Plasma-ashing of these units when the gate oxides are exposed often changes the electrical characteristics of the device. High-temperature annealing cycles will often bring the device

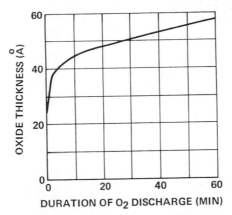

FIGURE 9.4. Growth rate of SiO_2 in oxygen plasma [8].

performance back into specifications. Plasma treatment does not affect MOS devices after the gate oxide has been metallized.

Silicon does oxidize slowly when exposed to an oxygen plasma as shown in Figure 9.4. The native oxide layer of 25 Å increases to 50 Å in 30 min. [8] This oxide layer is too thin to cause problems in most applications but can interfere with the metal–silicon electrical contact. A 3-sec dip in dilute HF completely removes the oxide.

Advantages of Plasma-Ashing

The inherent simplicity and reproducibility of plasma oxidation has made it easily amenable to automation. One automated unit can easily handle 80,000 wafers/month with a total chemical cost of under $10 for oxygen. The plasma is free from ionic impurities and when used with pure photoresists leaves no contamination on the wafer surface. Prolonged exposure to the oxygen plasma does not adversely affect most devices. This means that nonuniformity of photoresist thickness or ashing rate will not destroy a whole load of wafers. They can simply be left in the asher until all of the organic material has been removed. The plasma machines are completely shielded and offer no safety hazards for the personnel operating them. All three removal methods are compared in Table 9.1.

9.4. PLASMA-ETCHING

Fluorine-containing gases react with refractory metal oxides at high temperatures to form volatile metal fluorides. This type of reaction has been used to

TABLE 9.1. Comparison of Photoresist Removal Methods

Element of photoresist	Removal method		
	Organic solvent strippers	Oxidizing solutions	Plasma ashers
Time/100 wafers	15–20 min	15–20 min	20 min
Operator time/run	10–15 min	10–15 min	5 min
End-point determination	Operator	Operator	Automatic
Reactants	Proprietary mixtures	Electronic grade H_2O_2 or Cr_2O_3	O_2 gas
Chemical cost/10^5 wafers	$3000–4000	$3000–4000	$10
Equipment requirements	Hoods, deionized water rinse stations, safety masks, gloves, aprons, solvent disposal systems		Asher
Sources of contamination	Photoresist, solvent mixtures, rinsing baths, operator handling	Photoresist, oxidizing solutions, rinsing baths, operator handling	Photoresist, operator handling

etch sapphire substrates at 1450°C [9,10]. Beguin, et al. [11] observed the following reactions in a plasma torch:

$$CF_4 \rightarrow C\cdot + 4\, F\cdot$$

$$\rightarrow \,\colon CF_2 + 2\, F\cdot$$

$$4\, F\cdot + SiO_2 \rightarrow SiF_4 + 2\, O\cdot$$

The fluorine species in the plasma can also react with silicon nitride and silicon:

$$12\, F\cdot + Si_3N_4 \rightarrow 3\, SiF_4 + 2\, N_2$$

$$4\, F\cdot + Si \rightarrow SiF_4\cdot$$

This basic reaction scheme has been applied to the etching of silicon and its oxides and nitrides [12].

The etching reactors are very similar to the ashers shown in Figure 9.3. However, the etching procedure is somewhat different than that used for photoresist removal. Wafers covered with patterned photoresist are set vertically in quartz boats with their backs together to prevent attack of the unprotected silicon surfaces. The boats are placed in the reactor chamber which is evacuated all the way back to the CF_4 tank to exclude oxygen completely from the system. The CF_4 is introduced and allowed to reach equilibrium flow before the rf power is turned on. Etch times are determined empirically and set on the programmer. Great care must be taken to ensure the system is leak-tight since oxygen will remove the photoresist from the wafers.

The etch rates for several types of oxide and pyrolytic nitride are shown in Figure 9.5. Silicon etches approximately four times as fast as thermal oxide under the same conditions. This feature severely limits the applicability of CF_4-etching in device fabrication [13].

Any etch step which must go through SiO_2 or Si_3N_4 to expose the silicon surface, e.g., opening windows for diffusion or metallization, has to be monitored very closely to prevent extensive etching of the silicon. The presence of any exposed silicon in the reactor seems to retard the etching of Si_3N_4 or SiO_2 while the silicon etches faster than normal, indicating very selective attack by the fluorine atoms [13].

Figure 9.6a shows a step etched through SiO_2 and Si_3N_4 to a silicon surface. Note the extensive roughening of the Si. The fluorine does not attack aluminum or gallium arsenide phosphide. Figure 9.6b shows an etch cut through SiO_2 to aluminum metal. There is no attack on the metal. The walls of the etched opening are almost perfectly vertical.

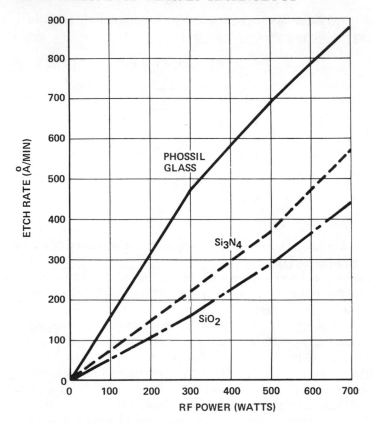

FIGURE 9.5.   Etch rate of dielectric films in a CF_4 plasma as a function of applied rf power [13].

Several very important processes lend themselves to plasma-etching. The first and most widely used is final passivation glass-etching to expose the aluminum wire-bonding pads and the silicon scribe grid. Extensive etching of the scribe grid areas is not critical since the wafers are broken into die along these lines.

The second etching operation involves the opening of diffusion windows in the fabrication of light emitting diodes on gallium arsenide or gallium arsenide phosphide wafers. Silicon nitride and silicon dioxide layers are deposited on the substrates as diffusion masks. The layers are only 500–1500 Å thick and require only a 3–10 min etch to open the windows.

Another limitation which must be considered when planning a plasma etch step is photoresist. Even with no oxygen in the reactor the photoresist is gradually eroded by the CF_4 plasma. The wafers often reach temperatures

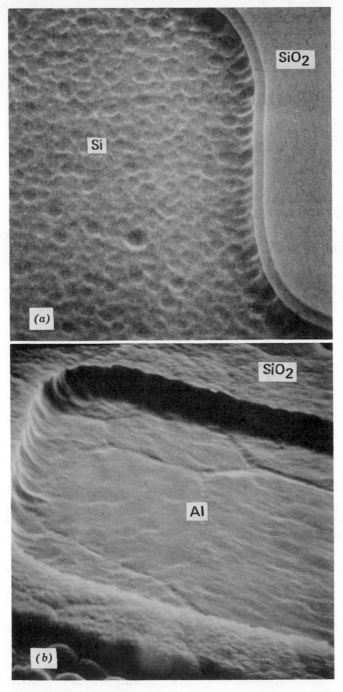

FIGURE 9.6. Steps etched through silicon oxide layers [13]: (*a*) To silicon substrate, note extension damage; (*b*) to aluminum.

in excess of 200°C during the etching cycle and any photoresist which survives the treatment is so extensively cross-linked that removal by conventional techniques is very difficult. Plasma oxidation, however, easily removes this film. A proprietary etch gas is now available which reduces the attack on the photoresist by 20–50%.

Advantages of Plasma-Etching

Reactors are commercially available which can etch SiO_2 from 50–60 wafers simultaneously at a rate of 300 Å/min. The equipment is fully automated and requires only 5 min of operator time to load each boat. The conventional etch techniques for silicon, silicon nitride, and silicon dioxide are shown in Table 9.2.

TABLE 9.2. Comparison of Etching Parameters

Etchant	Material		
	Thermal oxide	CVD oxide	Pyrolytic Si_3N_4
Conventional etchant	60°C; 4:1 buffered hydrofluoric acid	25°C; 4:1 buffered hydrofluoric acid	Refluxing 180°C; phosphoric acid
Etch rate	150 Å/sec	400 Å/sec	100 Å/min
Plasma etchant	CF_4	CF_4	CF_4
Etch rate	160 Å/min	480 Å/min	220 Å/min

The etch solutions suffer from the same drawbacks as the conventional resist strippers, i.e., they require costly chemicals, which are dangerous to use, require constant replenishment, and are expensive to dispose of. The plasma reactor is cleaner, cheaper, safer, faster, and requires less equipment and operator time. Unfortunately, the selective attack on silicon limits the applicability of plasma-etching to processes where no silicon is exposed or where the exposed silicon is not electrically important.

Sequential Plasma-Etch and Photoresist Removal

Plasma-etching and ashing are very similar operations. In each case a single reactant gas is activated by the plasma to form species which react with the exposed surface of the substrates to form volatile products which are pumped away. The reaction rate at any wafer depends on the local concentration of the activated species which is a function of local gas flow and local rf power. Reactors, therefore, are designed to give uniform gas flow in a uniform rf field. A reactor made for etching will work just as well for ashing with only a change in reactant gas and perhaps changes in vacuum seals to materials which are not attacked by either fluorine or oxygen plasmas. The automatic programmers have only to be modified to handle the additional switching.

Plasma-etcher strippers are now being used in production areas to select-ively etch final passivation glass and subsequently remove the photoresist. A comparison of the conventional and plasma processes explains the reason for plasma implementation. Both processes start after photoresist develop-ment.

Conventional	Plasma
Load in metal boat	Load in quartz boat
Bake at 165°C	Put boat in etcher stripper
Transfer to teflon carrier	Automatic etch strip
Etch in 4/1 buffered HF	Load in original carriers
Rinse in DI H_2O	
Dry	
Transfer to original carrier	
Take to cleaning station	
Transfer to cleaning carriers	
Photoresist removal	
Organic rinse DI H_2O	
Transfer to original carriers	
Total time: 98 min	Total time: 52 min
number of transfers: 5	Number of transfers: 2
Total operator time: 50 min	Total operator time: 15 min

The wafers have already been etched and the photoresist removed in the plasma process before the wafers even make it to the etch step in the con-ventional technique. The cost reduction due to chemical and equipment costs have already been discussed as has the yield improvement due to cleanliness and simplicity of the system. This process chart emphasizes another large advantage of the automated systems, i.e., a reduction of total operator time devoted to the process.

Equipment Considerations

The manufacturers of ashers and etchers have been able to transform a process from an experimental curiosity to a full-scale production method within 3 years. The production steps which now use plasma systems are the ones where a high degree of control is not necessary and the reaction-rate variations of $\pm 15\%$ in a reactor are not important. There are many more stages in production that would change over to plasma systems if the units could perform with $\pm 5\%$ variations throughout the reactor. The design parameters are straight forward: (*a*) Ensure uniform gas flow across the surface of each wafer; (*b*) develop an rf generator and coupling system which can be fine tuned to compensate for field variations within the reactor.

9.5. DEPOSITION AND GROWTH OF INORGANIC DIELECTRIC FILMS

Conventional Deposition Procedures

Inorganic dielectric films are used extensively for diffusion masks, as well as insulating and passivating layers. Silicon nitride and silicon dioxide are the primary materials used in diffusion, although recently interest has been generated in the use of silicon oxynitrides ($Si_xO_yN_z$). Silicon nitride is deposited from the reaction:

$$3\,SiH_4 + 4\,NH_3 \xrightarrow{\;700-900°C\;} Si_3N_4 + 12\,H_2$$

The films are excellent diffusion masks for all dopant atoms including Zn and Ga. The material is very unreactive. It can be etched only in 180°C refluxing phosphoric acid at 100 Å/min. The high deposition temperature and vigorous etch conditions preclude use of Si_3N_4 over aluminum metallization which alloys with silicon at 577°C and etches rapidly in phosphoric acid.

Silicon dioxide is most commonly used to mask boron and phosphorus diffusions but is ineffective for Ga and Zn. The oxide is grown on the silicon wafers at 900–1200°C in wet or dry oxygen. The material can be etched easily in dilute HF solutions. The reaction of tetraethylorthosilicate (TEOS) with oxygen at 750°C can be used to deposit silicon dioxide films. The glass is generally less clean than thermal oxide perhaps due to inclusion of organic decomposition products.

Layers of SiO_2 can be used as a final passivation when deposited from the reaction:

$$SiH_4 + O_2 \xrightarrow[N_2\;stream]{\;300-500°C\;} SiO_2 + 2\,H_2$$

The glass formed is compatible with aluminum and has good dielectric properties but is permeable to ionic contaminants which can completely degrade a semiconductor device. Addition of PH_3 to the reaction mixture gives a phosphosilicate glass which is a better ion barrier but still not thoroughly satisfactory. Silicon oxynitrides are formed by the reaction:

$$x\,SiH_4 + y\,NO + z\,NH_3 \xrightarrow{\;700-900°C\;} Si_xO_yN_z + H_2 + N_2 + H_2$$

At certain O/N ratios, these films exhibit diffusion-masking properties similar to Si_3N_4 but can be etched in dilute HF solutions much like SiO_2.

Glow-Discharge Deposition of Silicon Nitride

The glow-discharge reaction of SiH_4 and NH_3 to give Si_3N_4 has been extensively studied [14–18]. Silane and nitrogen also react in a glow discharge at temperatures above 100°C to give silicon nitride with properties very similar to silane–ammonia nitride [8,19]. The glow-discharge reactor used at Motorola to study the silane–ammonia reaction is shown in Figure 9.7. The substrate is placed on a radiantly heated aluminum block inside the evacuated quartz reaction chamber. The reactant gases are mixed prior to entry at the bottom of the chamber, and rf power is supplied through a two-turn coil surrounding the reactor immediately above the substrate holder.

The deposition procedure is simple. A clean substrate is placed on the sample holder. The chamber is evacuated to approximately 3×10^{-3} torr as the substrate is brought to the desired deposition temperature. An argon plasma (2–3 torr for 10 min) is then used to give a final clean-up prior to deposition. The reactant gases are introduced at the desired flow rates and

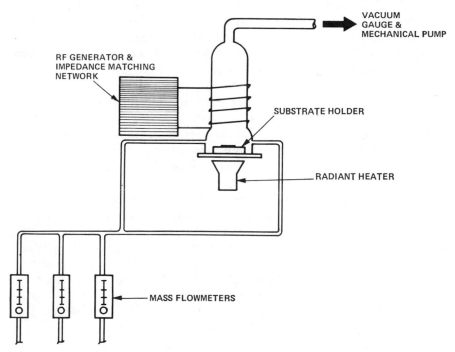

FIGURE 9.7.   System used in glow-discharge deposition of dielectric films.

the plasma is initiated. Proper cleaning of the substrates prior to deposition both chemically and in the plasma is necessary to ensure good adhesion and improve the electrical properties of the films.

The films were deposited from 100 to 500°C with NH_3/SiH_4 ratios of 9/1 to 3/7. The deposition rate of the films was independent of the substrate temperature but was strongly dependent on the rf power level and the silane content of the reactant mixture, ranging from 50 Å/min at low power and low silane, to 1000 Å/min at high power and high silane concentrations.

The resulting films were transparent and hard. Similar films showed pinhole densities of $10/m^2$ [8] and were shown to be amorphous by electron defraction [19]. The ir absorption spectra had broad absorptions from 10–14 μ due to silicon nitride stretching. The maxima of these absorptions shifted only slightly (11.6 $\pm$ 0.2 μ) throughout the entire range of ammonia/silane ratios. A small absorption due to N–H groups was observed at 3.0 μ in some spectra. This feature was not seen in the N_2 films [8]. Joyce, Sterling, and Alexander [14], used electron-diffraction radial-distribution functions to postulate that N/Si ratio of 1.33 in the films was achieved when the NH_3/SiH_4 ratio in the reactant gases was 3:1 and that free silicon was present in the material deposited at lower NH_3/SiH_4 ratios. Gereth and Scherber [8] confirmed the presence of excess silicon in the nitrogen-deposited films at high SiH_4/N_2 ratios by a backscatter technique. They observed Si_3N_4 film stoichiometry when the reactant SiH_4/N_2 ratio was 1.5×10^{-3}. The index of refraction of the films was insensitive to composition varying only from 1.95 to 2.01 throughout the entire range of silane–ammonia mixtures.

The etch rate of the deposited material is a function of deposition temperature and thermal history. Figure 9.8 shows the etch rate in room temperature buffered 4:1 HF of films deposited from a 1/4 silane–ammonia mixture as a function of deposition temperature. The etch characteristics of 400°C films make them the best suited for device fabrication. The films were very good Ga diffusion barriers [14].

Capacitance–voltage (CV) plots are used to determine the concentration of charges at the dielectric-silicon interface. The smaller the value of the flatband voltage (V_{fb}), the fewer surface charges there are. A shift in V_{fb} after any type of stress indicates that there are mobile charges in the material. A large hysteresis in a CV plot indicates the presence of charge trapping sites either in the dielectric or at the interface. The V_{fb} values for as deposited Si_3N_4 showed that surface charge densities of approximately $10^{11}/cm^2$ could be achieved with proper cleaning [14,15,18]. Swann, Mehta, and Gauge [15], could detect no V_{fb} shift after a sodium hydroxide rinse and a bias bake of 6 min at 300°C and 10^6 V/cm indicating very low permeability toward Na^+ ions. The dielectric strength of $5–11 \times 10^6$ with a dielectric

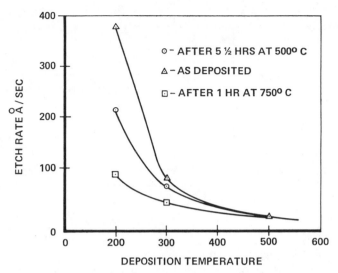

FIGURE 9.8. Etch rate of glow-discharge silicon nitride as a function of deposition temperature and thermal treatment.

constant of 6–10 agree well with values determined for pyrolytic Si_3N_4 (10^7 V/cm, 6.3) [20].

The Ar plasma-preclean proved to be a very important step in the processing. Without it flatband voltages of -20 V were not uncommon compared with -2 V using the method. Gereth and Scherber [8] studied the effects of changing the gas used in precleaning from Ar to O_2, N_2, or H_2. They found no appreciable difference in the resultant V_{fb} but a dramatic difference in the hysteresis behavior as well as the shift in flatband voltage under bias stress. Films deposited with Ar or N_2 plasma-precleaning or no plasma-precleaning showed high negative hysteresis and less than -2 V flatband voltage shifts after bias stress. Films formed after H_2 precleaning showed no hysteresis but had flatband voltage shifts of approximately -5V. Films deposited on oxygen plasma-precleaned wafers showed high-positive hysteresis and correspondingly high-positive flatband voltage shift after bias stress. Both the H_2 and O_2 plasmas appear to be reacting with the native oxide layer on the silicon, the hydrogen partially removing it and the oxygen adding to it.

Glow-Discharge Deposition of Silicon Oxynitride

Addition of nitric oxide to the silane–ammonia reaction mixture in a glow discharge gives silicon oxynitride films. The films were analyzed by Auger electron spectroscopy which showed that the amount of oxygen in the films

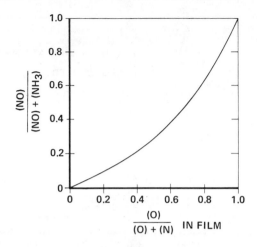

FIGURE 9.9. Composition of glow-discharge silicon oxynitride as a function of the nitric oxide content of the reactant mixture.

was related to the amount of NO in the reaction mixture as shown in Figure 9.9. The index of refraction varied from 1.5 with very high NO/NH_3 ratios to 2.0 without nitric oxide. The value of the index of refraction was a good measure of the oxygen content of the films. The index was reproducible to ± 0.01 for films made with the same reaction mixture and deposition temperature.

The etch rate of the films in buffered 4/1 HF at room temperature did not vary much with variation in oxygen content remaining near 30 Å/sec for films deposited at 400°C. The dielectric strength of all the films was approximately 10^7 V/cm. Further work will have to be done to determine whether the glow-discharge oxynitride films offer any real advantage over the nitride films.

Both materials can be etched in a CF_4 plasma at rates near 200 Å/min [13]. The good barrier properties and high dielectric strength coupled with easily controlled etching characteristics make glow-discharge silicon nitride and oxynitride good candidates for final passivation or insulation films.

Glow-Discharge Deposition of Silicon Oxide

TEOS reacts with oxygen in a glow discharge to give SiO_2. The deposition system is very similar to that shown in Figure 9.7 except the TEOS is bled into the reactor through a capillary tube under its own vapor pressure. The flow of TEOS is varied by controlling the temperature of the reservoir. Ing and Davern [21] deposited films at 75 and 200°C. Mukherjee and Evans [22] extended the study to above 600°C. Film deposition showed no induction period and was linear with time. The deposition rates increased linearly with increases in TEOS flow [21]. The highest rates observed were only 70 Å/min. Films deposited at rates above 50 Å/min left brown carbonaceous residues when etched in HF. The ir absorption spectra showed that considerable organic material was present in the films unless deposited above 500°C with high O_2 pressures. High (500°C) deposition temperatures were also necessary to minimize the number of OH groups in the films.

The films had dielectric constants of 4.5–5.5 and dielectric strengths of 5×10^6 to 10^7 V/cm. The high OH and organic content of the films deposited below 500°C make them unattractive as final passivation or dielectric layers.

The reaction

$$SiH_4 + 2 N_2O \rightarrow SiO_2 + 2 N_2 + 2 H_2$$

has been reported at temperatures from 25 to 1000°C [14]. The ratio of N_2O to SiH_4 was varied from 1/1 to 10/1. Deposition rates were 2–4 μ/hr. The ir absorption spectra for films made with high N_2O to SiH_4 ratios were very similar to those of quartz and vitreous silica [14]. Spectra of films deposited at 25°C had peaks at 2.3 and 3.0 μ indicating high OH content as well as a peak at 4.45 μ from Si–H groups. These bands became progressively smaller with higher deposition temperatures, disappearing above 300°C.

The etch rate of the films in "P" etch (15 part 49% HF, 10 part HNO_3, 300 part H_2O) as a function of deposition temperature is shown in Figure 9.10 for several N_2O/SiH_4 mixtures. The films deposited at 1000°C and $N_2O/SiH_4 = 2$ etch at the same rate as 1000°C thermal oxide grown on silicon.

Capacitance–voltage measurements were made on films deposited on very carefully cleaned wafers which were then plasma-precleaned prior to deposition. Films deposited at 450°C with $N_2O/SiH_4 = 6$ on these specially cleaned silicon substrates had surface state densities of $2 \times 10^{11}/cm^2$. Bias stress measurements conducted at 150°C and field strengths of 5×10^5 V/cm showed a V_{fb} drift of 1–2 V indicating that positive charges can drift through the material even at these low-stress values. Similar tests conducted on glow-discharge Si_3N_4 capacitors showed no V_{fb} shift. Addition of small amounts of phosphorous to the glass by adding PH_3 to the reaction mixture gave no stabilizing effects [15].

The oxide films deposited from the glow-discharge reaction of N_2O and SiH_4 are better suited for use in semiconductor device fabrication than those deposited from TEOS. They are cleaner and can be deposited with low OH content at temperatures of 450°C making them suitable for final passivation layers over aluminum.

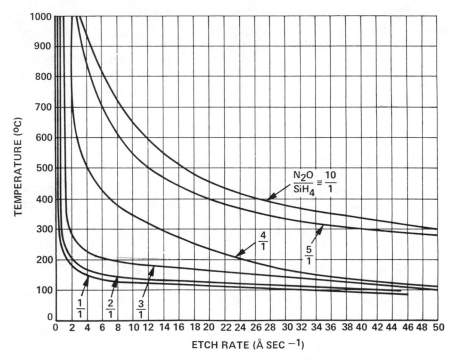

FIGURE 9.10.   Etch rate in "P" etch of glow-discharge silicon dixode [14].

Glow-Discharge Deposition of Aluminum Oxide

Aluminum oxide films have been deposited from the glow-discharge reaction of $AlCl_3$ and O_2 [23]. The films are of interest mainly for the fabrication of MOS devices with positive threshold voltages. The deposition apparatus is similar to Figure 9.7 except the oxygen is passed through an $AlCl_3$ evaporator which is maintained at a temperature selected to give a desired $AlCl_3$ vapor pressure. The deposition rate was independent of $AlCl_3$ vapor pressure above 0.3 torr and was independent of flow rate of oxygen at constant $AlCl_3$ pressure from 10 to 200 cm^3/min. The rate was directly related to the rf voltage applied. Deposition rates of 400 Å/min were easily achieved. Deposition was observed at 100°C, but the film exhibited better electrical properties at 480°C.

Infrared spectra of the films are very similar to those of films made from the thermal decomposition of aluminum triethoxide, each exhibiting only a broad absorption centered approximately at 15 μ. Electron diffraction shows the films to be amorphous as deposited but that partial crystallization occurs at temperatures above 800°C. The films deposited at 400°C and annealed at

700°C can be patterned using conventional photoresists in 70°C phosphoric acid.

The films have a dielectric constant of 8.5 and a dielectric strength of 5×10^6 V/cm. The dielectric strength degrades after heating to 800°C. The V_{fb} was always positive and increased with deposition rate. The films were polarizable, showing positive ΔV_{fb} of as much as 40 V under negative bias at 250°C. Annealing at 700°C made the negative polarization even more pronounced. The V_{fb} shift was not reversible. The electrical properties of these films are not well understood, but similar effects have been observed in the Al_2O_3 films formed by thermal decomposition of aluminum alkoxides. A more thorough understanding of the intrinsic electrical properties of Al_2O_3 will be necessary before these films are considered for use on production devices.

Plasma Anodization of Silicon

In 1965, Ligenza first reported the anodization of silicon in a microwave-induced oxygen plasma [24]. The apparatus used by Kraitchman [25] to study this reaction is shown in Figure 9.11. The quartz discharge tube has an

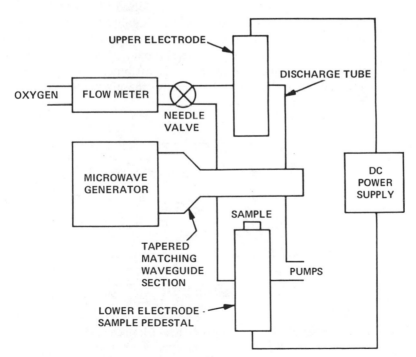

FIGURE 9.11. Plasma-anodization system [25].

inside diameter of 11 mm. The tapered waveguide section transmits the microwave power from the 500–1000 W generator to the discharge tube. The upper and lower electrodes are used primarily to increase the rate of oxidation. In this system O_2 gas was allowed to flow into the top of the reactor at a steady rate while being pumped out through the bottom. Ligenza [24] and Skelt and Howells [26] used static systems, in which an initial oxygen pressure was established and allowed to deplete during the reaction. Figure 9.12 shows the growth rate of oxide in the dynamic system at 6000 W with an O_2 flow rate of 50 cm³/min and sample temperature between 400–500°C. The curve A represents growth with no external voltage on the electrodes, curve B with 50 V on the bottom electrode and C is with 300 V. Ligenza found that the limiting thickness with no bias on the sample was 4400 Å [22]. This oxidation rate is much higher than that in a plasma asher as shown in Figure 9.4 due mainly to the much higher power density used here. Oxide growth rates as high as 7700 Å/min were observed. This is approximately equal to thermal oxide growth rate in steam at 1100°C and 1 atmosphere pressure. Kraitchman [25] reported that the etch rate in "P" etch, the ir spectrum, refractive index, resistivity, breakdown strength, and dielectric constant were indistinguishable from those of thermal oxide.

Capacitance–voltage plots made on plasma-anodized silicon MOS devices showed lower initial V_{fb} and a higher degree of stability than devices made using thermal oxide. Surprisingly the ΔV_{fb} shown by the anodized films was the opposite direction of that shown for thermal oxide. The negative shift in V_{fb} with positive bias shown by thermal oxide is attributed to the presence of mobile charged species in the glass. Apparently, the plasma-anodized films are free from mobile ionic impurities.

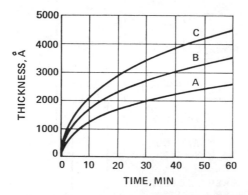

FIGURE 9.12.   Anodization rate of silicon [25]: (A) with no voltage applied to the sample, (B) with 50 V, (C) with 300 V.

The prospect of being able to grow ultraclean oxide layers on wafers at temperatures much too low to alter their diffusion profiles is exciting. However, the very high-power levels required will make the scale-up of this system to production size very difficult.

Equipment Requirements for Insulator Deposition

Glow-discharge–deposited dielectrics compare favorably with those formed by other processes. Microwave plasma anodization of silicon at 400°C gives films indistinguishable from oxides grown thermally at 1100°C in steam. The two reactions even have comparable growth rates. Silicon nitride and oxynitride can be deposited over aluminum. The dielectric properties of these films are as good as those of films formed at 750–900°C by conventional methods. The films can be patterned more easily using either wet chemical or plasma methods. The sodium-ion barrier properties of these films are better than for any other final passivation material compatible with aluminum.

The standard final passivation glass used in the semiconductor industry is phosphorous-doped silica called "phossil" glass. It is deposited from the gas phase reaction of silane, phosphine, and oxygen in a stream of nitrogen. The silicon, oxygen, and phosphorous levels in the glass are directly related to the gas phase composition during the reaction. A fluctuation in any of the flow rates will change the composition of the glass at that instant. Automatic flow controllers are available but expensive. The cost of installing three or four of them in each reactor is often prohibitive.

It has been shown that the composition of glow-discharge–deposited films is more sensitive to gas composition than in thermal reactions. However, the reactants used to deposit silicon dioxide, nitride, or oxynitride are all stable when mixed at room temperature. Once the optimum gas composition has been determined in laboratory-scale reactors, the starting gases can be ordered premixed in one standard cylinder. Only one gas flow controller would be necessary and any fluctuation in flow rate would result in a change in deposition rate rather than composition.

This very significant advantage is outweighed at present by the fact that no production-sized deposition systems are commercially available. Horsley [16] has reported making multilayer large-scale integrated circuit test patterns using glow-discharge SiO_2 and Si_3N_4. The crossover yields were as high as 99.99% using conventional etching techniques, but his reactor could deposit glass on only "several slices at once." The reactor shown in Figure 9.7 has been used at Motorola to deposit silicon nitride, oxynitride, and organic films but can hold only three 2 in. wafers per run.

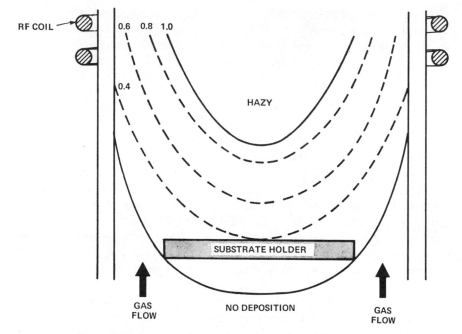

FIGURE 9.13.   Relative deposition rate as a function of position in the reactor.

The scale-up from laboratory to production scale is not trivial. Even assuming the gases can be premixed, the deposition rate and film composition depend on pressure, gas glow pattern, power density, and temperature not only in the system as a whole but immediately about each wafer. Figure 9.13 shows the relative deposition rates in the Motorola system as a function of substrate position. The temperature, gas composition, flow, and pressure as well as rf power were kept constant. Too low a position of the substrate holder relative to the rf coil gave no observable deposition; too high a position produced very rapid deposition but the resultant films were hazy. Intermediate positions gave clear films with the relative deposition rates indicated by the dotted lines. The deposition rate changes most drastically near the walls so that maximum uniformity occurs with a very small sample located near the center of the chamber.

Figure 9.14 shows scanning electron microscopy pictures of a clear and a hazy silicon nitride film. The surface of this hazy film was covered with nodules approximately 500 Å in diameter while the clear film had an almost featureless surface. The nodules are probably formed by the gas phase reaction of the silane and ammonia above the surface of the wafer. The small spheres of silicon nitride which form fall onto the surface of the wafer and

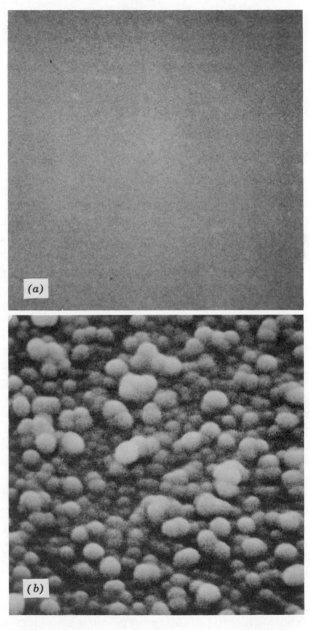

FIGURE 9.14. Scanning electron micrographs of glow-discharge silicon nitride films (20,000×): (*a*) Clear film, (*b*) hazy film.

become incorporated in the growing film. At very high-power levels the silane and ammonia react almost completely in the gas phase forming a fine white powder.

Particular care must be taken to design reactors in which all wafers see exactly the same gas flow, power level, and temperature conditions. For final passivation the uniformity of the reactor does not need to be better than 10% but the throughput should be in the 50–100 wafers/hr range.

9.6. DEPOSITION OF ORGANIC FILMS

Organic thin films are not frequently used in the wafer-processing stages of device fabrication, except as photoresists or etch masks. These films must be removed after use because they cannot stand the high temperatures used in passivation (450°C) or die-bonding (above 380°C). However, many devices which are to be encapsulated in plastic have to be coated with some sort of moisture or contamination barrier. At present, there is no material which has proven completely satisfactory in this role. Silicon polymers which cure at low temperatures are used but the moisture protection they provide is at best temporary. Several excellent reviews have been published on polymer dielectric films deposited by glow-discharge techniques [27–29]. They all agree that almost any simple organic material can be made to polymerize in a glow discharge. Bradley and Hammes [30] measured the electrical properties of polymers formed from 40 different monomers. They concluded that the films were generally more resistant to chemical attack, had higher melting points, lower solubilities, and greater thermal stability than conventional forms of polymers having the same composition. They also found that the electrical conductivity of the films was a function more of the carbon backbone than the nature of the monomer used in deposition. Therefore, the logical choice of monomer is the one which deposits a polymer at the highest rate. Bradley [31] observed that aromatic species gave very high polymer-deposition rates in the plasma but that such films tend to crack and peel away from the substrate when much thicker than 1 or 2 μ. Acrylonitrile, styrene, and perfluorinated olefins are less fragmented by the glow discharge, depositing films in which the original monomer units are essentially intact. The films are therefore less cross-linked and are more flexible with better adherence.

Vasile and Smolinsky [32] deposited organosilane films using vinyltri-methylsilane and hexamethyl disiloxane. The films were clear and smooth with a high order of cross-linking. They contained high concentrations of reactive radical sites as evidenced by a rapid increase in oxygen content upon exposure to air. Electron-spin resonance studies showed that a large number of unreacted sites remained in the polymer even after long storage in air. Denaro, Owens, and Crawshaw also observed a high radical concentration in plasma-deposited polystyrene films [33]. A study by Lee [34] showed that the presence of these radical sites does not adversely affect the electrical parameters of semiconductor devices. He coated a series of devices specifically designed to detect surface changes (REL CHIPS) [35] with polymers deposited from a series of fluorinated monomers. The devices contained MOS transistors and capacitors, components often degraded by plasma treatment. The devices showed no electrical changes after the coating, with the leakage currents remaining in the low picoamp range.

Work has yet to be published on the moisture and contamination barrier properties of these films. The good adhesion, chemical and moisture resistance of glow-discharge films coupled with the possibility of in situ plasma-pre-cleaning should make them attractive candidates for passivation just prior to plastic encapsulation. However, the economics of any process which must be performed on the units individually are different than for a process which can be performed at the wafer stage. A 2 in. wafer may contain 10,000 individual die. A plasma system which can simultaneously process 50–100 wafers or 500,000–1,000,000 die can handle only 100–1000 finished units at a time. Therefore, any operation which must be done on a unit basis has to be inexpensive to keep the cost per device as low as possible.

9.7. SUMMARY AND CONCLUSIONS

Plasma technology has been successfully applied in production-scale equipment to the etching of silicon compounds and to the removal of photoresist materials. These operations are very important in the production of semiconductor devices, and the application of plasma methods to them both individually or sequentially has resulted in substantial savings in time and money by reducing the amount of operator time, wafer handling, and chemical costs. The reactors used in each operation are simple units which are small by chemical industry standards but can hold enough wafers to make more than 1 million semiconductor devices.

Interactions with the substrates can limit the application of either plasma-ashers or etchers. The electrical parameters of certain devices can be drastically changed by exposure to the activated species in a plasma at certain stages in fabrication. A fluorine plasma will selectively attack exposed silicon, so that all etch steps have to be ones in which the silicon exposed is electrically unimportant. Processes can be modified to circumvent many of these effects, however, at present most production engineers are simply replacing conventional photoresist removal or etch steps with plasma systems without redesigning any of the processing sequences. When process designers become convinced that the economy, cleanliness, ease of operation, and possibility of automation now shown by plasma methods can be combined with reproducibility and uniformity of $\pm 5\%$, the use of plasma-etchers and ashers will become much more widespread.

The glow-discharge deposition of dielectrics has not been used in device production. The deposited materials are very good; silicon nitride and oxynitride can be deposited over aluminum metallization. The resultant films can be easily patterned using either conventional or plasma-etch techniques. The films are much better barriers against ionic contaminants than any conventional films used for final passivation. Silicon dioxide films grown by plasma anodization are indistinguishable from thermally grown oxide. However, at present, no production-scale equipment for film growth or deposition is commercially available. Until such equipment is produced, the plasma deposition of dielectrics will remain a laboratory curiosity.

ACKNOWLEDGMENTS

The author would like to thank several of the engineers at Motorola for their help on this chapter; notably, John Osborne, Marlo Cota, and Don McKinstry. A special thanks is due to my wife, Betty, for her help in preparing this manuscript.

REFERENCES

[1] H. R. HANSEN, J. V. PASCALE, T. DeBENEDICTIS, and P. M. RENTZEPIS, *J. Polymer Sci., Pt. A* 3, 2205 (1965).

[2] S. M. IRVING, *Kodak Photoresist Seminar* 2, 26 (1968).

[3] F. KAUFMAN and J. R. KELSO, *J. Chem. Phys.*, 32, 301 (1960).

[4] R. L. BERSIN, *Solid State Tech.* 39 (June 1970).

[5] H. HUGHES, W. L. HUNTER, and K. RITCHIE, *J. Electrochem. Soc.*, 120, 99 (1973).

[6] S. A. HARRELL, *Proc. 2nd Kodak Seminar Microminiaturization* 50 (1966).

[7] *International Plasma Bulletin*, International Plasma Corp., Hayward, Cal. Vol. 2, #6.

[8] R. GERETH and W. SCHERBER, *J. Electrochem. Soc.* 119, 1248 (1972).

[9] H. M. MANASEVIT and F. L. MORRITZ, *J. Electrochem. Soc.* 114, 204 (1967).

[10] H. M. MANASEVIT, *J. Electrochem. Soc.* 115, 435 (1968).

[11] C. P. BEGUIN, J. B. EZELL, A. SALVEMIN, J. C. THOMPSON, D. G. VICKROY, and J. L. MARGRAVE, "Chemical Synthesis in Radio-Frequency Plasma Torches," *in* R. F. BADDOUR, and R. S. TIMMINS, (eds.), *The Applications of Plasma to Chemical Processing*, MIT Press, Cambridge, Mass., 1967, p. 51.

[12] H. Abe, Y. Sonobe, and T. Enomoto, *Japan J. Appl. Phys.* **12**, 154, (1973).

[13] J. Osborne, Motorola Semiconductor Products Division, personal communication (1973).

[14] R. J. Joyce, H. F. Sterling, and J. H. Alexander, *Thin Solid Films* **1**, 481 (1968).

[15] R. C. G. Swann, R. R. Mehta, and T. P. Gauge, *J. Electrochem. Soc.* **114**, 713 (1967).

[16] A. W. Horsley, *Electronics* **84** (Jan. 20, 1969).

[17] H. F. Sterling and R. C. G. Swann, *Solid State Electronics* **8**, 653 (1965).

[18] R. W. Kirk and H. Gurev, Paper presented at Am. Inst. Chem. Eng. 64th Annual meeting (Nov. 1971).

[19] Y. Kuwano, *Japan J. Appl. Phys.* **7**, 88 (1968).

[20] V. Y. Doo, D. R. Nichols, and G. A. Silvey, *J. Electrochem. Soc.* **113**, 279 (1966).

[21] S. W. Ing, Jr., and W. Davern, *J. Electrochem. Soc.* **111**, 120 (1964).

[22] S. P. Mukherjee and P. E. Evans, *Thin Solid Films* **14**, 105 (1972).

[23] H. Kato and Y. Koga, *J. Electrochem. Soc.* **118**, 1619 (1971).

[24] J. R. Ligenza, *J. Appl. Phys.* **36**, 2703 (1965).

[25] J. Kraitchman, *J. Appl. Phys.* **38**, 4323 (1967).

[26] E. R. Skelt and G. M. Howells, *Surface Sci.* **7**, 490 (1967).

[27] L. V. Gregor, *IBM J.* **140** (March 1968).

[28] V. M. Kolotyrkin, A. B. Gil'Man, and A. K. Tsapuk, *Russ. Chem. Rev.* (*Eng. transl.*) **36**, 579 (1967).

[29] A. M. Mearns, *Thin Solid Films* **3**, 201 (1967).

[30] A. Bradley and J. P. Hammes, *J. Electrochem. Soc.* **110**, 15 (1963).

[31] A. Bradley, *J. Electrochem. Soc.* **119**, 1153 (1972).

[32] M. J. Vasile and G. Smolinsky, *J. Electrochem. Soc.* **119**, 451 (1972).

[33] A. R. Denaro, P. A. Owens, and A. Crawshaw, *Eur. Polymer J.* **4**, 93 (1968).

[34] S. M. Lee, *Insulation/Circuits* **33** (June 1971).

[35] S. M. Lee, J. J. Licari, and I. Litant, *Metal Trans.* **1**, 701 (1970).

Chapter 10

Engineering and Economic Aspects of Plasma Chemistry

Alexis T. Bell

10.1. REACTOR DESIGN AND SCALE-UP

Suitable procedures for the design and scale-up of plasma reactors have not yet been developed and are still a topic of active research. It is possible nevertheless to define which physical elements must be taken into account in order to provide a quantitative description of reactor performance. These elements will be discussed here together with an example illustrating their application to the design of a source of atomic oxygen.

The principal factors to be decided upon in the design of a reactor for a specified task are the reactor size and geometry. Closely coupled to this selection is the choice of operating conditions defined in terms of gas pressure, flow rate, discharge power, and source frequency. All of these elements can be varied independently at the discretion of the designer. Equally important but less amenable to independent control are the profiles of gas velocity and temperature within the reactor, both of which can affect the product yield. Before entering into a discussion of a reactor model which can be used for design purposes, it will be fruitful to consider, if only in a qualitative manner, the effects of source frequency and the profiles of gas velocity and temperature on the expected performance of a plasma reactor.

The choice of source frequency can place certain restrictions upon the size and geometry of the reactor. Since most plasma applications utilize electrodeless discharges, either a rf ($1–10^2$ MHz) or a microwave ($10^2–10^4$ MHz) source of power is chosen. The former offers a great deal of flexibility in the choice of reactor size and geometry since the external electrodes or coils used to supply energy can be made almost any size or shape desired. The major disadvantage of rf discharges is their tendency to give a nonuniform glow. This effect is illustrated in Figure 10.1 for a capacitively coupled discharge. As

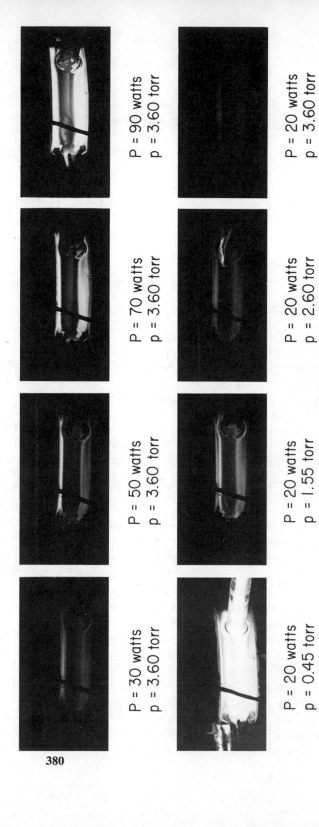

P = 90 watts
p = 3.60 torr

P = 20 watts
p = 3.60 torr

P = 70 watts
p = 3.60 torr

P = 20 watts
p = 2.60 torr

P = 50 watts
p = 3.60 torr

P = 20 watts
p = 1.55 torr

P = 30 watts
p = 3.60 torr

P = 20 watts
p = 0.45 torr

FIGURE 10.1. Photographic observations of the distribution of glow intensity in a rf discharge sustained in oxygen.

can be seen, the glow concentrates very near the sustaining electrodes. This effect is heightened with increasing discharge gap size and gas pressure. The causes for the nonuniformity have already been discussed in Section 1.5. By contrast, microwave discharges, because of the higher frequencies at which they operate, yield a more uniform discharge. If a resonant cavity is used to couple the microwave field to the discharge, then there will be certain limitations to the size of the discharge tube imposed by the dimensions of the cavity. Most of these limitations can be overcome, however, through the use of the slow wave structures discussed recently by Bosisio, Weissfloch, and Wertheimer [1].

The velocity profile of the gas as it flows through the reactor can have a very noticeable effect on product yield. As an example, let us consider the situation shown in Figure 10.2. Here we see a parallel plate discharge sustained by a rf power supply under conditions in which the glow is nonuniformly distributed. The gas flow occurring in the gap is shown with a parabolic velocity profile characteristic of laminar flow. At the low pressures used for most applications involving glow discharges, the Reynolds number is well below 2100 so that the flow is always laminar. As was shown in Figure 1.23, the rate of electron–molecule reactions in the regions where the glow is concentrated is considerably higher than at the center of the gap. These conditions lead to a nonuniform rate of production of active species. Coupled to this is the nonuniformity of the velocity profile. Those portions of the gas nearer the wall travel more slowly and thus have a longer residence time than portions of the gas moving along the center line. The combination of more intense rates of reaction and larger residence times in the vicinity of the containing walls contribute to higher yields of active species in these regions. A portion of this nonuniform distribution will, however, be smoothed out through the diffusion of active species from the wall regions toward the center of the gap.

A somewhat different effect of velocity distribution is illustrated in Figure 10.3. In this instance, we are considering a reactor of uniform diameter fitted with inlet and outlet tubes of a smaller diameter. The anticipated velocity profiles are shown at several cross-sections along the reactor. Because of the

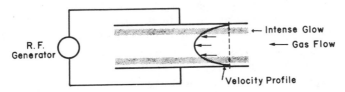

FIGURE 10.2. Illustration of flow-velocity distribution in a rf discharge.

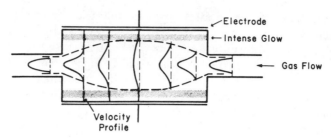

FIGURE 10.3.   Illustration of bypassing in a rf discharge.

small diameter inlet and outlet tubes, the gas flow tends to occur along the center line. Thus the regions near the cylindrical walls of the reactor will contain essentially stagnant gas and are poorly ventilated by the flow. For such a configuration, the gas flows through the reactor in less time than that anticipated from a calculation based on the total reactor volume. A reasonable estimate of the effective reaction volume is shown by the dashed curves. As a result of the bypassing of the flow through parts of the reactor the conversions obtained will be less than that expected [2].

The temperature profile within the reactor will also affect the local rates of reactions and it is thus desirable to consider the form of this profile for a given set of operating conditions. Due to the complexity of all the processes involved, it is not possible presently to give an accurate prediction of the gas temperature in a glow discharge as a function of the operating conditions and as a result one must rely on empirical observations. This point may be appreciated if one considers that in order to perform such calculations it is necessary to know what fraction of the dissipated electrical power is converted into molecular kinetic energy through the action of elastic electron–molecule collisions, self-absorption of radiation released by the plasma, and exothermic reactions occurring both within the gas phase and at the container walls. Balancing these sources of thermal energy will be heat losses, which occur principally by conduction through the gas and by infrared radiation.

To give an illustration of the dependence of gas temperature on operating variables we shall consider the measurements of Bell and Kwong [2] shown in Figures 10.4, 10.5, and 10.6. As may be seen in Figure 10.4, the temperature remains fairly uniform throughout most of the discharge but decreases in the vicinity of the cylindrical wall. The effects of power and pressure on the gas temperature are shown in Figures 10.5 and 10.6 for a point 1.5 cm from the center of the discharge.

If the effects of spatial variations in the gas velocity and temperature can be excluded, it is then possible to examine the problem of reactor design in terms of a relatively simple model. As an example of this approach, we will

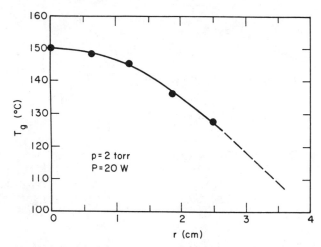

FIGURE 10.4. Gas temperature as a function of position in a rf discharge in oxygen.

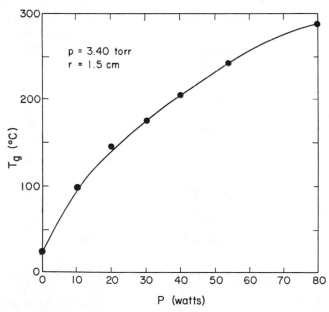

FIGURE 10.5. Gas temperature as a function of power in a rf discharge in oxygen.

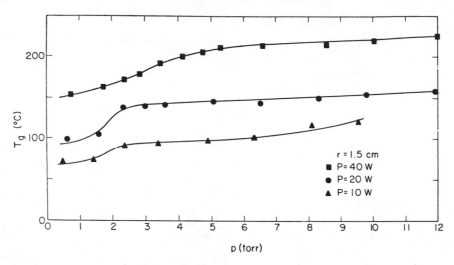

FIGURE 10.6. Gas temperature as a function of pressure in a rf discharge in oxygen.

utilize the model developed in Section 1.6 to describe the formation of atomic oxygen in a microwave discharge. Recall that from the point of view adopted there the plasma was regarded as a cylindrical volume characterized by an average electron density and a spatially independent value of E_e/p. Under these assumptions the overall kinetics for the formation of atomic oxygen can be expressed as

$$\frac{4FN}{(2N - n_1)^2}\frac{dn_1}{dV} = 2k_1\langle n_e\rangle(N - n_1) - \frac{1}{2R}n_1v_r\gamma + 2k_2n_1^2(N - n_1)$$

$$- 2k_3n_1^3 - 2k_4n_1(N - n_1)^2 \quad (10.1)$$

where n_1 is the concentration of atomic oxygen, N is the total gas concentration, $\langle n_e\rangle$ is the volume-averaged electron density, F is the molar flow rate of molecular oxygen entering the reactor, V is the volume of the reactor, R is the radius of the discharge tube, v_r is the random velocity of oxygen atoms, γ is the wall recombination coefficient, and k_1–k_4 are the reaction rate constants for the dissociation of molecular oxygen and the homogeneous recombination of atomic oxygen. The term on the left-hand side of (10.1) describes the loss of atomic oxygen by convection. On the right-hand side, the first term describes the rate of atomic oxygen formation via dissociation of molecular oxygen; the second term describes the rate of atomic oxygen

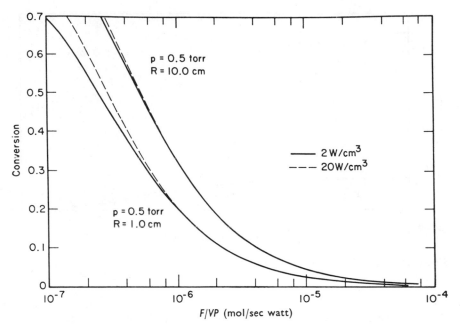

FIGURE 10.7. Conversion as a function $F/V\bar{P}$.

recombination on the reactor walls; and the remaining three terms describe the rate of atomic oxygen recombination via homogeneous gas phase reactions. Values for the dissociation rate constant k_1 and the electron density can be obtained from Figures 1.26 and 1.19. The remaining rate constants and the wall recombination coefficient γ may be found in Table 1.1.

Integration of (10.1) leads to curves of conversion and yield versus the ratio of the molar flow rate F to the reactor volume V. By plotting the results as a function of $F/\bar{P}V$ in which $\bar{P}$ is the power density the specific effects of power density can be reduced. Figures 10.7 and 10.8 are therefore plotted in this manner. The values of k_1 and n_e for the curves shown in these figures are given in Table 10.1. As may be seen in Figure 10.7 the curves for 2 and 20 W/cm³ either overlap or diverge to only a slight degree. This behavior will be observed as long as the first term on the right-hand side of (10.1) is dominant. By contrast, Figure 10.8 shows that a parametric effect of power density can be identified when the dissociation rate term in (10.1) is not dominant.

Figures 10.7 and 10.8 show that an increase in $F/\bar{P}V$ produces a decrease in the extent of conversion. This may be viewed as the result of a reduction in the amount of energy dissipated in each mole of oxygen entering the reactor.

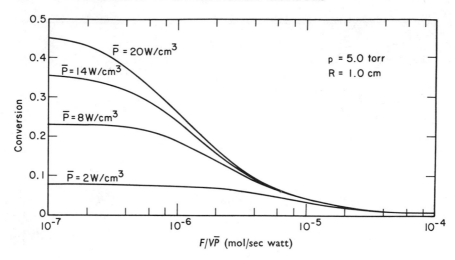

FIGURE 10.8.   Conversion as a function $F/V\bar{P}$.

An increase in pressure for a fixed tube size leads to a reduction in the conversion due to a decrease in the values of k_1 and n_e caused by the increase in the group $p\Lambda$. Also contributing to the decreased conversion is the enhancement of the rates of the three-body homogeneous recombination reactions. Finally, it can be observed that, for a fixed pressure, an increase in tube size produces an increase in the conversion. This effect is due to the reduction of the wall surface area per unit volume of the reactor which reduces the significance of the wall recombination reaction.

In a number of applications of plasma chemistry, it is desireable to produce reactive species in the discharge and then transfer them downstream to a point where they can be used. For such applications it is important to transfer the species with a minimum of loss. An example of this type of operation would be the production of a stream of atomic oxygen to be used in the surface treatment of a polymer which is located outside of the active discharge zone.

TABLE 10.1.   Values of k_1 and n_e

p (torr)	R (cm)	k_1 (cm^3/sec)	$n_e/\bar{P}$ (W^{-1})
0.5	10	6.7×10^{-11}	1.97×10^{11}
0.5	1	1.0×10^{-10}	3.58×10^{10}
5.0	1	6.7×10^{-11}	1.97×10^{10}

The decay in atomic oxygen concentration as it flows downstream from the discharge can be expressed by the following equation

$$\frac{D_{12}}{r}\frac{\partial}{\partial r}\left(r\frac{\partial n_1}{\partial r}\right) - \frac{4FN}{(2N - n_1^2)\pi R^2}\frac{\partial n_1}{\partial z} - 2k_2 n_1^2(N - n_1) - 2k_3 n_1^3$$

$$- 2k_4 n_1(N - n_1)^2 = 0 \quad (10.2)$$

where D_{12} is the diffusivity of atomic oxygen through molecular oxygen. The first term on the right-hand side of (10.2) describes the loss of atomic oxygen by diffusion to the tube walls and the remaining terms carry the same identity as in (10.1). The importance of wall-recombination process is accounted for through the boundary condition used at $r = R$,

$$-D_{12}\left(\frac{\partial n_1}{\partial r}\right)_{r=R} = \frac{1}{4}n_1 v_r \gamma \quad (10.3)$$

Equation (10.3) states that the radial diffusion flux of atomic oxygen evaluated at the tube wall must equal the rate at which the atoms recombine.

When the value of γ is relatively small as it is for glass surfaces ($\gamma \simeq 10^{-4}$), the necessary radial flux can be supplied without producing a large radial concentration gradient. Under these circumstances the concentration of atomic oxygen is essentially constant in the radial direction and (10.2) can be rewritten in a simplified form as

$$\frac{4FN}{(2N - n_1)^2 \pi R^2}\frac{dn_1}{dz} = -\frac{1}{2R}n_1 v_r \gamma - 2k_2 n_1^2(N - n_1) - 2k_3 n_1^3 - 2k_4 n_1(N - n_1)^2$$

$$(10.4)$$

It may be recognized that (10.4) differs from (10.1) only in the absence of a term describing the rate of dissociation.

Figure 10.9 illustrates the axial concentration profile which is expected in the reactor as well as downstream. These profiles were obtained from solutions to (10.1) and (10.4) and the requirement that at $z = L$ the solutions to both equations agree. The concentration of atomic oxygen downstream from the discharge is governed by the following factors: (a) The concentration of atoms at the discharge exit, (b) the gas-flow rate, (c) the wall-recombination coefficient, and (d) the gas pressure. The first of these four factors has been discussed above. The influence of gas-flow rate is easy to discern. The higher the flow rate, the lower will be the residence time in a given volume and hence the opportunity for recombination either on the wall or in the gas phase. The magnitude of the wall-recombination coefficient can have a

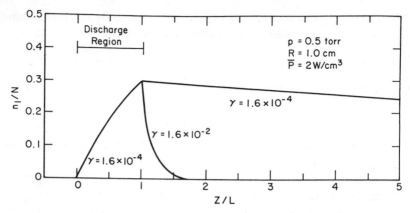

FIGURE 10.9. Atomic oxygen concentration in and downstream of an oxygen discharge.

pronounced effect on the attenuation of the atomic oxygen stream. For example, glass walls at room temperature have a coefficient of about 10^{-4} while metals have a coefficient of about 10^{-1}. Finally we must consider the effect of gas pressure. As the gas pressure is increased, the rates of the three-body recombination reactions are enhanced. Since these processes have a cubic dependence on pressure they can produce a strong influence on the loss of atomic oxygen when the corresponding terms in (10.4) become significant in magnitude.

The design of a discharge reactor will now be illustrated through the application of the models developed above. As a first example, we shall consider the design of a reactor to provide a stream containing 20% of atomic oxygen at a flow rate of 1.49×10^{-4} mole/sec (2×10^{2} cm³/min at STP). The reactor tube radius is chosen to be 1.0 cm and the gas pressure is to be 0.5 torr. From Figure 10.7 we may conclude that for a 20% concentration of atomic oxygen, which corresponds to an 11.11% conversion, $F/\bar{P}V = 2.03 \times 10^{-6}$ mole/W sec. The molar flow rate of molecular oxygen required is 6.70×10^{-4} mole/sec. As a result $\bar{P}V = 34.0$ W. From Figure 10.7 we notice that there is no additional influence of $\bar{P}$ beyond its appearance in the group $F/\bar{P}V$. Consequently $\bar{P}$ and V can be chosen in any convenient manner as long as the total dissipated power is 34.0 W.

As a second example, let us consider the production of 10% of atomic oxygen at a pressure of 5.0 torr. The flow rate of atomic oxygen is to be 1.49×10^{-4} mole/sec and the discharge tube radius is 1.0 cm. The molar flow rate of molecular oxygen is in this case 1.41×10^{-3} mole/sec (1900 cm³/min at STP). If the power density is chosen as 2 W/cm³, the value of $F/\bar{P}V$ taken from Figure 10.8 is 4.86×10^{-6} and the total power required will be 290 W.

On the other hand, if the power density is chosen as 10 W/cm³, the value of $F/\bar{P}V$ is 7.74 × 10⁻⁶ and the total power is 182 W. In this case we may conclude that the use of a higher power density reduces the total power required to produce the required supply of atomic oxygen.

The scale-up of a plasma reactor represents a problem which parallels the design problem. Thus let us, for example, determine the power required to give a gas stream whose composition and linear velocity are the same as in the first example. The discharge tube radius is to be increased tenfold while the pressure remains at 0.5 torr. We may begin by noting that the conversion remains the same as in the previous example. Consequently, from Figure 10.7, we conclude that $F/\bar{P}V = 3.71 \times 10^{-6}$ mole/W sec. The molar flow rate of molecular oxygen must increase by a hundredfold in order to maintain the same linear velocity. Consequently, the total power required for the large discharge tube will be 2 × 10³ W.

10.2. ECONOMICS OF USING PLASMA CHEMISTRY

It is difficult to discuss the economics of applying plasma chemistry to a commercial process because the detailed considerations differ widely from case to case. Consequently, we shall limit our discussion to a survey of the major costs which include capital equipment, labor, chemicals, and utilities in an effort to assess their level and to compare them to those for more conventional processing techniques. The benefits to be derived from using plasma chemistry will also be discussed.

Because the presence of electrodes within the plasma zone is usually objectionable, microwave and rf generators are used most frequently to supply the discharge power. In terms of costs generators operating in either frequency range may be obtained at comparable prices. For low-power generators delivering less than 500 W, the prices range between $4 and 5/W. For intermediate powers in the vicinity of 1 kW, the price decreases to $2–3/W. Finally, above 10 kW the price drops yet further to about $1–2/W. These prices are not intended to be accurate but rather to serve as a rough guide for estimating costs. A further observation which can be made is that the price of a generator alone does not vary greatly from one manufacturer to another.

The cost of reactors and ancillary glassware is more difficult to estimate since this will vary from application to application. One may observe, however, that these costs are almost always minor compared to the cost of the generator.

If one considers the existing and proposed plasma chemistry processes as examples, it is possible to conclude that this type of process does not demand a large amount of labor. The principal reason for this is that the plasma-generating equipment is stable and reliable and can be designed for automatic control or programming of the operating variables. It is important to note that one of the most attractive situations for the application of a plasma process is that in which the existing process involves an extensive series of steps. Such situations occur quite frequently in the semiconductor industry and in the processing of polymers to alter their surface characteristics.

A classic example of how a plasma-chemical process can reduce manufacturing steps and costs can be represented by the etching of compound semiconductors such as gallium arsenide phosphide. Materials such as these cost in the range of $20–15/in.^{2} and hence must be handled with extreme care. A comparison of the steps involved in etching these materials by conventional and plasma chemical means is illustrated in Table 10.2. As may be seen, the plasma-chemical approach greatly reduces the number of steps and the consequent chances for breakage. By eliminating all wet chemistry the plasma-chemical approach also reduces the chances of product contamination by ionic impurities.

TABLE 10.2. Comparison of Conventional and Plasma Chemical Approaches for Etching Compound Semiconductors

Conventional approach	Plasma chemical approach
1. Deposit Si_3N_4 film	1. Deposit Si_3N_4 film
2. Deposit SiO_2 film over Si_3N_4	2. Apply photoresist over Si_3N_4
3. Apply photoresist over SiO_2	3. Mask and expose photoresist
4. Mask and expose photoresist	4. Dissolve unexposed photoresist
5. Dissolve unexposed photoresist	5. Etch Si_3N_4 in CF_4 plasma
6. Etch SiO_2 in HF solution	6. Remove photoresist in O_2 plasma
7. Rinse	
8. Dry	
9. Remove photoresist	
10. Rinse	
11. Etch Si_3N_4 in boiling phosphoric acid	
12. Rinse	
13. Dry	

Most of the plasma processes presently considered for industrial application do not utilize large amounts of chemicals. As a result, the attendant costs are not significant. Furthermore, there are cases in which a plasma process has replaced an existing one which made use of corrosive and toxic chemicals or chemicals which needed to be of high purity. A case in point is the use of a helium discharge to replace sodium dissolved in liquid ammonia for the treatment of Teflon surfaces in order to make them bondable by adhesives. A second example is the use of a carbon tetrafluoride discharge to replace boiling reagent grade phosphoric acid for the etching of silicon nitride. In this case not only are the chemicals costs reduced but the costs of waste disposal are totally eliminated.

The major utility consumed by a plasma process is electricity. In order to compute utilities costs it is necessary to know the efficiency with which line power is converted to high-frequency power and the efficiency of coupling the latter to the discharge. Present rf and microwave equipment has a line power conversion efficiency which lies in the range of 50–80%. With proper impedance matching circuits, it is possible to couple nearly 100% of the high-frequency power in the discharge. As a result the actual line power consumed can be determined by dividing the discharge power by the line power conversion efficiency.

Electrical power costs, however, are rarely a significant part of the total cost of operation. The reason for this is that one is usually involved in depositing only a small amount of material or making a chemical change to only a small depth in an existing material. This cost can become significant, however, if one considers the manufacture of a chemical compound on a bulk scale.

The advantages of using plasma chemistry depend on how it is used. Thus in semiconductor applications the significant advantages are the reduction in the number of process steps and the attendant handling and breakage, the elimination of hazardous chemicals, and the introduction of a relatively simple and "clean" process. In the case of wool shrinkproofing the major advantages are the production of a unique product, the elimination of costly resins, and the simplicity of the process. If one is to generalize, it is possible to say that the principal payoffs are found in one of the following categories: (*a*) uniqueness of product, (*b*) improved product quality, (*c*) reduced labor and handling, and (*d*) reduced utilization of hazardous or costly chemicals.

REFERENCES

[1] R. G. BOSISIO, M. R. WERTHEIMER, and C. F. WEISSFLOCH, *J. Phys. E* **6**, 628 (1973).

[2] A. T. BELL and K. KWONG, *AIChE J.* **18**, 990 (1972).

Design of Radio-Frequency Equipment Used to Sustain Electric Discharges†

A.1. INTRODUCTION

Low-pressure glow discharges are most conveniently sustained by using high-frequency electric fields produced by either a radio-frequency or a microwave power supply. Most of the work described in this book has been carried out using rf generators. The following reasons can be given for this selection:

1. Commercial availability of rf equipment specifically designed for plasma chemical studies
2. Ease of adapting radio transmitters for plasma studies
3. Relatively low cost of measuring equipment
4. Ease of which rf power can be coupled to a wide variety of reactor designs

For these same reasons it is believed that rf equipment will continue to be used in future studies of plasma chemistry. The purpose of this Appendix is to provide a guideline for researchers desiring to develop their own rf generators or to adapt existing radio equipment for the purpose of sustaining a discharge.

A.2. DESIGN CONSIDERATIONS

A complete rf system for producing a discharge consists of a generator, an impedance matching network, means for applying the rf field to the reactor, a reactor, and means for measuring the rf power supplied to the discharge. These components are illustrated in Figure A.1. A brief discussion of each of these components will now be given.

Radio-Frequency Generator

An rf generator which is to be used for sustaining a discharge must meet two primary criteria. First, it must comply with the Federal Communications

† Prepared with the assistance of Georges Gorin, Tegal Corporation.

393

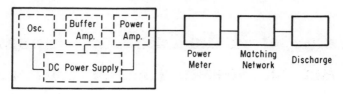

FIGURE A.1.   Block diagram of components used to sustain an rf discharge.

Commission regulations covering the operation of equipment in the frequency bands licensed for commercial use. Second, the generator must be capable of withstanding large variations in the load impedance. Whereas radio communications transmitters meet the first criterion, they are usually designed to operate into a fixed impedance and hence are not well suited for producing a gaseous discharge unless modified.

The combinations of individual stages in the block labeled rf generator shown in Figure A.1 are particularly convenient for generating rf power. The first stage is a crystal-controlled rf oscillator which most frequently operates at 13.56 MHz and serves as the source of the rf signal. The signal from the oscillator is then amplified to a level of about 5 to 10 W using either solid-state or tube amplifiers. Figure A.2 illustrates a typical design for an oscillator stage.

Final amplification is accomplished by the power amplifier. This stage is designed around a power tube rather than a solid-state device in order to accommodate large variations in load impedance. The safest type of tube to use for the power amplifier is a power triode since its only real limitation is its plate dissipation rating. This rating can easily be met by limiting the

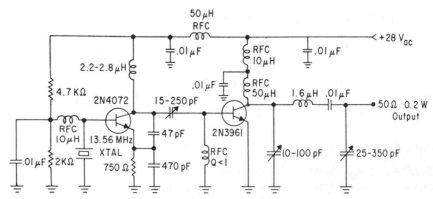

FIGURE A.2.   Schematic for an rf oscillator and buffer amplifier.

amount of dc power available from the dc power supply. The main disadvantage of a triode is that it requires a large amount of driving power and for this reason tetrodes or pentodes are often used. If such tubes are used, the power stage should be protected not only against excessive plate dissipation but also and more critically against screen grid dissipation.

Due to its high efficiency, the power tube is chosen in such a way that the total plate dissipation will be equal to

$$P_d = P_{in} = P_{out}/0.65 \tag{A.1}$$

where P_d is the power dissipated at the plate, P_{in} is the dc power supplied to the tube, and P_{out} is the maximum rf power delivered. The factor 0.65 is used to describe a plate efficiency of 65%, a level which is easily attainable. If a high rf power is required, it is often less expensive to let

$$P_d = P_{in} - P_{out} \tag{A.2}$$

and to incorporate protective devices in the circuit design. Figure A.3 illustrates a typical design for a power amplifier stage.

Matching Network

The purpose of the matching network is to achieve a match between the impedance of the generator and that of the discharge. When such a match is attained the power transfer from the generator to the discharge is at peak efficiency. This point can readily be illustrated by assuming that both the generator and the discharge impedances have only resistive components. Then with the aid of Figure A.4 and the relationship

$$P_{load} = I^2 R_{load} \tag{A.3}$$

we can deduce

$$P_{load} = \frac{V_{out}^2}{2(R_{out} + R_{load})^2} \tag{A.4}$$

In (A.3) and (A.4), P_{load} is the power dissipated in the discharge, R_{load} is resistance of the discharge, R_{out} is the resistance of the generator at its output, I is the magnitude of the rf current flowing in the circuit, and V_{out} is the magnitude of rf voltage supplied by the generator. From (A.4) we can deduce that for a given V_{out} the value of P_{load} is maximized when $R_{out} = R_{load}$.

Since the impedance of a discharge changes with its operating conditions and its exact value is usually unknown, it is easier to fix the output impedance

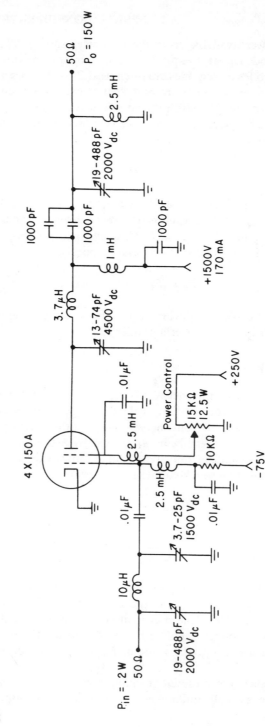

FIGURE A.3. Schematic for a power amplifier.

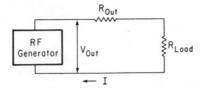

FIGURE A.4. Simplified schematic
to illustrate the relationship between
generator and load impedances.

of the generator at a known value such as 50 ohms (Ω) and then transform
the impedance of the discharge (about 5000 to 15,000 Ω) down to 50 Ω.
This procedure also makes it possible to use conventional test equipment and
rf cables, both of which are designed to be compatible with a 50 Ω output
impedance. Figure A.5 illustrates a typical pi-type of matching network.

Power Measurement

The power delivered by the generator is easily measured by placing a 50 Ω
through-line wattmeter between the output of the rf generator and the input
of the matching network as shown in Figure A.6. When the matching network
is properly tuned, its input impedance will be equal to 50 Ω, for which the
through-line wattmeter will indicate zero reflected power and the value read
for forward power will be the amount of rf power delivered to the discharge.

Coupling of Radio-Frequency Power

Coupling of rf power from the output of the matching network to the discharge is accomplished by means of either a coil or a set of capacitor plates.
These two forms of coupling are referred to as inductive and capacitive
coupling respectively and are shown in Figure A.7. The type of coupling to
be used depends upon the size and shape of the reaction chamber. If the
ratio of reactor diameter to length exceeds 1 to 1.5, either type of coupling
can be used. When the ratio is less than 1 then only capacitive coupling can
be used.

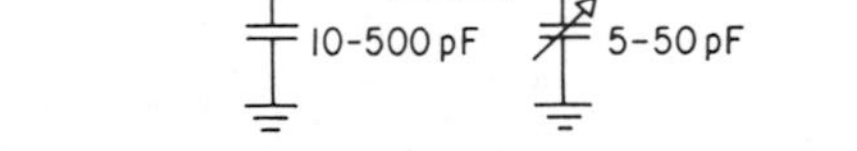

FIGURE A.5. Schematic for a pi-type
matching network.

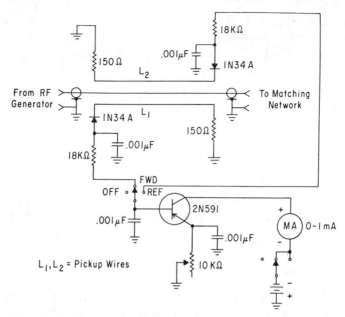

FIGURE A.6   Schematic for a 50 Ω through-line wattmeter (taken from "The Radio Amateur's Handbook," American Radio Relay League, Newington, Conn., p. 556).

In order to facilitate matching of the discharge to the generator, the values of the coupling element should fall below a specified limit. As an example, for capacitive coupling the shunt capacitance of the electrodes should be less than 40 pF and for inductive coupling the inductance of the coil should be less than 3 μH. In both instances, it has been assumed that the source frequency is 13.56 MHz.

When a discharge is present within the reactor the net load impedance is comprised of the impedance of the coupling elements plus the impedance of

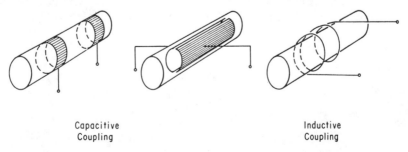

Capacitive
Coupling

Inductive
Coupling

FIGURE A.7.   Examples of capacitive and inductive coupling.

the discharge. The magnitude of the load impedance will depend upon the size of the reaction chamber, the size and configuration of the coupling element, the type of gas used, and the power and pressure at which the discharge is sustained. As a result the magnitude of the load impedance can vary considerably.

A.3. REFERENCES FOR CIRCUIT DESIGN

The following references should be consulted for additional details regarding the design of specific circuits.

"The Radio Amateur's Handbook," American Radio Relay League, Newington, Conn.

"RCA Power Circuits—DC to Microwave," Radio Corporation of America, Harrison, N.J.

"Care and Feeding of Power Grid Tubes," Eimac Division of Varian, San Carlos, Ca., 1967.

"RCA Transmitting Tubes," Radio Corporation of America, Harrison, N.J.

"The Semiconductor Data Library," Motorola Semiconductor Products, Inc., Phoenix, Ariz.

Index